MW00634590

SLURRY TRANSPORT USING CENTRIFUGAL PUMPS

Third Edition

SLURRY TRANSPORT USING CENTRIFUGAL PUMPS

Third Edition

K. C. WILSON
Queen's University
Kingston, Ontario, Canada

G. R. ADDIE
GIW Industries Inc.
Grovetown, Georgia, USA

A. SELLGREN
Luleå University of Technology
Luleå, Sweden

R. CLIFT
University of Surrey
Guildford, Surrey, UK

Library of Congress Cataloging-in-Publication Data

A C.I.P. Catalogue record for this book is available
from the Library of Congress.

ISBN-10: 0-387-23262-1
ISBN-13: 9780387232621

ISBN-10: 0-387-23263-X (e-book)
ISBN-13: 9780387232638

Printed on acid-free paper.

Printed in the United States of America.

9 8 7 6 5 4 3 2 1 SPIN 11327295

springeronline.com

Acknowledgements

This book is the Third Edition of a text that emerged from a series of annual courses on slurry transport held by GIW Industries at Augusta and Grovetown, Georgia over the past 28 years. The authors wish to acknowledge those who have been involved with the GIW slurry course and who have contributed to the book. In particular, we would like to thank Dr. M.R. Carstens for his hard work in the early years and to acknowledge the late Dr. R. Alan Duckworth for his enthusiastic participation in later years. We also wish to acknowledge the early contributions to the course made by Dr. M. Roco, and the later contributions by Dr. K. Pagalthivarthi, and Dr. H. Tian. Mr. Reab Berry participated with enthusiasm in the organization of the course and contributed some of the practical examples. A series of test engineers generated data of the highest quality under conditions that were often hectic, and our highest thanks go to the present test engineer, Mr. Lee Whitlock.

We particularly wish to thank Mr. R. Visintainer for his contribution to the course and text, and to the members of the GIW staff who prepared the figures. Ms. Anne Davis typed the original course notes, and deserves our special thanks for the organisational skills that she devoted to the course. As with the previous editions, Ms. Karen MacIntyre merits our warmest thanks for her skilful processing of the text. Thanks for text preparation also go to Mrs. Phyllis Korman and Mr. David Nielsen.

Last but not least, the authors wish to express their gratitude to Mr. Thomas W. Hagler Jr. of GIW Industries (now a subsidiary of KSB Aktiengesellschaft, Frankenthal, Germany). Throughout his tenure as Chief Executive Officer of GIW, Mr. Hagler has been tireless in his enthusiastic support for the laboratory, the course, and the book; and this support has been continued by the present CEO, Mr Dennis Ziegler.

Contents

Chapter 1

INTRODUCTION

1.1 Applications of Slurry Transport

Vast tonnages are pumped every year in the form of solid-liquid mixtures, known as slurries. The application which involves the largest quantities is the dredging industry, continually maintaining navigation in harbours and rivers, altering coastlines and winning material for landfill and construction purposes. As a single dredge may be required to maintain a throughput of 7000 tonnes of slurry per hour or more, very large centrifugal pumps are used. Figures 1·1 and 1·2 show, respectively, an exterior view of this type of pump, and a view of a large dredge-pump impeller (Addie & Helmley, 1989).

The manufacture of fertiliser is another process involving massive slurry-transport operations. In Florida, phosphate matrix is recovered by huge draglines in open-pit mining operations. It is then slurried, and pumped to the wash plants through pipelines with a typical length of about 10 kilometres. Each year some 34 million tonnes of matrix are transported in this manner. This industry employs centrifugal pumps that are generally smaller than those used in large dredges, but impeller diameters up to 1.4 m are common, and drive capacity is often in excess of 1000 kW. The transport distance is typically longer than for dredging applications, and

Figure 1.1. Testing a dredge pump at the GIW Hydraulic Laboratory

Figure 1.2. Impeller for large dredge pump

hence a series of pumping stations is often used. Figure 1·3 shows a booster-pump installation in a phosphate pipeline.

Many other types of open-pit mining use slurry transport, and the number of such applications is increasing as it becomes clear that, for many short-haul and medium-haul applications, slurry transport is more cost-effective than transport by truck or conveyor belt. Partially-processed material from mining and metallurgical operations and other industries is often already in slurry form, facilitating pipe transport. Much of this is carried out using relatively small lines. For example newly-mixed concrete is a slurry and is sometimes piped from the mixing plant to other parts of a construction site.

Figure 1.3. Booster pump in a phosphate-matrix pipeline

As this slurry has a high resistance to flow, and as considerable static lifts are also common in construction, the pumps employed are usually of the positive-displacement type. Such pumps are also used in other high-head applications; and the best-known of these is the Black Mesa pipeline which transports a partially-processed coal slurry from the mine to a electrical generating station more than 400 km distant. This line, which involves heads at each pumping section of 75 atmospheres or more, began operation in 1970. Since that time many similar, but larger, coal pipelines have been proposed, but none has yet been built.

Recent decades have seen a great increase in the transport of waste materials, in slurry form, to suitable deposit sites. The concept is an old one and the earliest known use was by Hercules, who removed a decade's accumulation of animal droppings from King Augeas' stables by diverting two rivers to form an open-channel slurry transport system. Waste-disposal problems are even more severe at present, and maintenance of the environment requires that wastes be conveyed to dedicated and monitored disposal sites, either underground or on the surface. This requirement can often be satisfied by backfilling mines (either deep or open-pit), and slurry transport is the favoured placing method. Such backfilling operations are usually characterised by significant tonnages with short to medium hauls, and hence centrifugal pumps are the natural choice for these applications.

Large backfilling operations are found in Alberta, where petroleum is extracted from oil sands which are obtained by open-pit mining operations. After the extraction operations are completed, the sand (actually a mixture of sand, fine particles and some residual organics) is used as backfill in areas previously mined. An example of a high-pressure pump in an oil-sands tailings pipeline (Addie *et al.*, 1995) is shown on Figure 1.4.

Figure 1.4. High-pressure pump for a tar-sands tailings pipeline

1.2 Topics Covered in this Book

Much of the material presented in this book has been developed in connection with a short course presented annually since the early 1970's under the joint sponsorship of GIW Industries and Augusta State University. Three of the authors of the book have been associated with the course since its inception, and the fourth has been involved for many years. The course, and thus this book, also draws from the work of many other contributors, and only the major contributions could be recognized in the Acknowledgements.

Over the years the course has laid stress on the need for experimentation, and the students have always participated in slurry testing at the GIW Hydraulic Laboratory. The data obtained from the experiments carried out for the course are at state-of-the-art level and have often been included in papers presented at international conferences or published in technical journals (from Clift *et al.*, 1982; to Whitlock *et al.*, 2004). Points arising during the course also led to other research (for example Clift & Clift, 1981; Sellgren & Addie, 1989; Wilson, 2004). Such results were incorporated into subsequent offerings of the course, and also in the successive editions of the present text.

The book falls into two basic sections. The first half, up to the end of Chapter 8, is concerned primarily with various sorts of slurry behaviour including scale-up from one pipe size to another and prediction of particle deposition and pressure gradients. The second portion of the book is concerned with the behaviour of centrifugal pumps handling slurries, and with how pumps and pipelines interact as a system.

Within the first half of the book, Chapter 2 leads into the main topics by providing a review of the basic principles of fluid mechanics. This is needed not only as a ready reference source for points on which some readers may wish to brush up, but also to establish a common basis for readers of different backgrounds, e.g., Civil, Mechanical and Chemical engineers. (A common list of symbols is also required, of course. This appears at the end of Chapter 1). Chapter 2 also deals with the settling of solid particles in liquids, which forms a necessary background for slurry flow analysis. As mentioned above, engineers dealing with slurry transport come from various backgrounds. These reflect not only educational discipline, but also different experiences with design or operation of slurry transport systems. The effect of this wide difference in backgrounds has been noted since the earliest offerings of the short course. In fact, although the technical content of the course has advanced considerably over the years, one point has remained constant, and that is the invocation of the old tale of the blind men and the elephant (Saxe, 1872). In this fable six blind men, who had always wanted to know what elephants were like, were finally allowed to spend a few minutes in contact with one of these beasts. Since each man had touched a different part of the animal, they later found themselves in greater disagreement than before, comparing the elephant to everything from a house (based on its side) to a rope (based on its tail); see Figure 1.5.

In order to avoid the mistakes of the blind men it is necessary to elucidate the different aspects of slurry flow. Chapter 3 deals with homogeneous flow of non-Newtonian slurries. It considers both laminar and turbulent flow, mentioning various rheological models and presenting the laminar scaling laws based on the Rabinowitsch-Mooney transform. Equivalent affinity laws are derived for turbulent flow. The effects of solids concentration are outlined, and the laminar-turbulent transition is investigated. This chapter closes with a case study which demonstrates many of the points discussed.

Figure 1.5. The elephant

Examination of the various sorts of slurry flow of settling particles begins in Chapter 4, where the basic distinction between pseudo-homogeneous flow and settling-slurry flow is introduced, and the underlying physical mechanisms are outlined. Significant aspects of the literature are also reviewed here, to provide background for the following four chapters, which present recent work on the analysis of slurry flow.

Chapters 5 and 6 deal with settling slurries in horizontal pipes. In line with the analysis of Chapter 4, the distinction is made between particles supported by granular contact and those supported by the fluid, i.e. contact load and suspended load. Chapter 5 is concerned with contact load, which is analysed by setting up a force balance. The pressure gradient provides the driving force which tends to move the solids, while the resisting force, which tends to hold the particles in place as a deposit, is associated with the submerged weight of the contact-load particles. The force-balance analysis is used to determine the velocity at the limit of stationary deposition (a condition that must be known so that deposition can be avoided). A nomographic chart based on the analysis provides a convenient method of estimating the deposition point, and this is complemented by equations which show the effect of solids concentration and other variables (a variable not included here is pipe slope; it is considered later, in Chapter 8). The force balance of Chapter 5 is also applied to cases where all the particles are in motion, giving rise to a method of calculating pressure drop for coarse-particle transport. This method is illustrated by a case study of large clay balls moving through the pipeline from a dredge.

The heterogeneous flow dealt with in Chapter 6 involves both contact load and suspended load. The fraction of contact load, indicated by the stratification ratio, is found to be very important in determining hydraulic gradients and specific energy consumption for flows of this type. Using a method introduced previously, it is shown how the stratification ratio can be obtained from experimental tests of slurry behaviour and scaled to prototype pipelines. If results of flow tests are not available for the slurry of interest, the parameters in the scaling relations can be estimated from fluid and particle properties, on the basis of recent developments in the understanding of particle support mechanisms. Case studies demonstrate the application of the methods developed in this chapter to pipeline flow of slurries of sand and coal.

Chapter 7 deals with the flow of complex slurries. It first presents an analysis of slurries with very broad particle grading, and then considers the commercially-important case of coarse particles travelling in a non-Newtonian carrier medium, and develops scaling relations for cases of this type.

Chapter 8 considers slurry flow in pipes which are not horizontal. The simplest case to analyse is vertical flow, which is employed in a U-tube instrument for determining solids concentration, and also has direct applications, particularly in the mining industry. In addition to vertical flow, the flow in inclined pipes is covered in this chapter, which includes sections pertaining to the effect of inclination angle on deposition limit and on friction loss. These effects can be very important in inclined suction pipes, as shown in the case study of a suction dredge.

The second half of the book begins by shifting the focus from the slurry to other components of the pipeline transport system, beginning in Chapter 9 with the performance and testing of centrifugal pumps. This chapter discusses characteristic curves, affinity laws, specific speed and suction performance. The basics of hydraulic and mechanical design are considered, indicating some of the differences between slurry pumps and water pumps. Finally, performance testing is discussed, including layout and instrumentation of test loops, and test procedures for determining pump characteristics for total dynamic head, efficiency, and net positive suction head.

Chapter 10 continues the study of pumps by describing the way in which the presence of solid particles influences the head produced by a pump, and its efficiency. These effects tend to become more severe for higher solids concentrations and for larger or denser particles, indicating the basic similarity between the influence of solids on pump performance and on pipeline friction. Scaling equations for pumps show how the solids effect diminishes with increasing machine size. All of these relations are brought together in a generalised solids-effect diagram, and illustrated by worked examples.

Many of the mechanisms by which the solids decrease pump performance also act to promote erosive wear, a subject which is considered in Chapter 11. This chapter introduces the two major wear mechanisms of sliding abrasion and particle impact, showing how they are influenced by features of fluid and particle motion which were considered in previous chapters. From wear mechanisms the discussion turns to wear-resistant materials (hard metals, elastomers and ceramics) and then to experimental methods for wear testing. The numerical modelling of flow and wear patterns is dealt with next, introducing computer techniques and evaluating some of their strengths and weaknesses. The results of practical experience with slurry system design and operation are then presented, and particle attrition is also considered.

Chapter 12 introduces other components of slurry transport systems. The configuration of suction piping merits considerable attention, since difficulties with cavitation can usually be eliminated by careful design of this component. Pump-house layout and spacing are also considered, along with pump drive trains. Accurate instrumentation is required to provide the information needed for system control, and thus the instruments used for monitoring slurry flow are considered in some detail.

Chapter 13 shows how the pump and pipeline characteristics interact. It considers the general form of the characteristics, and the question of the stability of their intercept, which defines system operability, and thus determines pump selection and other aspects of system design. Both the pump characteristic and that for the pipeline are affected by particle size and concentration, and it is shown how these quantities affect system operability, and also how temporal variation of the characteristics of slurry entering the system can cause operating difficulties.

Chapter 14 deals with pump selection and cost considerations, bringing together materials, wear and capital and maintenance costs in terms of the concepts of Life-Cycle Cost and Total Cost of Ownership. Chapter 15 uses the knowledge developed in the previous chapters to deal with practical problems in the design and operation of slurry systems, including both series and parallel pumping configurations. A case-study approach is used in dealing with these problems, together with others involving difficulties with pump suction, start-up transients, and water hammer.

Chapter 16 discusses the environmental aspects of slurry systems. There is a brief introduction to the role of hydraulic transport in mining and mineral processing systems: integration of operations from mine to mill output; comparison of hydraulic conveying with belt conveying and trucks, in terms of energy consumption and other environmental impacts. Current emphasis on tailings management as a way of reducing the environmental impact of mining and mineral processing emphasizes the importance of hydraulic transport at high concentrations and forms the conclusion of the book.

A number of worked examples are included in the text. These require only a hand calculator. The case studies are presented both in SI units and in traditional U.S. units. The SI versions are located in the text, and the versions in U.S. Customary Units are gathered together in the Appendix.

1.3 List of Symbols

The principal symbols used in this book are listed here, together with brief definitions. The basic units of each quantity in terms of mass [M] length [L] and time [T] are given in square brackets. Dimensionless quantities are indicated by [-].

a	celerity of transient disturbance $[LT^{-1}]$
A'	coefficient in Eq. 4.3 [-]. The case where $A' = 1.0$ is the equivalent-fluid model
A_p	maximum projected area of a particle, see Eq. 2.66 $[L^2]$
b	breadth between pump shrouds [L]
B	coefficient in Eq. 4.6, dependent on flow stratification [-]
c_m	absolute fluid velocity in meridional direction $[LT^{-1}]$
c_r	absolute fluid velocity in radial direction $[LT^{-1}]$
c_t	absolute fluid velocity in tangential direction $[LT^{-1}]$
C_D	particle drag coefficient, see Eq. 2.50 [-]
C_r	relative volumetric concentration of solids, C_{vd}/C_{vb} [-]
C_{rm}	value of C_r at which deposition velocity is a maximum [-]
C_v	volumetric concentration (volume fraction) of solids [-]
C_{vb}	volumetric solids concentration in loose-poured bed [-]
C_{vd}	delivered volumetric solids concentration, see Section 2.4 [-]
C_{vi}	*in situ* volumetric solids concentration, see Section 2.4 [-]
C_W	solids concentration by weight (or mass) [-]

d	particle diameter [L]
d*	dimensionless particle diameter, see Eq. 2.57 [-]
d_a	area-equivalent particle diameter, see Eq. 2.66 [L]
d_{50}	mass-median particle diameter [L]
d_{85}	diameter for which 85% (by mass) of the particles are finer [L]
D	diameter (internal diameter of a pipe, diameter of a pump impeller) [L]
D_s	pump suction diameter [L]
E	voltage [V]
E_i	energy coefficient for specific impact wear (J/m^3)
E_{sp}	energy coefficient for specific wear (J/m^3)
f	Moody friction factor, see Eq. 2.29 [-]
f_f	value of f for equal volumetric flow of fluid [-]
f_w	value of f for equal volumetric flow of water [-]
f_{12}	value of f at stratified-flow interface [-]
fn( )	a function
F_D	drag force [MLT^{-2}]
F_L	lift force [MLT^{-2}]
F_N	normal intergranular force against pipe wall [MLT^{-2}]
F_W	immersed weight force [MLT^{-2}]
g	gravitational acceleration [LT^{-2}]
G	coefficient in Eq. 3.18
h	height [L]
h_a	atmospheric head (in height of liquid being pumped) [L]
h_v	vapour-pressure head [L]
H	head, see Eq. 2.9 [L]
H_A, H_B	head at locations, A, B. [L]
H_i	theoretical head, see Eq. 9.10 [L]
H_r	head ratio, see Section 10.1 [-]
H_s	suction head [L]
i	hydraulic gradient [height of water/length of pipe] see Eq. 2.44
i_f	value of i for equal volumetric flow of fluid [height of water/length of pipe]
i_m	value of i for flow of mixture [height of water/length of pipe]
i_{mh}	value of i for flow of homogeneous mixture [height of water/length of pipe]
i_{pg}	hydraulic gradient for plug flow, see Eq. 5.8 [height of water/length of pipe]
$\bar{i}_v$	average hydraulic gradient in ascending and descending vertical flows [height of water/length of pipe]

i_w	value of i for equal volumetric flow of water [height of water/length of pipe]
i_w'	hydraulic gradient in return water line [height of water/length of pipe]
I	electric current [A]
j_m	friction gradient in height of mixture/length of pipe, see Eq. 2.45
k	coefficient in power-law rheological model
K	volumetric shape factor of particles (Eq. 2.67) [-] *or* McElvain's, head-reduction coefficient (Eq. 10.3) [-]
L	length, measured along pipe [L]
m	exponent in Eq. 3.18 [-]
M	exponent in stratification-ratio equation [-]
n	angular speed in revolutions per second $[T^{-1}]$ *or* Richardson-Zaki exponent, see Eq. 2.69 [-] *or* exponent in power-law rheological model [-]
n'	Rabinowitsch-Mooney exponent [-]
n_s	dimensionless specific speed, see Eq. 9.14 [-]
N	angular speed in revolutions per minute
N_s	specific speed in customary units, see Eq. 9.15
NPSH	net positive suction head, see Eq. 9.16 [L]
p	pressure $[ML^{-1}T^{-2}]$
p_A, p_B	pressure at locations A, B. $[ML^{-1}T^{-2}]$
p_v	vapour pressure $[ML^{-1}T^{-2}]$
P	power $[ML^2T^{-3}]$
Q	volumetric flow rate $[L^3T^{-1}]$
Q_{BEP}	flow rate at best efficiency point $[L^3T^{-1}]$
Q_f	volumetric flow rate of fluid $[L^3T^{-1}]$
Q_m	volumetric flow rate of mixture $[L^3T^{-1}]$
Q_s	volumetric flow rate of solids $[L^3T^{-1}]$
r	radius [L]
R_H	head reduction factor, see Eq. 10.3 [-]
R_{T3}	representative shell radius, see Fig. 11.4 [L]
$R\eta$	efficiency reduction factor, see Eq. 10.3 [-]
Re	Reynolds number VD/v [-]
Re_*	shear Reynolds number U_*D/v [-]
Re_p	particle Reynolds number v_td/v [-]
S_f	relative density of fluid [-]
S_m	relative density of mixture [-]
S_{md}	delivered relative density of mixture [-]
S_{mi}	*in situ* relative density of mixture [-]
S_s	relative density of solids [-]

SDEC	specific delivered energy consumption, i.e. electrical energy drawn from the grid (kWh/tonne-mile)
SEC	specific energy consumption, see Eq. 4.9 (kWh/tonne-km)
t	time [T]
T	torque $[ML^2T^{-2}]$
TDH	total dynamic head, see Eq. 2.10 [L]
u	local velocity or velocity of moving part $[LT^{-1}]$
u_{max}	maximum local velocity (at pipe centreline) $[LT^{-1}]$
u_1	reference value of local velocity $[LT^{-1}]$
u^+	dimensionless velocity U/U_* [-]
U	velocity $[LT^{-1}]$
U_*	shear velocity, $\sqrt{(\tau_o/\rho)}$ $[LT^{-1}]$
U_{*f}	shear velocity for equivalent flow of fluid $[LT^{-1}]$
U_1	mean velocity in upper portion of stratified flow
U_2	mean velocity in lower portion of stratified flow
v_t	terminal settling velocity of a single particle $[LT^{-1}]$
v_t'	hindered settling velocity $[LT^{-1}]$
v_t^*	dimensionless settling velocity [-]
v_{ts}	settling velocity of spherical particle $[LT^{-1}]$
V	mean velocity, $4Q/(\pi D^2)$ $[LT^{-1}]$
V_A, V_B	values of V at locations A, B $[LT^{-1}]$
V_f	mean velocity of fluid $[LT^{-1}]$
V_m	mean velocity of mixture $[LT^{-1}]$
V_N	value of V_m for equivalent Newtonian flow $[LT^{-1}]$
V_r	relative velocity V_m/V_{sm} [-]
V_s	value of V_m at limit of deposition $[LT^{-1}]$
V_{sm}	maximum value of V_s $[LT^{-1}]$
V_T	value of V_m at laminar-turbulent transition $[LT^{-1}]$
V_{50}	value of V_m at which 50% of solids are suspended by fluid $[LT^{-1}]$
w	particle-associated velocity, see Eq. 6.2 $[LT^{-1}]$ *or* velocity of fluid relative to impeller vane $[LT^{-1}]$$w_{50}$ value of particle-associated velocity for d_{50} $[LT^{-1}]$
w_{85}	value of particle associated velocity for d_{85} $[LT^{-1}]$
W_c	wear coefficient, inverse of E_{sp} (m^3/J)
x	distance, usually measured along pipe [L]
X_f	fraction of fine solids, see Section 7.2
X_h	fraction of heterogeneous solids, see Section 7.2
X_p	fraction of pseudo-homogeneous solids, see Section 7.2
X_s	fraction of stratified solids, see Section 7.2
y	distance measured perpendicular to flow boundary [L]

y^+ dimensionless form of y, see Eq. 2.26 [-]
Y specific value of y [L]
z distance measured vertically [L]
z_A, z_B values of z at locations A, B [L]

α angle defining position in pipe, see Fig. 5.1 [-]
β angle defining interface in stratified flow, see Fig. 5.1 [-] *or* vane angle of pump [-]
γ exponent used in Section 8.5 *or* exponent used in Section 10.2
δ sub-layer thickness [L]
ε roughness height [L]
ζ relative excess pressure gradient, see Eq. 5.12 [-]
η efficiency of a machine, see Eq. 9.5 [-]
η_B slope of Bingham rheological model [$ML^{-1}T^{-1}$]
η_D efficiency of drive train [-]
η_P efficiency of pump [-]
η_r efficiency ratio, See Section 10.1 [-]
η_t 'tangent viscosity' of non-Newtonian [$ML^{-1}T^{-1}$]
θ angle of inclination of pipe [-] *or* electrical phase angle [-]
κ von Karman's coefficient for mixing length [-]
λ exponent used in Section 10.4 [-]
Λ lag ratio of solids, see Eq. 2.42 [-]
μ shear viscosity (dynamic viscosity) of fluid [$ML^{-1}T^{-1}$]
μ_{eq} equivalent turbulent-flow viscosity, see Eq. 3.6 [$ML^{-1}T^{-1}$]
μ_s mechanical friction coefficient of solids against pipe wall [-]
ν kinematic viscosity of fluid [L^2T^{-1}]
ξ ratio v_t/v_{ts}, see Eq. 2.68 [-]
ρ density [ML^{-3}]
ρ_f density of fluid [ML^{-3}]
ρ_m density of mixture [ML^{-3}]
ρ_{md} density of mixture delivered [ML^{-3}]
ρ_{mi} density of mixture *in situ* [ML^{-3}]
ρ_s density of solids [ML^{-3}]
ρ_w density of water [ML^{-3}]
σ cavitation parameter for hydraulic machinery [-]
σ_f component of σ_{10} due to fluid turbulence, see Eq. 6.6 [-]
σ_s normal stress of solids (granular pressure), see Section 5.2 [$ML^{-1}T^{-2}$] *or* component of σ_{10} associated with particle grading, see Eq. 6.6 [-]
σ_{10} standard deviation of log (V_m/V_{50}), see Eq. 6.6 [-]
τ shear stress [$ML^{-1}T^{-2}$]

τ_o	shear stress at flow boundary [$ML^{-1}T^{2}$]
τ_s	intergranular component of shear stress [$ML^{-1}T^{-2}$]
τ_y	yield-point shear stress [$ML^{-1}T^{-2}$]
τ_{12}	shear stress at stratified-flow interface [$ML^{-1}T^{-2}$]
φ	angle of internal friction of particles (static) [-]
φ'	angle of internal friction of particles (dynamic) [-]
Φ	Durand variable, see Eq. 4.4 [-]
ψ	Head coefficient for a pump [-]
Ψ	Durand variable, see Eq. 4.5 [-]
ω	angular velocity in radians per second [T^{-1}]

REFERENCES

Addie, G.R. & Helmly, F.W. (1989). Recent improvements in dredge pump efficiencies and suction performances. *Proc. Europort Dredging Seminar*, Central Dredging Association.

Addie, G., Dunens, E. & Mosher, R. (1995). Performance of high pressure slurry pumps in oil sand tailing application, *Proc. 8th International Freight Pipeline Society Symposium,* Pittsburg, PA.

Clift, R., & Clift, D.H.M. (1981). Continuous measurement of the density of flowing slurries. *Int. J. Multiphase Flow*, Vol. 7, No. 5, pp. 555-561.

Clift, R., Wilson, K.C., Addie, G.R., & Carstens, M.R. (1982). A mechanistically-based method for scaling pipeline tests for settling slurries. *Proc. Hydrotransport 8*, BHRA Fluid Engineering, Cranfield, UK, pp. 91-101.

Saxe, J.G. (1872). The blind men and the elephant. in *Fables and Legends of Many Countries Rendered in Rhyme.* Republished in *The Home Book of Verse* (Ed. B.E. Stevenson) Holt, Reinhart and Winston, New York, 9th Ed. (1953) Vol. 1 pp. 1877-1879.

Sellgren, A. & Addie, G.R. (1989). Effect of Solids on large centrifugal pump head and efficiency. Paper presented at *CEDA Dredging Day*, Amsterdam, The Netherlands.

Whitlock, L., Wilson, K.C. & Sellgren, A. (2004). Effect of near-wall lift on frictional characteristics of sand slurries. *Proc. Hydrotransport 16*, BHR Group, Cranfield, UK, pp. 443-454.

Wilson, K.C. (2004). Energy consumption for highly-concentrated particulate slurries, *Proc. 12th Int'l Conf. On Trannsport & Sedimentation of Solid Particles,* Prague, Sept. 20-24.

Chapter 2

REVIEW OF FLUID AND PARTICLE MECHANICS

2.1 Introduction

Before considering the flow of mixtures of liquids and solid particles, we must first treat the flow of single-phase liquids and the motion of particles in liquids. These topics also provide an introduction to concepts, terminology and notation used in later chapters. The treatment in this chapter is at the level of a review, intended primarily to reinforce knowledge which readers will have encountered in their undergraduate engineering curriculum, but may have not utilized in the interim. As in other parts of this book, the level of presentation is directed to practical engineering application. This chapter is not intended as an introduction to fluid mechanics in general, to turbulence, or to the micromechanics of particle-fluid systems. All these subjects have significance for fundamental research, but they are not required for an engineering treatment of slurry flow.

To characterise a simple fluid such as water only two material properties are required: density and viscosity. The density, denoted by ρ, represents the mass of fluid per unit volume. The viscosity is a measure of the resistance of the fluid to deformation by shearing, and is best illustrated by reference to a conceptual experiment illustrated in Fig. 2.1. The fluid fills a gap of thickness y between two flat parallel plates. The plate forming the y = 0 plane is kept stationary, while the other plate (at y = Y) is moved parallel to

the first at a steady velocity U. The fluid immediately in contact with each solid plate keeps the same velocity as the plate; this is known as the 'non-slip boundary condition'. Thus the velocity of the fluid at y = 0 is zero and at y = Y is U. The fluid velocity, u, at any intermediate position is yU/Y. The quantity U/Y is the velocity gradient in the fluid, denoted by du/dy. Its significance is that it represents the rate of shear deformation of the fluid; it is therefore known as the 'rate of shear strain' or, more simply, the 'shear rate'.

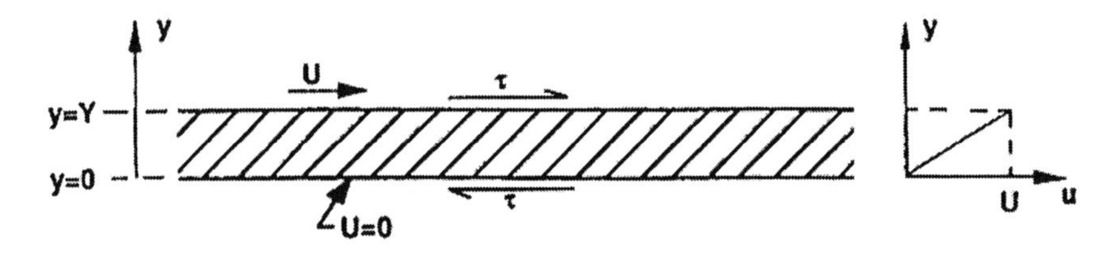

Figure 2.1. Shear deformation of a fluid (schematic)

To maintain the steady motion in Fig. 2.1, it is necessary to apply a force to the moving plate and an equal and opposite restraining force to the stationary plate. These forces are parallel to the plates, and in the direction of their relative motion. The force on one plate per unit area is known as the 'shear stress', denoted by τ. As shown in Fig. 2.1, it represents the shear stress exerted by each plate on the fluid in the gap. Simple Newtonian mechanics dictates that the fluid exerts an equal and opposite shear stress on the plate, and that the shear stress within the fluid at any plane parallel to the plates is also τ.

Conceptually, an experiment like that in Fig. 2.1 could be used to measure τ as a function of du/dy. In practice, flat plates are inconvenient; practicable ways to measure viscosity are introduced in Chapter 3. For a simple fluid like water, the relationship between τ and shear rate takes the form shown in Fig. 2.2, with τ linearly proportional to du/dy. This type of relationship defines a 'Newtonian fluid', and the constant of proportionality is known as the 'shear viscosity' (or simply the 'viscosity') and will be denoted by μ. Thus

$$\tau = \mu \frac{du}{dy} \tag{2.1}$$

Equation 2.1 is the 'constitutive equation' of a Newtonian fluid. The ratio μ/ρ is known as the 'kinematic viscosity' of the fluid, denoted by ν.

Values for the density and viscosity of water are given in Table 2.1. For practical purposes, over the range of conditions encountered in the industries using hydraulic transport, liquids are incompressible so that ρ and μ can be taken as independent of pressure. However they both depend on temperature. Specifically, the viscosity of water decreases significantly with increasing temperature.

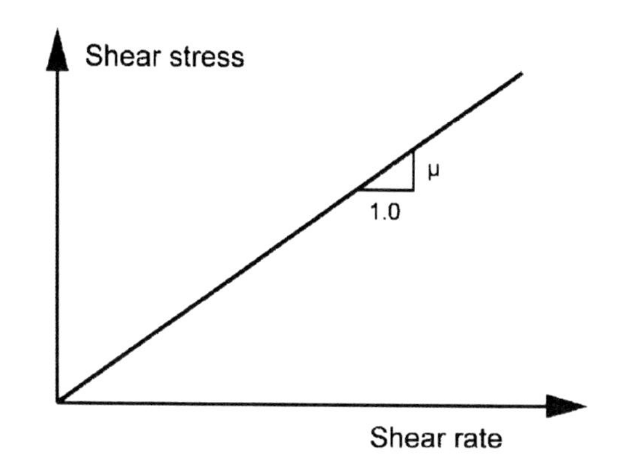

Figure 2.2. Characterisation of a Newtonian fluid (schematic)

Table 2.1. Properties of Water

Temperature (°C)	Density, ρ (kg/m^3)	Viscosity, $\mu \times 10^3$ (Pa.s)	Kinematic viscosity, $\nu \times 10^6$ (m^2/s)	Vapour pressure, ρ_v (kPa)
0	999.8	1.781	1.785	0.61
5	1000.0	1.518	1.519	0.87
10	999.7	1.307	1.306	1.23
15	999.1	1.139	1.139	1.70
20	998.2	1.002	1.003	2.34
25	997.0	0.890	0.893	3.17
30	995.7	0.798	0.800	4.24
40	992.2	0.653	0.658	7.38
50	988.0	0.547	0.553	12.33
60	983.2	0.466	0.474	19.92
70	977.8	0.404	0.413	31.16
80	971.8	0.354	0.364	47.34
90	965.3	0.315	0.326	70.10
100	958.4	0.282	0.294	101.33

2.2 Basic Relations for Flow of Simple Fluids

Much of this book is concerned with steady motion, in which the mean velocity at any point does not change with time (although it may vary with location). Whether the flow is steady or unsteady, analysis is based on three fundamental laws: the continuity balance (or conservation of matter); linear or angular momentum balances (which amount to the application of Newton's laws to fluids); and the mechanical energy balance (which is essentially the first law of thermodynamics applied to fluids). Some simple applications of these laws to solid-liquid mixtures are considered in Section 2.4. For steady flow of an incompressible fluid in a pipe or conduit, the continuity equation states simply that the volumetric flow is the same through each section across the pipe. Consider a pipe in which the diameter changes between sections A and B, as shown schematically in Fig. 2.3. If the total volumetric flowrate in the pipe is Q, and the pipe is taken to be 'running full', then the mean velocity at section A is

$$V_A = 4Q/\pi D_A^2 \tag{2.2}$$

where D_A is the internal pipe diameter at A, so that the pipe's cross-sectional area is $\pi D_A^2/4$. Similarly, at section B the mean velocity is

$$V_B = 4Q/\pi D_B^2 \tag{2.3}$$

The equation of continuity for this section of pipe then takes the form

$$Q = \pi D_A^2 V_A/4 = \pi D_B^2 V_B/4 \tag{2.4}$$

For cases in which it is necessary to consider the local velocity in the pipe at distance r from the axis, i.e. u(r), then the total volumetric flow is evaluated from the integral

$$Q = \int_0^{D/2} u2\pi r dr = 2\pi \int_0^{D/2} u r dr \tag{2.5}$$

in which the area of the element from r to (r+dr) is $2\pi r dr$ and the velocity through it is u. Specific applications of Eq. 2.5 are considered below, and in Chapter 3.

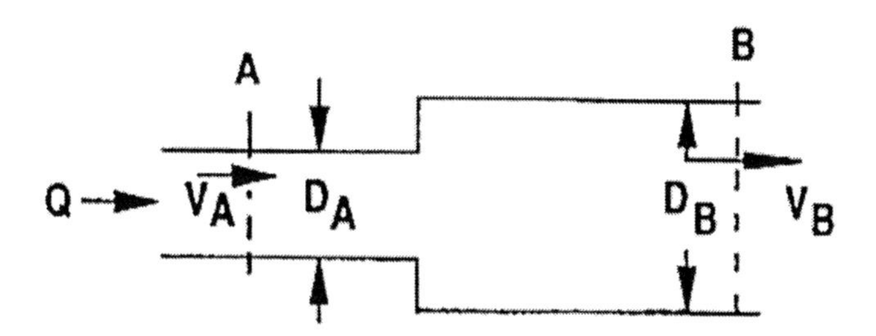

Figure 2.3. Flow in a pipe with change in diameter (schematic)

We next turn to the linear momentum equation, considering the simple case illustrated by Fig. 2.4: a fluid in steady motion through a straight horizontal pipe of constant diameter. Flow is 'fully developed', i.e. conditions do not vary between positions along the pipe. Therefore the shear stress exerted by the pipe walls on the fluid, τ_o, is the same at all sections. Between two sections A and B, a distance L apart, the total area of pipe wall is πDL so that the total force exerted by the pipe walls on the fluid is $\pi DL\tau_o$. The linear momentum equation applied to this case of steady uniform fully-developed flow states that the total force on the fluid between sections A and B must be zero, (because the momentum flux across section A is equal to that across section B), giving

$$\frac{\pi D^2}{4}(p_A - p_B) + \pi DL\tau_o = 0 \qquad (2.6)$$

where p_A and p_B are the (static) pressures in the fluid at the two sections. Rearranging Eq. 2.6

$$-\frac{dp}{dL} = -\frac{(p_A - p_B)}{L} = \frac{4\tau_o}{D} \qquad (2.7)$$

or

$$\tau_o = -\frac{(p_A - p_B)D}{4L} \tag{2.8}$$

Equations 2.7 and 2.8 apply whether or not the fluid is Newtonian (i.e. obeys Eq. 2.1).

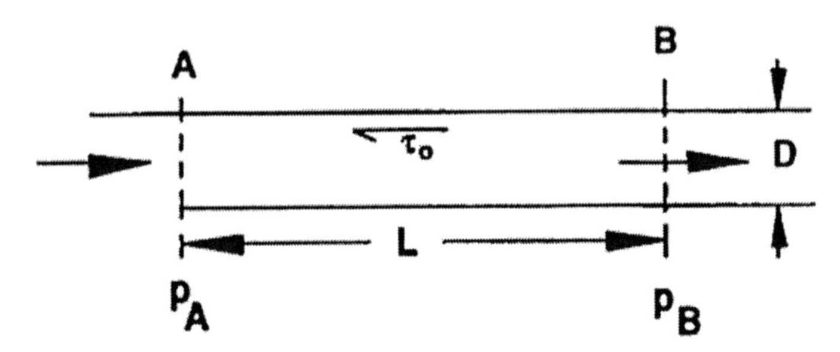

Figure 2.4. Flow in a straight horizontal pipe of constant diameter (schematic)

Another basic equation, for angular momentum, is needed for the analysis of centrifugal pumps. It will be presented in that context in Chapter 9. The energy equation is the next to be dealt with here. The mechanical energy balance for a flowing fluid is usually written in the form known as 'Bernoulli's equation'. It will be given here in terms of the 'head' of the fluid. 'Head' is a concept used extensively by Civil and Mining Engineers, and much of the literature on hydraulic conveying is written in terms of head and 'hydraulic gradient' (see below). Head is a measure of the mechanical energy of a flowing fluid per unit mass. It indicates the height by which the fluid would rise if the energy were converted to potential energy, and therefore has the units of length. The 'total dynamic head' of a fluid of density ρ flowing at velocity V in a pipe at elevation z above a reference level and at pressure p is

$$H = \frac{V^2}{2g} + \frac{p}{\rho g} + z \tag{2.9}$$

where the first term in this expression represents the kinetic energy of the fluid ('velocity head'), the second results from the static pressure in the fluid ('pressure head'), and the final term is the elevation ('static head').

Now consider a liquid propelled by a centrifugal pump, as shown in Fig. 2.5. Upstream of the pump, at section A on the pump suction, the total dynamic head of the fluid is H_A. Similarly, at section B on the pump discharge, the total dynamic head is H_B. The increase in head across the pump, which is a simple measure of the energy imparted to the fluid by the pump, is known as the 'total developed head', TDH; i.e.

$$TDH = H_B - H_A = \frac{V_B^2 - V_A^2}{2g} + \frac{p_B - p_A}{\rho g} + (z_B - z_A) \tag{2.10}$$

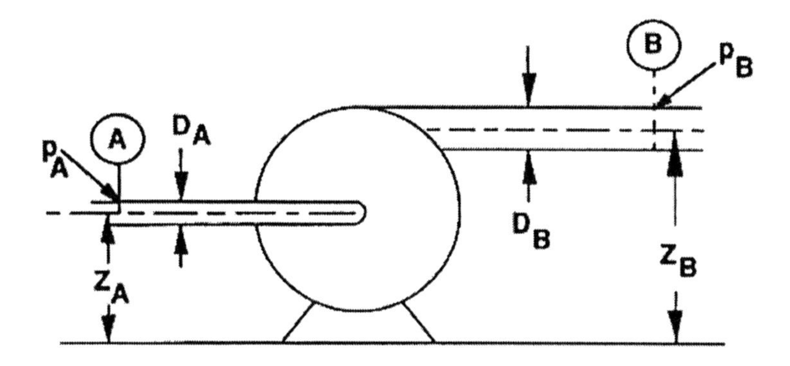

Figure 2.5. Fluid passing through a centrifugal pump (schematic)

In terms of the total volumetric flow rate, Q, and the pipe diameters at the two sections, D_A and D_B, (Eq. 2.2 and 2.3), the TDH equation becomes

$$TDH = \frac{8Q^2}{\pi^2 g}\left(\frac{1}{D_B^4} - \frac{1}{D_A^4}\right) + \frac{(p_B - p_A)}{\rho g} + (z_B - z_A) \tag{2.11}$$

The second term in Eq. 2.11, the 'pressure head', is normally by far the largest. The 'static head' term (z_B - z_A) is usually relatively small, and depends on the pump geometry and dimensions. Centrifugal pumps are sometimes made with different suction and discharge diameters (see Chapter 8), so that the first or 'velocity head' term can become significant at large flowrates.

For a fluid flowing in a straight horizontal pipe of constant diameter, the only head term which varies along the pipe is that arising from changes in pressure. For the flow shown in Fig. 2.4, the reduction in head per unit length of pipe is known as the 'hydraulic gradient':

$$i = -\frac{(p_A - p_B)}{\rho g L} = \frac{4\tau_o}{\rho g D} \tag{2.12}$$

Here i has been related to the wall shear stress, τ_o, using Eq. 2.7. The units of the hydraulic gradient are (m head lost)/(m pipe run), or feet of head per foot of pipe. Thus the numerical value of i is independent of the system of units used. However, i is not strictly a dimensionless number: for example, if the flow took place on the moon or in any other environment of changed gravity, the value of i would be different even though τ_o and the pressure gradient were unchanged.

A simple illustration of the significance of i is given by Fig. 2.6. If we imagine that 'sight glasses' are attached to the pipe - i.e. vertical open-ended transparent tubes - then the height to which the fluid rises in each sight glass shows the pressure head inside the pipe at that point. For steady fully-developed pipe flow, with constant hydraulic gradient and pressure gradient, the levels in sight glasses set along the pipe will lie on a straight line. This is known as the 'hydraulic grade line', and its inclination is the hydraulic gradient, i.

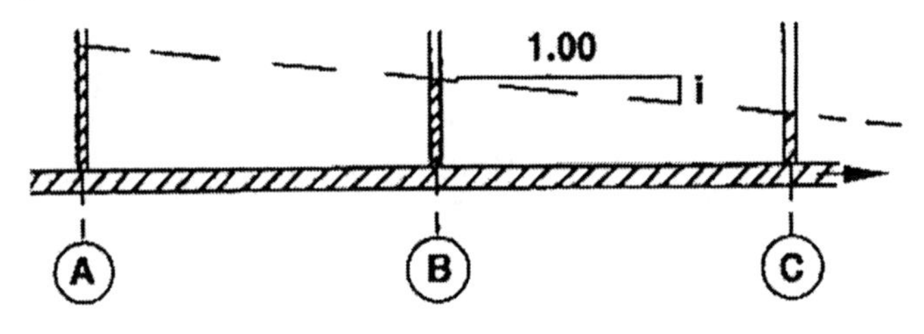

Figure 2.6. Significance of hydraulic gradient (schematic)

Consider now a pipe at an angle to the horizontal, as shown in Fig. 2.7. The pressure difference is measured between two sections, A and B, by connecting the pipe at these points to a manometer or a differential pressure sensor. Point B is Δz above A. The sensor is h above A, and therefore (Δz-h) below B, and the connections from the sensor to the pipe are filled with the same fluid as in the pipe, of density ρ. If the pressure in the pipe at A is p_A, then the pressure on the upstream side of the sensor is lower by the hydrostatic column h:

$$p_A' = p_A - \rho g h \tag{2.13}$$

Similarly, the pressure on the other side of the sensor is

$$p_B' = p_B + \rho g(\Delta z - h) \qquad (2.14)$$

Thus the pressure difference actually measured will be

$$\Delta p = p_A' - p_B' = p_A - (p_B + \rho g \Delta z) \qquad (2.15)$$

i.e. the sensor will measure the change in the combination of pressure head and static head between A and B. For the case where the pipe is of constant section this is equal to the change in total head, i.e. $i\rho gL$ where L is the distance from A to B. Thus a simple manometric measurement such as that illustrated in Fig. 2.7 serves to measure the hydraulic gradient.

Standard methods for predicting i for Newtonian fluids are given in the following Section. For slurry flow very careful definitions of hydraulic gradient are required. These will be introduced in Section 2.4.

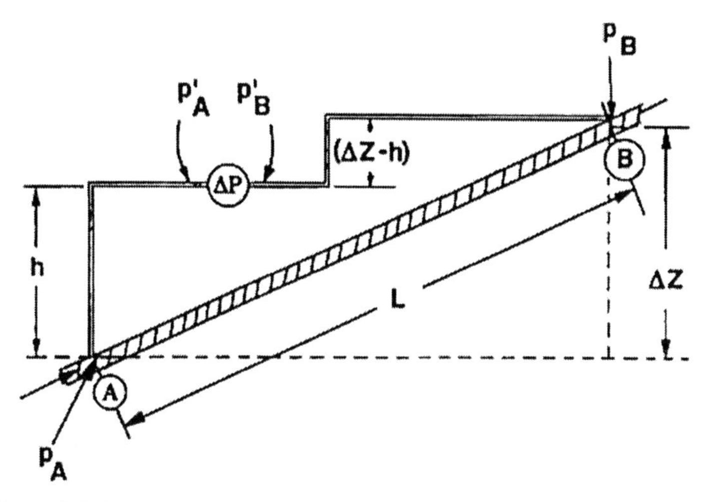

Figure 2.7. Measurement of hydraulic gradient in an inclined pipe (schematic)

2.3 Friction in Laminar and Turbulent Flow of Simple Fluids

Figure 2.8 shows a fluid flowing along a straight horizontal cylindrical pipe under the action of the pressure gradient dp/dx, where x is distance measured along the pipe. From Eq. 2.8, the shear stress at the pipe wall is

$$\tau_o = \frac{D}{4}\left(-\frac{dp}{dx}\right) \tag{2.16}$$

A similar force balance on the fluid within the cylinder of radius r coaxial with the pipe shows that the shear stress varies across the pipe section as shown in Fig. 2.8, and at radius r the shear stress is

$$\tau = \frac{r}{2}\left(-\frac{dp}{dx}\right) \tag{2.17}$$

It is now necessary to introduce an important distinction between two modes of flow of a fluid. In *laminar* flow, each element of the fluid moves on a steady path; in the case of flow in a pipe, all these paths are straight and parallel to the axis. In general, this type of motion occurs when viscous effects in the fluid predominate. In *turbulent* flow, elements of fluid follow irregular fluctuating paths caused by moving eddies. Thus, although the average or 'mean' velocity at any point within the fluid is parallel to the wall, the instantaneous velocity fluctuates in both magnitude and direction. In general, turbulent flow occurs when inertial effects predominate. For water in pipes of industrial scale, the flow is invariably turbulent. However, laminar flow can be important for non-settling slurries, which are discussed in Chapter 3.

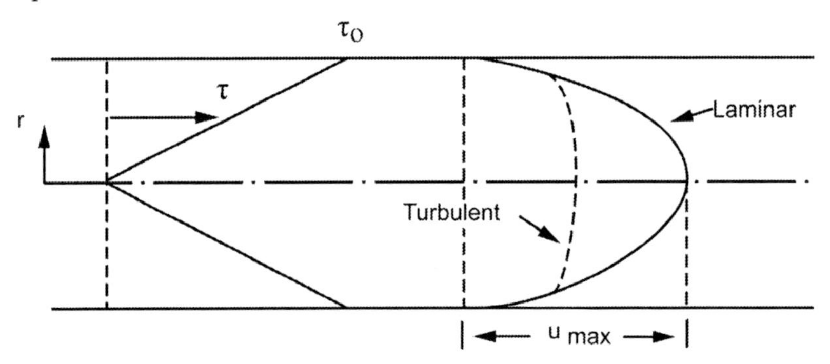

Figure 2.8. Stress and velocity distributions in pipe flow

Equation 2.17 applies whether the flow is laminar or turbulent. However, the resulting velocity profile in the pipe differs between the two types of flow. Consider first a Newtonian fluid in laminar flow, for which Eq. 2.1 applies in the form

$$\tau = \mu \left(-\frac{du}{dr} \right) \tag{2.18}$$

where u is the velocity at radius r. From Eq. 2.17 and 2.18,

$$\frac{du}{dr} = \frac{r}{2\mu} \left(\frac{dp}{dx} \right) \tag{2.19}$$

To obtain the velocity profile, Eq. 2.19 is integrated with the two conditions:

i. by symmetry, the velocity gradient is zero on the pipe axis, i.e. du/dr = 0 at r = 0;

ii. from the no-slip condition, the velocity is zero at the wall, i.e. u = 0 at r = D/2.

The resulting velocity profile takes the characteristic parabolic form, shown in Fig. 2.8 and given by

$$u = u_{\max} \left(1 - \frac{4r^2}{D^2} \right) \tag{2.20}$$

Fluids with other than Newtonian properties have slightly different velocity profiles, to be mentioned in Chapter 3. In Eq. 2.20, u_{max} is the maximum velocity in the pipe, which occurs on the axis and is given by

$$u_{\max} = \frac{D \tau_o}{4\mu} = \frac{D^2}{16\mu} \left(-\frac{dp}{dx} \right) \tag{2.21}$$

Using Eq. 2.5, the total flowrate in the pipe is obtained as

$$Q = \frac{\pi D^2}{8} u_{\max} \tag{2.22}$$

Thus the mean velocity in the pipe, as defined in Eq. 2.2, is

$$V = 4Q/\pi D^2 = u_{\max}/2 \tag{2.23}$$

i.e. for a Newtonian fluid in laminar flow in a cylindrical pipe, the maximum velocity is twice the mean velocity. Finally, from Eq. 2.21 and 2.22

$$Q = \frac{\pi D^4}{128\mu}\left(-\frac{dp}{dx}\right) \tag{2.24}$$

and

$$\tau_0 = \mu\left(\frac{8V}{D}\right) \tag{2.25}$$

Equation 2.25 will be generalised in Chapter 3 for non-Newtonian fluids.

If the fluid is in turbulent flow, Eq. 2.18 no longer applies, because the fluctuations exchange slow- and fast-moving fluid across surfaces within the flow. The effect of this momentum transfer is to set up stresses, known as 'Reynolds stresses', which dominate over the purely viscous stresses except near the walls. As a result, the velocity profile takes the form shown schematically in Fig. 2.8, rather flat in the central core of the flow but with a large velocity gradient in the wall region.

A basic parameter in turbulent flow is the group $\sqrt{\tau_o/\rho}$ which has the dimensions of velocity. It is known as the 'shear velocity' and is denoted by U_*. The velocity fluctuations associated with turbulent eddies have the same order of magnitude as the shear velocity. In fully turbulent flow the velocity gradient du/dy is directly proportional to U_* and inversely proportional to the 'mixing length'. This length is related to the size of the turbulent eddies, and for turbulent flow near a pipe wall the mixing length is evaluated as κy where y is the distance from the wall and κ is von Karman's coefficient. The value $\kappa = 0.4$ is often employed, and will be used here, giving the velocity gradient as $2.5U_*/y$.

The local velocity, u, obtained by integration, varies with the logarithm of y. If the pipe wall is 'hydraulically smooth', the velocity distribution is given by

$$\frac{u}{U_*} = 2.5\ell n\left[\frac{yU_*}{\nu}\right] + 5.5 \tag{2.26}$$

where ln indicates the natural logarithm. The left hand side of this equation is a dimensionless velocity, denoted u^+, while the ratio yU_*/ν is a dimensionless distance from the wall, denoted y^+. At small values of y^+ the logarithmic velocity law of Eq. 2.26 will not apply, because immediately adjacent to a smooth wall there is a 'sub-layer' where viscous stresses are more important than Reynolds stresses and the flow can be considered to be laminar. As viscosity is dominant here, du/dy must equal τ_o/μ, by Eq. 2.1. This relation gives a linear variation of u with y, equivalent to the statement that u^+ equals y^+ at small values of y^+.

This velocity relation extends from the smooth wall at $y^+=0$ out to about $y^+ = 5$ (see Fig. 2.9). Here turbulence first begins to be felt. The 'buffer layer' (from $y^+ = 5$ to a value of y^+ defined by various authors as between 25 and 50) is characterised by a gradual shift to fully turbulent flow, so that Eq. 2.26 becomes strictly valid only for y^+ greater than the buffer-layer limit. Additional information on velocity profiles may be found in Kay & Nedderman (1985) and Reynolds (1974). Although mathematical analysis of the buffer layer may be required in sophisticated treatments of turbulence, for most engineering purposes it is sufficient to employ a simplified treatment in which the transition to turbulent behaviour is assumed to occur abruptly at the point where Eq. 2.26 intersects the linear velocity relation applicable near the wall. At this point both y^+ and u^+ equal 11.6. In effect, the sub-layer is assumed to extend from the wall to $y^+ = 11.6$, and Eq. 2.26 is used for all larger values of y^+. The thickness δ of this sub-layer (referred to as the viscous sub-layer) is given by

$$\delta = \frac{11.6\,\mu}{(\rho_f \tau_0{}^{1/2})} = \frac{11.6\,\mu}{\rho_f V}\left(\frac{8}{f}\right)^{1/2} \tag{2.27}$$

where f is the friction factor defined in Eq. 2.29 below. Thus the sub-layer becomes thinner as the flow in the pipe is increased.

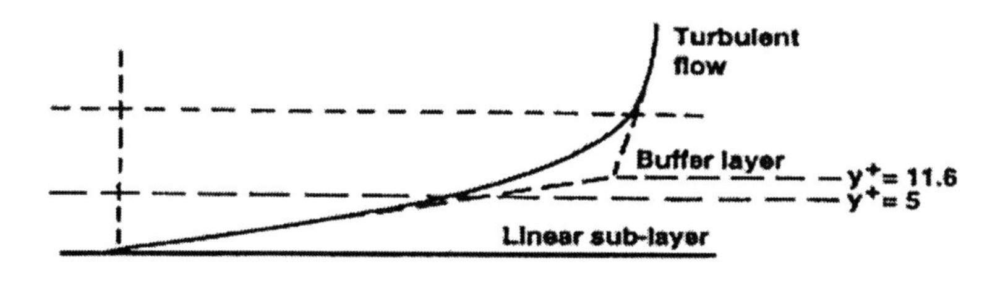

Figure 2.9. Velocity profile near wall in turbulent pipe flow

As the axis of the pipe is approached, the observed velocity profile diverges somewhat from the logarithmic law of Eq. 2.26. Detailed models of turbulent flow take this divergence into account, but for present purposes the logarithmic law is adequate for all of the flow except the viscous sub-layer. As this layer usually occupies a very small portion of the pipe area (See Example 2A, below), the average velocity V (the discharge divided by the section area $\pi D^2/4$) can be obtained by integrating Eq. 2.26. The result may be written

$$\frac{V}{U_*} = 2.5\ell n\left(\frac{U_* D}{\nu}\right) \tag{2.28}$$

Both dimensionless groups in this equation merit careful consideration. The ratio V/U_* can be expressed directly in terms of the dimensionless group known as the 'friction factor'. In this book we use the Moody form of the friction factor, defined as

$$f = 8\tau_o/\rho V^2 \tag{2.29}$$

This is the definition commonly used by Civil and Mechanical engineers, whereas Chemical engineers may be more familiar with the Fanning friction factor, equal to f/4. From Eq. 2.29, it follows that $V/U_* = (8/f)^{1/2}$, or

$$U_* = V(f/8)^{1/2} \tag{2.30}$$

The dimensionless quantity U_*D/ν found in Eq. 2.28 is called the 'shear Reynolds number' Re_* which, like the better known 'pipe Reynolds number' Re (i.e. VD/ν), gives an indication of the state of flow. For cases where the pressure gradient, pipe diameter and fluid properties are known, U_* and Re_* can be determined immediately, and the mean velocity V is found by substituting these quantities into Eq. 2.28.

In cases where V and Re are known and U_* (and the pressure gradient) are required, use is made of the relation

$$Re_* = \frac{U_* D}{\nu} = \left(\frac{f}{8}\right)^{1/2}\frac{VD}{\nu} = Re\left(\frac{f}{8}\right)^{1/2} \tag{2.31}$$

On this basis Eq. 2.28 takes the form

$$(8/f)^{1/2} = 2.5\ell n\left(Re\,(f/8)^{1/2}\right) \tag{2.32}$$

With Re known, this equation can be solved for f. Although iteration is required, the range of f is small and the solution can be obtained quickly.

It should be noted that Eq. 2.32 (like Eq. 2.28, from which it is derived) applies only to turbulent flow with 'hydraulically smooth' pipe walls. This type of behaviour does not require asperities to be completely absent from the pipe wall, merely that the size, ε, of the typical roughness asperity is too small to penetrate the laminar sub layer and influence the turbulent portion of the flow. For larger values of ε the relative roughness, ε/D, is a significant parameter influencing pipe friction. For 'fully rough' pipes the viscous sub- layer is hidden between the asperities on the pipe wall, so that the roughness interacts directly with the turbulent flow. In this case viscosity is no longer important and the friction equation depends on ln(D/ε) instead of $\ln(Re_*)$. An appropriate transition function which incorporates both 'smooth' and 'rough' behaviour as limiting cases is given by the Colebrook-White equation. Rearranged somewhat (Streeter and Wylie, 1979) it may be expressed as

$$\frac{V}{U_*} = (8/f)^{1/2} = -\,2.43\ell n\left[\frac{\varepsilon/D}{3.7} + \frac{2.51}{Re(f)^{1/2}}\right] \tag{2.33}$$

Here the inverse of the von Karman coefficient is assigned the value of 2.43, not significantly different from the 2.5 used in other friction equations. As was the case for Eq. 2.28, if the pressure gradient is known (together with D, ε, and ν) Eq. 2.33 allows the mean velocity to be calculated directly. If Re is known, together with ε/D, the equation can readily be iterated for f. The same result can be obtained directly from a graph of the relationship given by Eq. 2.33. This plot is often known as the 'Moody diagram' or the 'Stanton-Moody diagram'. The portion of the diagram that is of interest here is displayed on Fig. 2.10.

Various regions can be distinguished on the Moody diagram. For Re < 2,000, f is independent of roughness and is given by

$$f = 64/Re \tag{2.34}$$

This range corresponds to laminar flow, and Eq. 2.34 is simply Eq. 2.25 written in terms of the dimensionless groups introduced above.

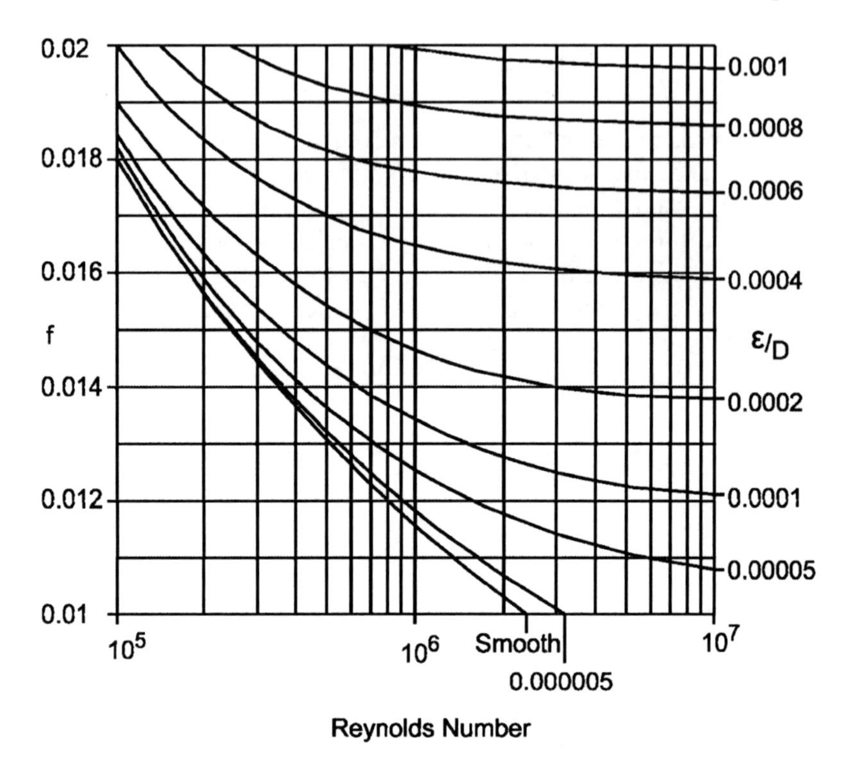

Figure 2.10. Pipe friction factor in normal operating range

For Re between about 2,000 and 3,000, flow can be laminar or turbulent; in industrial practice, it will almost always be turbulent. For Re > 3,000, flow is turbulent and f depends on both Re and ε/D: the curves on Fig. 2.10 are each drawn for one value of the relative roughness. In general, the friction factor decreases as Re increases and as roughness decreases. However, for sufficiently large values of Re, the laminar sub-layer becomes thinner than the asperities so that the range indicated as 'fully rough' is entered: the horizontal curves show that f now depends on ε/D but is independent of Re. Even in the 'transitional rough' range, the dependence of f on Re is weak. Thus, for turbulent flow in a given pipe with fixed relative roughness, it is frequently sufficient to treat the friction factor as a constant, characteristic of the pipe.

Detailed characterisation of roughness is a subject in itself, because the microscopic geometry will differ from one rough surface to another. However, commercial pipes are commonly characterised in terms of the 'equivalent sand-grain roughness'; i.e. their friction characteristics are compared with systematic measurements obtained by Nikuradse (1933) by gluing sand grains to the walls of test sections of pipes. For example, the equivalent sand-grain roughness for commercial steel pipe is about 46 μm. It must be remembered that the effective roughness can change in service: corrosion can increase ε, while the polishing action of the particles in a slurry can reduce ε (and possibly also increase D if erosion is severe).

For a given flow of a given fluid in a given pipe, it is possible to calculate Re and ε/D, and hence obtain f from Fig. 2.10 or Eq. 2.33. From Eq. 2.12 and Eq. 2.29, the hydraulic friction gradient follows as

$$i = f\frac{V^2}{2gD} \tag{2.35}$$

As f is approximately constant for turbulent flow in a given pipe, Eq. 2.35 shows that the hydraulic gradient varies roughly as V^2(or as Q^2).

The approach to estimating hydraulic gradient i summarised above and used throughout this book gives results essentially equivalent to those of other, more obviously empirical methods. An example of such methods is the 'C-factor' of Hazen and Williams which is still sometimes used in the mining industry. However, the use of f, with its dependence on Reynolds number and relative roughness, is preferred because it gives an indication of flow conditions and is more readily extended to interpretation of slurry flows.

For turbulent flow only, the loss of head associated with fittings such as bends and valves is usually estimated by multiplying the velocity head by a loss coefficient The most widely-used loss coefficients are 0.5 for a standard 90-degree elbow (0.2 for a long-radius elbow) and 0.8 for an abrupt (unrounded) entry. At a pipe exit the full velocity head is lost, equivalent to an exit loss coefficient of 1.0. in Further information on fitting losses is given in the McGraw-Hill Pump Handbook (Karassik et al., 2001).

The energy or head loss per unit length of pipe provides a measure of the hydraulic power required to deliver a flowrate Q through a horizontal pipe of length L, i.e.

$$P = \rho g Q i L \tag{2.36}$$

which can be rearranged, using Eq. 2.2 and 2.35, as

$$P = \frac{8\rho Q^3 f}{\pi^2 D^5} L \tag{2.37}$$

The very strong dependence on pipe diameter is worth noting. If power consumption is a major consideration, the economic incentive is towards using pipes of large diameter. The effect of this consideration on slurry system design will be evaluated in later chapters.

Example 2.A - Flow of Water in a Pipe

Water at room temperature of 20°C flows at 0.12 m^3/s through a standard 8-inch steel pipe. Calculate:

(a) the mean velocity and pipe Reynolds number;
(b) the Moody friction factor;
(c) the hydraulic gradient;
(d) the thickness of the sub-layer;
(e) the shear velocity
(f) the hydraulic power required.

(a) Standard 'schedule 20' 8-inch pipe has an internal diameter of 0.2064 m. Therefore the pipe cross-sectional area is

$$\pi \times (0.2064)^2 / 4 = 0.03346\ m^2$$

and so the mean velocity V is

$$0.12/0.03346 = 3.59\ m/s$$

This mean velocity is towards the low end of the range typically used for settling slurries. For water at 20°C, ρ = 998.2 kg/m^3 and μ = 1.002 x 10^{-3} Pa.s (Table 2.1). Therefore the Reynolds number is

$$Re = \frac{\rho V D}{\mu} = \frac{998.2\,(3.59)\,(0.2064)}{1.002 \times 10^{-3}} = 7.38 \times 10^5$$

Because Re is dimensionless, the same value is obtained whatever system of units is used. Note that Re is high and the flow is well into the turbulent range.

(b) For commercial steel pipe, the equivalent sand-grain roughness is typically 4.6 x 10^{-5}m, as noted previously. Therefore the relative roughness is

$$\varepsilon/D = 4.6 \times 10^{-5} / 0.2064 = 2.23 \times 10^{-4} = 0.000223$$

Referring to Fig. 2.10, flow conditions are in the 'transitional rough' range where the friction factor depends on the relative roughness and (weakly) on Re. From Fig. 2.10 or Eq. 2.33

$$f = 0.0152$$

Values for the Moody friction factor in the range 0.01 to 0.02 will prove to be fairly typical for water alone pumped at conditions commonly used for settling slurries.

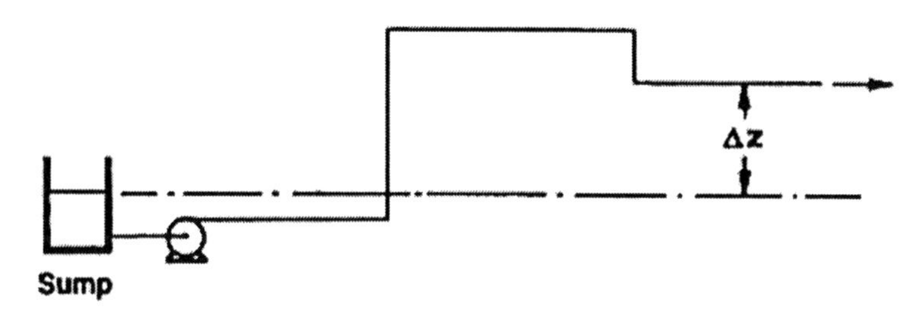

Figure 2.11. Simple piping system

(c) From Eq. 2.35, the hydraulic gradient is

$$i = \frac{f V^2}{2gD} = \frac{0.0152(3.59)^2}{2(9.81)(0.2064)} = 4.84 x 10^{-2} \text{ m. head per m. pipe}$$

Values of the order of a few metres head per hundred metres of pipe are again typical of water pumped at conditions appropriate for a settling slurry.

(d) From Eq. 2.29, the wall shear stress is

$$\tau_o = \rho f V^2 / 8 = 998.2 (3.59)^2 (0.0152)/8 = 24.4 \text{ Nm}^{-2}$$

The thickness of the viscous sublayer is estimated from Eq. 2.27 as

$$11.6 \mu/(\rho\tau_0)^{1/2} = 11.6 \ (1.002 \times 10^{-3})/(998.2 \times 24.4)^{1/2} = 7.4 \times 10^{-5} \text{ m}$$

This value, i.e. 74 μm, can be compared with the equivalent sand-grain roughness, of 46 μm. As expected for a flow well into the 'transitional rough' range, the sub-layer thickness and the equivalent roughness are of comparable magnitude. The value for sub-layer thickness, a few tens of microns, is typical and is worth noting.

(e) From Eq. 2.30, the shear velocity is

$$U_* = V_m (f/8)^{1/2} = 3.59(0.0152/8)^{1/2} = 0.16 \ m/s$$

and

$$U_* / V = (f/8)^{1/2} = 0.044$$

This value of U_*, of the order of V/20, is typical.

(f) The power required from the pumps is given by Eq. 2.37 as

$$\frac{P}{L} = \frac{8 \rho Q^3 f}{\pi^2 D^5} = \frac{8\,(998.2)(0.12)^3 (0.0152)}{\pi^2 (0.2064)^5} = 57 \ W/m$$

i.e. 57 kW/km, which again gives a typical order of magnitude for water alone.

In later chapters, we will examine the operability of piping systems using centrifugal pumps, by matching the 'system characteristic' with the 'pump characteristic'. Figures 2.11 and 2.12 illustrate this idea for a system conveying a liquid alone. For a simple piping system shown schematically in Fig. 2.11, the total head required varies with flowrate as shown by curve 1 in Fig. 2.12. The 'static lift' term, Δz, from the surface in the sump to the pipe discharge, does not depend on flow rate. The flow-dependent terms include the friction losses in the pipe and fittings, and also the 'velocity head' at the pipe discharge because this is not recovered as pressure. Thus, for a pump with head-discharge characteristic (see Chapter 9) given by curve 2, the system will operate at the discharge corresponding to the intersection of the characteristics at point A. Operation is stable, because the pump characteristic is falling while the system characteristic is rising. Thus a slight decrease in flow reduces the head demand of the system but increases the head delivered by the pump, to return operation to point A. Similarly, increases in flow are automatically returned to A. Closing a valve, for example on the pump discharge, reduces the flow by increasing the resistance of the system, for example by moving the system characteristic to curve 3, so that the operating point moves to B. However, for a simple liquid with a rising system characteristic, operation remains stable.

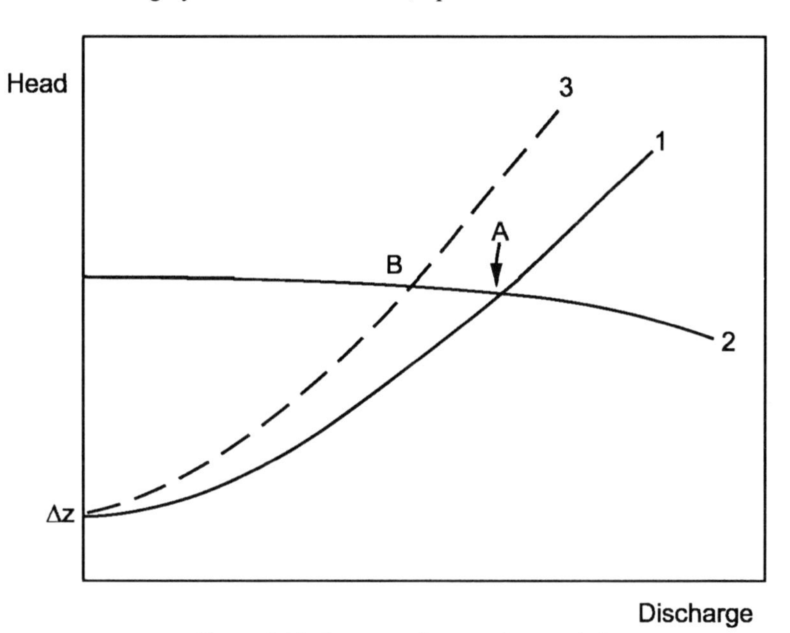

Figure 2.12. System and pump characteristics

2.4 Basic Relations for Slurry Flow

When we turn from flow of a simple liquid to that of a slurry, i.e. a mixture of solid particles in a carrier fluid, the need immediately arises for a more precise system of nomenclature. For example, instead of a single density, ρ, several densities must now be distinguished. These include the density of the fluid, ρ_f, that of the solid particles, ρ_s, and that of the mixture ρ_m. For a very large number of slurries, the carrier fluid is water, with density of approximately 1000 kg/m^3. The symbol ρ_w is employed in this instance, using ρ_f for fluids of other densities. The value of ρ_w forms the basis for expressing the relative density or 'specific gravity' of other materials. For example there is a wide range of applications when the solids being conveyed have a density around 2650 kg/m^3, a typical value for sand, and in this case the relative density, S_s or ρ_s/ρ_w, is 2.65. Although in many cases the fluid is standard-density water, this is not always the case. For example in marine dredging operations, the carrier fluid is sea water, for which the relative density, denoted S_f, varies from place to place but has a typical value of about 1.03.

For a mixture of solids and fluid, the relative density S_m (i.e. the mean specific gravity of the mixture) is given by the general formula

$$S_m = S_f + (S_s - S_f)\,C_v \tag{2.38}$$

where C_v is the volumetric concentration, i.e. the fraction of the mixture volume which is occupied by the solids. When the fluid is water of standard density, S_f is unity and the equation for S_m becomes

$$S_m = 1 + (S_s - 1)\,C_v \tag{2.39}$$

The volumetric concentration, C_v, is employed in this book, but it should be noted that a different measure, the weight concentration C_W, is commonly used in some industries. If the weight concentration C_w is known, C_v can be calculated from: $S_f C_w/[S_s - (S_s - S_f)C_w]$. Conversely, C_w is given by the expression $S_s C_v/[S_f + (S_s - S_f)C_v]$. As we are concerned here with general approaches applicable to any slurry, we will work throughout in terms of *volume* concentration. This parameter provides a much clearer indication of slurry consistency, applicable whatever the solid density.

In specifying concentrations, care is required to distinguish between delivered and *in situ* values. The delivered concentration is the fraction of solids delivered from (or fed to) the conveying system. If the slurry discharged from the system is collected in a tank, then the volume fraction of solids for the mixture in this tank is the delivered volumetric concentration, denoted C_{vd}. On the other hand, the resident or *in situ* concentration is the average concentration present in the system, or in some part of it, such as a certain length of pipe. If, say, this length of pipe were isolated by suddenly closing valves at both ends (ignoring the effects of water-hammer), then the volumetric fraction of solids in the isolated pipe is the resident or *in situ* volumetric concentration, denoted C_{vi}.

Although it might appear at first sight that these two measures of concentration should give the same value, this is only the case for truly non-settling slurries for which there is no tendency for the two components to segregate. The values of C_{vd} and C_{vi} differ when the average velocity of the solids (V_s) is not the same as that of the fluid (V_f), and this is typically the case in stratified flows (see Chapter 3). Although it is not necessary to have stationary solids in the pipe for C_{vd} and C_{vi} to differ, an extreme example is given by the case where a deposit of stationary solids fills, say, the lower half of the pipe. Water flows through the upper part of the pipe, but moves only a few solid particles, which tend to roll along the top of the bed. The *in situ* concentration C_{vi} is quite large, but as most of the solids are not moving at all, the *average* velocity of the solids is small, much less than that of the water. As a result, in this case the concentration of solids in the delivered mixture, C_{vd}, is much smaller than C_{vi}.

The volumetric flowrate of liquid, Q_f, is the product of V_f and the cross-sectional area occupied by the fluid, i.e. $(1-C_{vi})\pi D^2/4$. Similarly, the volumetric flowrate of solids is $V_s C_{vi} \pi D^2/4$. The total flowrate of the mixture, Q_m, is given by the sum of the fluid and solids flowrates, and is also equal to $\pi D^2/4$ times the mean velocity of the mixture. Thus

$$\frac{4 Q_m}{\pi D^2} = V_m = V_f (1 - C_{vi}) + V_s C_{vi} \tag{2.40}$$

The delivered volumetric concentration, C_{vd}, represents Q_s/Q_m, i.e.

$$C_{vd} = \frac{Q_s}{Q_m} = \frac{V_s}{V_m} C_{vi} \tag{2.41}$$

Equation 2.41 shows directly that the delivered concentration must be less than the *in situ* value provided V_s is less than V_m. This condition is described as 'lag', 'hold up' (or, less accurately, 'slip') of the solids. The 'lag' or 'slip' is the velocity difference $V_m - V_s$, and the lag ratio Λ is obtained by dividing this quantity by the mean velocity, i.e.

$$\Lambda = \frac{V_m - V_s}{V_m} \tag{2.42}$$

On this basis Eq. 2.42 is re-written

$$C_{vi} = \frac{1}{1 - \Lambda} C_{vd} \tag{2.43}$$

Thus, when there is hold-up of the solids relative to the liquid the *in situ* concentration is greater than the delivered concentration, and the difference between the two concentrations increases when the lag ratio increases.

This conclusion has a number of far-reaching implications. An obvious corollary is that measuring the *in situ* solids concentration, for example by a radiation technique (see Chapter 12), does not indicate the delivered concentration. A further corollary concerns the analysis of the transport of settling slurries, for which the hold-up effect is significant. As will be seen, the *in situ* concentration is most important in determining friction losses. However, design methods must be based on the delivered concentration, with the *in situ* concentration either inferred or not estimated explicitly. Furthermore, slurry testing must normally use closed-loop systems in which the inventory and therefore the resident concentration is fixed, and thus the delivered concentration will vary as V_m is changed. Hence the simple approach suitable for single-phase fluids cannot be applied to slurries.

Another area in which slurries require a more careful treatment than single-phase fluids is that of the friction gradient or energy gradient. In the form of the pressure gradient dp/dx, the friction loss associated with the flow of a slurry in a pipe is unambiguous. However, the expression of pressure loss as a 'hydraulic gradient' (such as m water per m length of pipe) is so common in slurry pipelining that it cannot be avoided. Only precise definitions of all quantities will prevent ambiguity. As expressed in Eq. 2.12, for a single fluid the hydraulic gradient is $(-dp/dx)/\rho g$. The possible ambiguity arises from the density be used in this expression. For flows of water alone the appropriate density is ρ_w (approximately 1000 kg/m^3), so that the corresponding hydraulic gradient is

$$i = \frac{1}{\rho_w g}\left(-\frac{dp}{dx}\right) \tag{2.44}$$

giving the friction loss as metres head of water per metre of pipe.

Following a proposal made by Dr. M.R. Carstens in the early years of the GIW Slurry Course, the symbol i is used in this book to denote 'hydraulic gradients' given by Eq. 2.44 and expressed in height of clear water per length of pipe. As both ρ_w and g are constants, this definition is equivalent to stating that, for a horizontal pipe, i is simply the pressure gradient divided by a constant (9810 N/m^3 in S.I. units). The set of subscripts introduced above in connection with densities will also be employed here. Thus i_m represents the pressure gradient for a mixture, but expressed in height of water per length of line. Similarly i_f expresses the pressure gradient for a fluid (if different from water) in terms of height of water, and i_w applies when the fluid is water. (The similarity to the use of water as a reference fluid for specific gravity will be readily apparent.) In evaluating the extra friction loss caused by the conveyed solids, we use the 'solids effect' (i_m - i_f), where i_f is the friction gradient for carrier liquid alone at flowrate equal to the mixture flowrate Q_m. The evaluation of this quantity will be discussed extensively in Chapters 4, 5 and 6.

An alternative definition of the mixture hydraulic gradient is based on the mean density of the delivered slurry, ρ_{md} (and the associated relative density S_{md}). For clarity, this will be noted by j rather than i, giving

$$j_m = \frac{1}{\rho_{md}\, g}\left(-\frac{dp}{dx}\right) = \frac{1}{S_{md}\,\rho_w\, g}\left(-\frac{dp}{dx}\right) \tag{2.45}$$

This measure of the hydraulic gradient is more useful for some purposes, such as matching system and pump characteristics (see Chapter 13). From Eq. 2.44 and 2.45, the two are related by

$$i_m = S_{md}\, j_m \tag{2.46}$$

Under some circumstances, to be examined in later chapters, it can happen that the additional pressure gradient attributable to the solids is proportional to the increase in slurry density, i.e. the 'equivalent fluid' model applies with

$$i_m = S_{md}\, i_w \quad \text{and} \quad j_m = i_w \tag{2.47}$$

A slurry of this type will be termed an 'equivalent fluid' because, in effect, it behaves as a single-phase liquid with the density of the delivered slurry.

2.5 Settling of Solids in Liquids

The properties of slurries depend very strongly on the tendency of the particles to settle out from the conveying liquid. For transport of settling slurries, an important parameter is the *terminal velocity*, v_t , i.e. the velocity at which a single particle settles through a large volume of quiescent liquid. The terminal velocity depends on the liquid properties (ρ_f and μ) on the particle diameter (d) and its density (ρ_s) and, to a lesser extent, on its shape. For vertical flow of settling slurries the *hindered settling velocity* is also of importance. When particles have fully settled, their concentration, achieved without compacting or vibrating the sediment, is referred to in later chapters as the 'loose packed' volume fraction, denoted C_{vb}.

Particle sizes are commonly reported as 'screen size', i.e. the opening in a standard sieve or screen. Particle size distributions are then reported as the fraction (by mass or weight) passing through one screen in the series but retained on the next. Standard screen series have often been expressed as 'mesh size' based on the number of openings per inch, but openings in mm (or μm) are now in common use.

In general, particles in slurries are not spherical, but the sphere represents a convenient reference case in the analysis. Figure 2.13 shows the forces acting on a rigid sphere settling through a fluid. The weight of the particle is partially reduced by the buoyancy of the surrounding fluid. When the spherical particle is moving steadily at its terminal velocity v_{ts}, the resulting 'immersed weight', 'submerged weight' or 'net weight' is balanced by the drag of the fluid:

$$F_D = \frac{\pi d^3}{6}(\rho_s - \rho_f)g \tag{2.48}$$

Calculation of the terminal velocity therefore depends on estimating the velocity at which Eq. 2.48 is satisfied. Since there is no general theoretical result which enables this equation to be solved for v_{ts}, we resort to dimensional analysis. In general, with fn denoting some function,

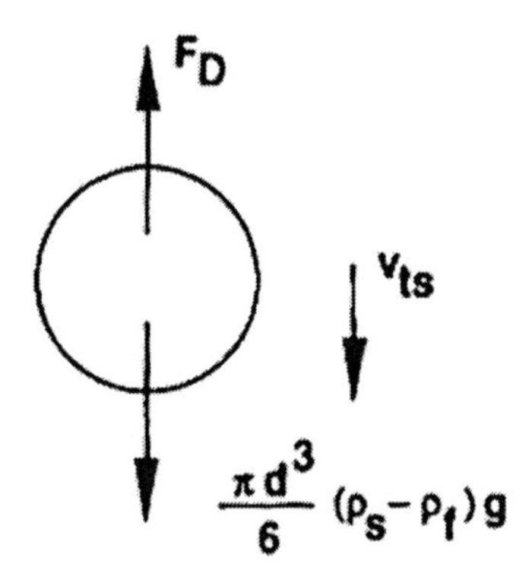

Figure 2.13. Single spherical particle falling at its terminal velocity

$$F_D \equiv fn(\rho_f, \mu, d, v_{ts}) \tag{2.49}$$

where v_{ts} refers specifically to the terminal velocity of a spherical particle. Given that there are five parameters and the usual three basic dimensions of mass, length and time, Buckingham's theorem shows that two dimensionless groups suffice to express Eq. 2.49 in general form. Most commonly, the two dimensionless groups selected are:

$$\textit{drag coefficient}: C_D = \frac{8 F_D}{\pi d^2 v_{ts}^2 \rho_f} \tag{2.50}$$

$$\textit{particle Reynolds number}: Re_p = \rho_f v_{ts} d / \mu \tag{2.51}$$

so that Eq. 2.49 is written in general form as

$$C_D \equiv fn(Re_p) \tag{2.52}$$

The function indicated in this equation has been fitted to the many determinations of drag or terminal velocity, to give the 'standard drag curve' shown in Fig. 2.14. For $Re_p < 3 \times 10^5$, which amply covers the range encountered in slurries, the curve is approximated well by an expression given by Turton & Levenspiel (1986)

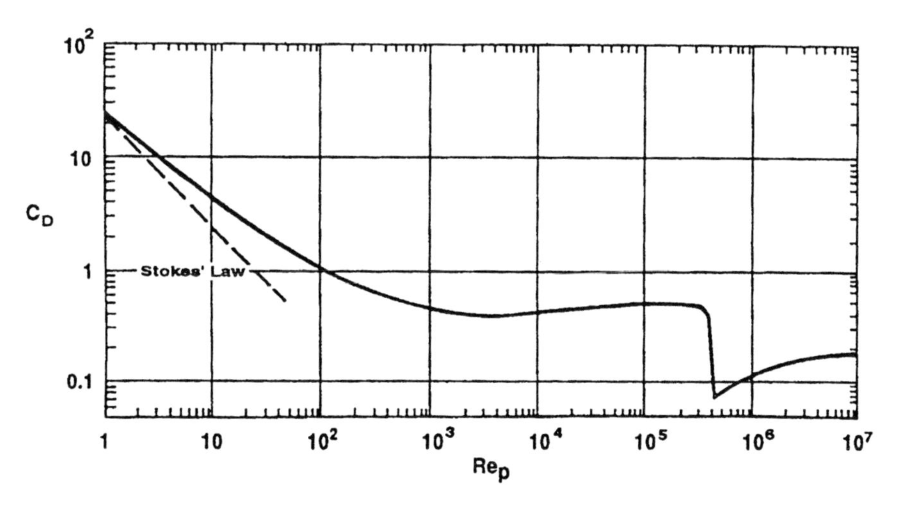

Figure 2.14. Standard drag coefficient curve for spheres

$$C_D = \frac{24}{Re_p}(1 + 0.173\, Re_p^{0.657}) + \frac{0.413}{1 + 1.63x\,10^4\, Re_p^{-1.09}} \tag{2.53}$$

The terminal velocity can then be determined from Eq. 2.48 and 2.53, but the procedure is iterative, with successive estimates for v_{ts} updated to converge on the solution.

The reasons for the form of the curve in Fig. 2.14 are discussed in detail by Clift *et al.* (1978). For low particle Reynolds numbers, say $Re_p < 0.1$, Stokes' law applies, with the drag force given by the theoretical result:

$$F_D = 3\,\pi\,\mu\,v_{ts}\,d \tag{2.54}$$

The terminal velocity follows as

$$v_{ts} = \frac{d^2(\rho_s - \rho_f)\,g}{18\,\mu} \tag{2.55}$$

At much larger Reynolds numbers, in the approximate range $750 < Re < 3 \times 10^5$, the drag coefficient is roughly constant and close to 0.445. This is known as the 'Newton's law' range, based on Newton's experiments with falling objects (e.g. inflated pigs' bladders falling within the dome of St. Paul's cathedral, see Newton, 1726). In this range the terminal velocity of a sphere falling through water can be calculated using

$$v_{ts} = 1.73\sqrt{gd\,(S_s - 1)} \tag{2.56}$$

As a general guide for sand-density particles in water, Stokes' law applies for particles smaller than about 50 μm , while Newton's law applies for particles larger than about 2mm.

As noted above, two independent dimensionless groups are needed to express empirical drag results in general form, However, these groups can be selected for convenience. Various possibilities and their uses are reviewed by Clift *et al.* (1978). Following the suggestion of Grace (1986), it is found convenient to define a dimensionless particle diameter d* (equal to the cube root of what is often called the Archimides number)

$$d^* = \left[3\,C_D\,Re_p^2/4\right]^{1/3} = d\left[\frac{\rho_f(\rho_s - \rho_f)g}{\mu^2}\right]^{1/3} \tag{2.57}$$

and a dimensionless terminal velocity

$$v_{ts}^* = Re_p / d^* = \left[\frac{4\,Re_p}{3\,C_D}\right]^{1/3} = v_{ts}\left[\frac{\rho_f^2}{\mu(\rho_s - \rho_f)g}\right]^{1/3} \tag{2.58}$$

so that Eq. 2.49 becomes in general dimensionless form

$$v_{ts}^* = fn(d^*) \tag{2.59}$$

and calculation of v_{ts}^*, and hence v_{ts}, requires no iteration. This functional relation has been worked out in considerable detail, and is entirely suitable for particles falling in Newtonian fluids. However, for the analogous case in non-Newtonian fluids (dealt with in Chapter 7) difficulties arise because the viscosity is included in both dimensionless variables. To cover both Newtonian and non-Newtonian cases, an alternative method has been worked out, as described by Wilson *et al.* (2003) and Wilson & Horsley (2004). This method is based on a pair of dimensionless variables that employ concepts developed in the pipe-flow analysis of Prandtl (1933) and Colebrook (1939).

The method expresses the velocity ratio (mean velocity to shear velocity) as a function of the shear Reynolds number (based on shear velocity rather than mean velocity). In pipe flow the shear velocity V* is the square root of the ratio of the shear stress at the pipe wall (uniform for a circular pipe) to the fluid density. The shear stress set up on the surface of a spherical particle is non-uniform, but the mean surficial shear stress, denoted $\bar{\tau}$, forms a useful basis for analysis. This stress is given by the submerged weight force divided by the surface area of the sphere, which is πd^2, where d is the sphere diameter. The submerged weight force is the product of the sphere volume $\pi d^3/6$ and $(\rho_s - \rho_f)g$, where g is gravitational acceleration and ρ_s and ρ_f are the densities of the solid and fluid phases, respectively. Thus the mean surficial shear stress is given by

$$\bar{\tau} = (\rho_s - \rho_f)g\,d/6 \tag{2.60}$$

For the falling-particle case, the shear velocity is based on $\overline{\tau}$. Thus

$$V^* = \sqrt{(\rho_s - \rho_f)gd/6\rho_f} \tag{2.61}$$

or, with S denoting ρ_s/ρ_f

$$V^* = \sqrt{(S-1)gd/6} \tag{2.62}$$

For pipe flow, the velocity ratio is the mean velocity divided by V^*; and for falling particles the analogous ratio is based on the terminal fall velocity of the particle V_t, giving V_t/V^* as the velocity ratio. The shear Reynolds number Re^* has the form:

$$Re^* = \rho_f V^* d/\mu \tag{2.63}$$

Here d is the particle diameter (analogous to the pipe diameter for the pipe-flow case) and, for a Newtonian fluid, μ is the viscosity.

Next, it is appropriate to investigate the form of the settling curve for Newtonian fluids for the V_t/V^* and Re^* axes. In terms of these variables, the drag coefficient equals $8/[V_t/V^*]^2$ and the conventional Reynolds number Re equals the product of Re* and V_t/V^*. For Re larger than about 1100 the drag coefficient can be taken as effectively constant at 0.445, equivalent to $V_t/V^*=4.24$ for $Re^*>260$. At the other end of the Reynolds number range, settling obeys Stokes law which can be expressed as $V_t/V^*=Re^*/3$. In the intermediate region, the coordinates of individual points on the C_D-Re curve can be transformed into V_t/V^* and Re^* values and plotted on the new curve, which is found to have the shape shown on Fig. 2.15.

In developing Fig. 2.15, a series of points on the C_D-Re curve were calculated using the equation of Turton & Levenspiel (1986). These points were then transformed to V_t/V^* and Re^* co-ordinates, and equations were fitted to the transformed points. For $Re^* \leq 10$, the fit equation is:

$$V_t/V^* = Re^*/[3(1+0.08\,Re^{*1.2})] + 2.80/[1+3.0(10^4)(Re^{*-3.2})] \tag{2.64}$$

In the range $10 < Re^* < 260$ a different fit equation is used, based on $x = \log(Re^*/10)$ and $y = \log(V_t/V^*)$. It is:

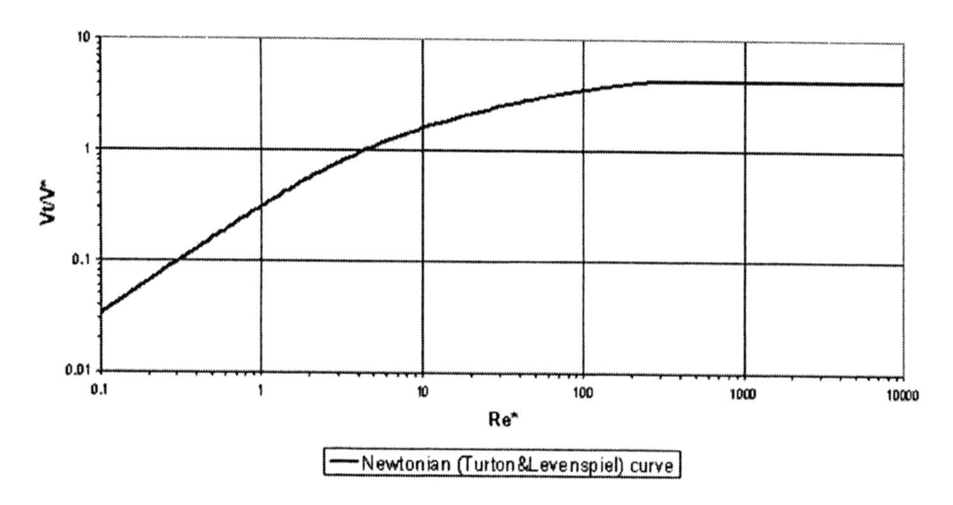

Figure 2.15. Curve of relative fall velocity versus shear Reynolds number (from Wilson et al., 2003)

$$y = 0.2069 + 0.500\,x - 0.158\,x^{1.72} \tag{2.65}$$

The curve represented by these equations is shown, on logarithmic coordinates, on Fig. 2.15. As mentioned above, for $Re^* > 260$, $V_t/V^* = 4.24$.

Approaches to estimating terminal velocities for non-spherical particles have been reviewed by Clift *et al.* (1978). For the conditions of interest in hydraulic conveying, the most suitable approach is based on Heywood's 'volumetric shape factor'. In the range of particle Reynolds numbers from roughly unity to of order 100 - which is the range of interest here - a particle orients itself during settling so as to maximise the drag. Generally, this means that an oblate or lenticular particle, i.e. a shape with one dimension smaller than the other two, will settle with its maximum area horizontal. The drag of the fluid on the particle then depends most critically on this area, A_p. This is also the area seen if the particle lies in a stable position on a flat surface, for example a microscope slide. Therefore, for estimation of drag, the non-spherical particle is characterised by the 'area-equivalent diameter', i.e. the diameter of the sphere with the same projected area:

$$d_a = \sqrt{4\,A_p/\pi} \tag{2.66}$$

For particles whose sizes are determined by sieving rather than microscopic analysis, d_a is slightly smaller than the mesh size. However, unless the particles are needle-shaped, the difference between d_a and the screen opening is relatively small, generally less than 20%. The shape of the particle is described by the 'volumetric shape factor' defined as

$$K = (volume\ of\ particle)/\ d_a^3 \tag{2.67}$$

so that K is 0.524 for a sphere. Representative values for various mineral particle are given in Table 2.2. In general, the more angular or flakey the particle, the lower is the value of K.

Table 2.2. Typical values of volumetric shape factor for mineral particles

Mineral Particles	Typical K
Sand	0.26
Sillimanite	0.23
Bituminous coal	0.23
Blast furnace slag	0.19
Limestone	0.16
Talc	0.16
Plumbago	0.16
Gypsum	0.13
Flake graphite	0.023
Mica	0.003

The procedure for calculation of terminal velocity is first to use the method presented above to calculate v_{ts} for the sphere of diameter d_a and the same density as the particle of interest. The value for the non-spherical particle is then given by

$$v_t = \xi\, v_{ts} \tag{2.68}$$

where the velocity ratio ξ is a function of the shape factor, K, and a weak function of the dimensionless diameter d^*. Values for ξ are obtained from the curves in Fig. 2.16. It will be seen that, for common particles like sand and coal, the terminal velocity is typically 50-60% of the value for the equivalent sphere. The value of ξ allows for the lower volume (and therefore lower immersed weight) of the particle compared to the sphere, and also for differences in drag.

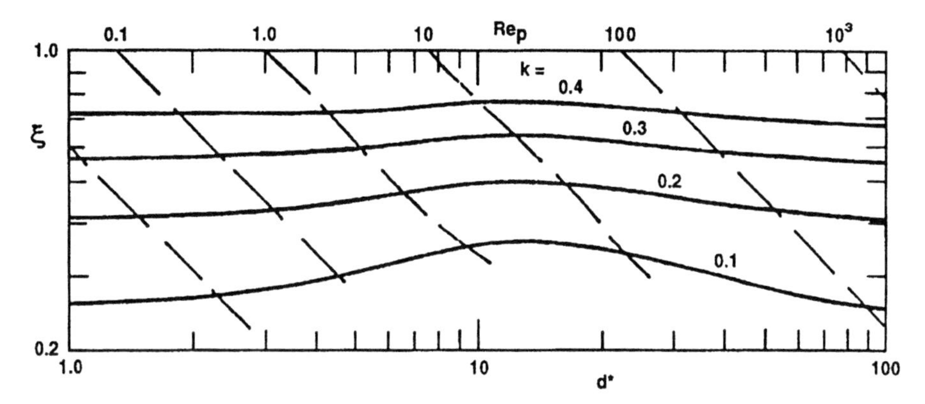

Figure 2.16. Ratio of terminal velocity of non-spherical particle to value for sphere, ξ, as function of dimensionless diameter, d^*

The hindered settling velocity, v_t', is normally less than v_t and is strongly dependent on the volume concentration of solids. For fine particles in the 'clay' range, the hindered settling behaviour is dominated by the forces between the particles. These forces depend on the chemical nature of the particle surfaces and of the liquid, and specifically on acidity, electrolyte concentration, and trace concentrations of surface-active agents. At present, these effects cannot be predicted reliably. Exactly the same remarks apply to the rheological properties of slurries of these particles, and this leads to the need for the 'testing and scale-up' approach for non-settling slurries set out in Chapter 3. In general, the effect of interparticle forces reduces v_t' even more than purely hydrodynamic effects. However, if the particles flocculate or agglomerate, then the hindered settling velocity is increased.

For larger particles – those which behave as individual grains – the hindered settling velocity can be estimated reasonably reliably by the correlation of Richardson & Zaki (1954):

$$v_t' = v_t (1 - C_v)^n \tag{2.69}$$

Equation 2.69 allows for two phenomena which reduce v_t' below v_t: the increased drag caused by the proximity of other particles, and the upflow of liquid as it is displaced by the descending particles. The index n depends on d^*, and is larger for particles settling in the range of Stokes' law (n = 4.6) and smaller in the range of Newton's law (n = 2.4).

Example 2.B - Calculation of Terminal and Hindered Settling Velocities

Estimate the terminal velocity and hindered settling velocities of sand particles with shape factor 0.26 in water at room temperature. Consider particles of sizes 0.2, 0.5, 1.0 and 2.0 mm. (It will be shown in Chapter 6 that particle settling is of greatest significance for heterogeneous flow, which applies to particles in this size range.)

(a) Terminal Velocities of Equivalent Spheres.
For water at 20°C, $\rho_w = 998.2$ kg/m^3 and $\mu = 1.002 \times 10^{-3}$ Pa.s (from Table 2.1). The density of quartz sand is typically 2650 kg/m^3. Therefore the shear velocity based on the mean surficial shear stress (Eq. 2.62) is:

$$V^* = \sqrt{(S_s - 1)gd} \text{ or } \sqrt{(1.65)9.81d}$$

with d in metres. Table 2.3 shows the values that were obtained for this quantity for the various particle sizes. Also shown in the table are values of Re*, i.e. $\rho_f V^* d/\mu$. This quantity is used to calculate v_{ts}/V^*, using Eq. 2.64 for the 0.2 mm particle (for which Re* < 10) and Eq. 2.65 for the other particles, based on their shear Reynolds numbers. Multiplying the ratio v_{ts}/V^* by V^* gives the fall velocity for a spherical particle, v_{ts}, which is shown in Table 2.3. Note that the 0.2 mm particles lie well beyond the Stokes' law range, while the 2 mm particles are not quite into the Newton's law range.

(b) Terminal Velocities of Sand Grains
To obtain the terminal fall velocity of the sand particles, the values of v_{ts} must be multiplied by the velocity ratio ξ, which, in turn, depends on the dimensionless particle size d*, given by Eq. 2.57

$$d^* = \left[\frac{\rho_f(\rho_s - \rho_f)g}{\mu^2} \right]^{1/3} d$$

or

$$d^* = \left[\frac{998.2\,(2650 - 998.2)\,(9.81)}{(1.002 \times 10^{-3})^2} \right]^{1/3} d = 2.526 \times 10^{+4}\, d$$

The values of the velocity ratio, ξ, are obtained from Fig. 2.16, by interpolation between the contours for K = 0.2 and 0.3 to K = 0.26, at the appropriate value of d^* in each case. Using the values for ξ and v_{ts} from part (a) gives the results shown in Table 2.3. Note that the variation of ξ over the range of particle sizes is rather small.

Table 2.3. Settling velocity calculations

d (m)	2×10^{-4}	5×10^{-4}	1×10^{-3}	2×10^{-3}
V* (m/s)	0.0232	0.0367	0.0519	0.0735
Re*	4.64	18.4	51.9	147
v_{ts}/V^*	1.041	2.105	2.990	3.841
v_{ts} (m/s)	0.0241	0.0772	0.1552	0.2823
d^*	5.05	12.6	25.3	50.5
Ξ	0.55	0.58	0.55	0.52
v_t (m/s)	0.013	0.045	0.085	0.147

REFERENCES

Clift, R., Grace, J.R. & Weber, M.E. (1978). *Bubbles, Drops and Particles,* Academic Press, New York.

Grace, J.R. (1986). Contacting modes and behaviour classification of gas-solid and other two-phase suspensions. *Can. J.Chem.Eng.* Vol. 64, pp. 353-363.

Karassik, I.J., Messina, J.P., Cooper, P. & Heald, C.C.(2001) *Pump Handbook Third Edition,* McGraw- Hill.

Kay, J.M. & Nedderman, R.M. (1985). *Fluid Mechanics and Transfer Processes,* Cambridge University Press.

Newton, I. (1726). *Principia Mathematica Philosophiae Naturalis.* 3rd Ed., Book II Scholium to Proposition XL, Royal Society, London. Reprinted by University of Glasgow, 1871. See also *Newton's Principia, Motte's Translation Revised,* Vol. 1, University of California Press, Berkeley, CA, 1962.

Nikuradse, J. (1933). Strömungsgesetze in rauhen Rohren, *Forschungsheft* - Verein deutsche Ingenieure, No. 361.

Reynolds, A.J. (1974). *Turbulent Flows in Engineering.* Wiley, London.

Richardson, J.F. & Zaki, W.M. (1954). Sedimentation and fluidisation: Part I. *Trans. Instn. Chem. Engrs. Vol. 32*, pp. 35-53.

Streeter, V.L., & Wylie, E.B. (1975). *Fluid Mechanics.* McGraw-Hill, New York.

Turton, R. & Levenspiel, O. (1986). A short note on the drag correlation for spheres. *Powder Technol. Vol. 47,* pp. 83-86.

Wilson, K.C., Horsley, R.R., Kealy, T., Reizes, J.C. & Horsley, M. (2003). Direct prediction of fall velocities in non-Newtonian materials. *Int'l J. Mineral Proc.,* Vol. 71/1-4 pp. 17-30.

Wilson, K.C. & Horsley, R.R. (2004). Calculating fall velocities in non-Newtonian (and Newtonian) fluids: a new view. *Proc. Hydrotransport 16,* BHR Group, Cranfield, UK.

Chapter 3

FLOW OF NON-SETTLING SLURRIES

3.1 Introduction

For slurries of very fine particles — say with hindered settling velocity less than about 1.5 x 10^{-3} m/s — then the tendency for the particles to settle out from the liquid can be neglected. As a result, the slurry can be treated for most purposes as a single-phase fluid. The particles move at the local fluid velocity. Therefore the delivered and *in situ* concentrations are identical (*cf.* Section 2.4) and, provided that the line is not left full of stagnant slurry for an extended period, a stationary bed of solids does not form. Set against these simplifications is the fact that the slurry usually shows non-Newtonian behaviour.

As a result of these general properties, the friction gradient for a non-settling slurry varies with mixture velocity and solids concentration, as shown schematically on Figure 3.1. The friction gradient, j_m, is given in terms of the head loss of mixture. The rising curve to the right of the figure represents turbulent flow. If the friction factor is constant, the turbulent curve is a single line with j_m proportional to V_m^2, as shown on the figure (*cf.* Section 2.3). If the friction factor varies with volumetric solids concentration C_v, turbulent flow will be represented by a bundle of curves rather than a single one. This bundle, normally quite narrow, has not been shown on Fig. 3.1. Discussion of this point will be deferred to Sections 3.3 and 3.7.

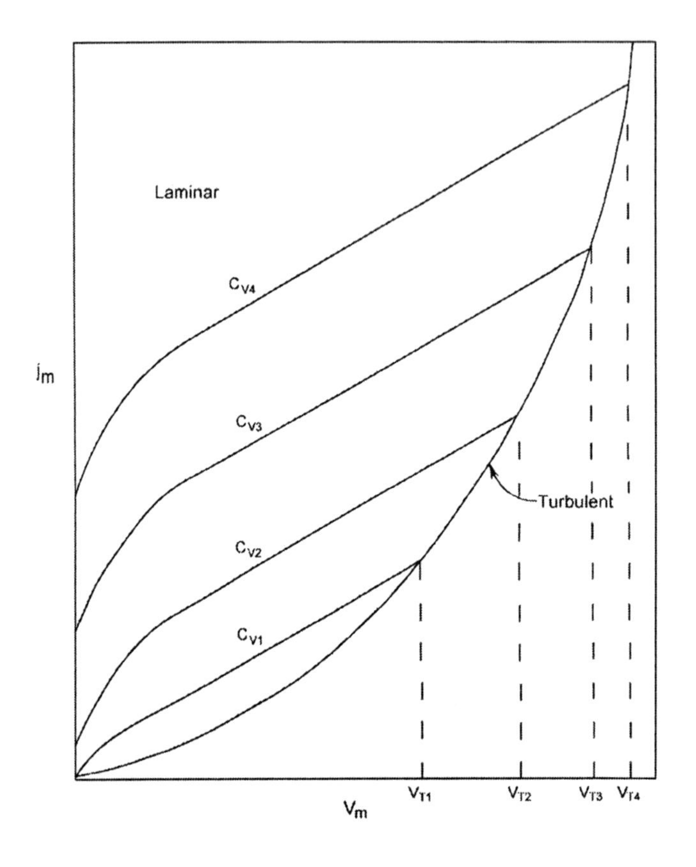

Figure 3.1. Effect of concentration on friction gradient for homogeneous slurries (schematic)

The behaviour in laminar flow depends strongly on solids concentration, *via* the rheological characteristics of the slurry. Figure 3.1 shows laminar flow lines for a series of concentrations, increasing from C_{v1} to C_{v4}. The first of these, C_{v1}, represents a relatively low solids concentration. This curve for laminar flow passes through or close to the origin, corresponding to zero or negligible yield stress (see Section 3.2). The friction gradient typically increases less than proportionately to V_m, a point to be examined in detail in Sections 3.2 and 3.4. At mixture velocity V_{T1} the laminar flow characteristic intersects the characteristic for turbulent flow. As a first approximation this intercept can be taken as the transition between laminar and turbulent flow.

In practice, the transition may be less sharp than suggested by the intercepts on Fig. 3.1. The laminar/turbulent transition in homogeneous slurries is discussed further in Sections 3.3, 3.5 and 3.7, and also in Case Study 3.1.

When concentration is increased from C_{V1} to C_{V2}, the slurry shows a yield stress (Section 3.2), and therefore j_m does not go to zero for zero V_m. The higher frictional resistance increases the laminar/turbulent transition to the range around V_{T2}. Further increases in concentration increase both the yield stress and the transition velocity. It is also useful to display plots of j_m *versus* V_m on logarithmic co-ordinates. An example is shown on Fig. 3.2, using data for red mud in an 81 mm pipe.

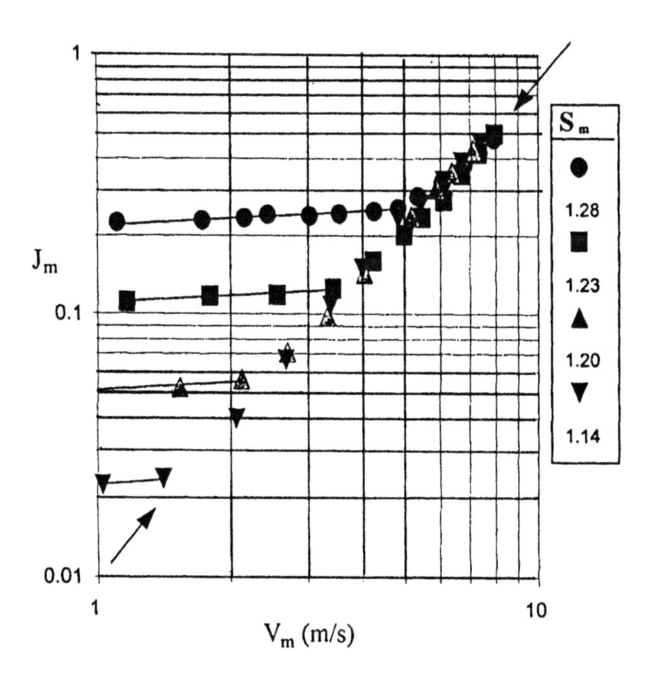

Figure 3.2. Logarithmic friction-gradient plot for red mud in an 81 mm pipe, after Wilson (1999)

Fully predictive design for a homogeneous slurry requires the relationship between shear stress and strain rate. At present, this cannot be obtained reliably from first principles. The approach developed here therefore employs scale-up, using data from laboratory tests or small-scale pipe trials. The basic concepts are developed in Sections 3.2 and 3.3, and the scale-up techniques for laminar and turbulent flow are explained in Sections 3.4 and 3.5. Section 3.6 then returns to the concepts illustrated in

Fig. 3.1, giving examples of empirical equations which have been derived from test data to predict the effect of varying solids concentration. The transition from laminar to turbulent flow is then re-visited in Section 3.7.

In addition to the applications discussed above, non-Newtonian homogeneous slurries are sometimes employed as carrier fluids for coarse particles. These complex flows are discussed in Chapter 7.

3.2 Rheometry and Rheological Models

The relationship between shear stress (τ) and strain rate (du/dy) for a homogeneous slurry is known as the rheogram. As noted in Chapter 2, for a Newtonian fluid the relation between these two quantities is simple:

$$\tau = \mu \, du/dy \tag{3.1}$$

where μ is the (dynamic) viscosity of the fluid. It follows that the rheogram of a Newtonian fluid is a straight line which passes through the origin and has slope μ (see Fig. 2.2). Rheograms for non-Newtonians are not so simple. For example, Fig. 3.3 shows the rheogram for a fine-particle slurry tested at the GIW Hydraulic Laboratory. It can readily be seen that this rheogram does not pass through the origin (the strain rate remains zero until a certain yield stress τ_y is exceeded) and is not straight (although in this case the behaviour is approximately linear at sufficiently high values of strain rate).

The definition of viscosity given by Eq. 3.1 can be retained, but this viscosity no longer represents the slope of the rheogram itself, but rather that of secant line shown on Fig. 3.3, and hence μ as defined by Eq. 3.1 can also be called the 'secant' viscosity. The tangent to the rheogram, though less meaningful physically, appears in some mathematical formulations (to be mentioned below). This 'tangent' viscosity will be denoted η_t. For a curve such as that displayed on Fig. 3.3, it is obvious that both μ and η_t vary with position, and hence they both depend on du/dy (or, alternatively, on τ). The stippling on Fig. 3.3 indicates the area beneath the leftward portion of the rheogram. The area ratio α (which will be used later) is the ratio between the stippled area and the triangular area below the secant line. As with μ and η_t, both the area beneath the rheogram and the area ratio α depend on du/dy (or on τ). Specific mathematical functions are often fitted to the experimentally-determined points plotted on the rheograms. These functions define rheological models, some of which will be presented below.

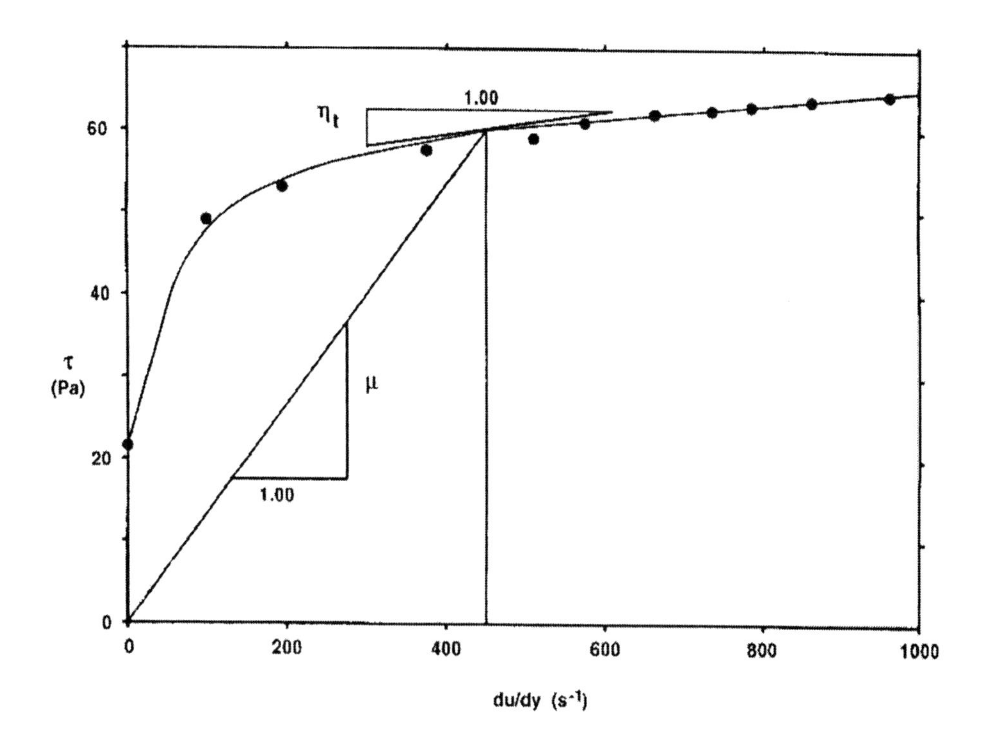

Figure 3.3. Rheogram for phosphate slimes tested at GIW Hydraulic Laboratory, after Wilson (1986)

The data points required for plotting rheograms like Fig. 3.3 can be obtained from instruments of two basic types: rotational viscometers and tube viscometers. Rotational viscometers are widely available and require only a small sample of slurry for testing, but they introduce much larger uncertainties than do tube viscometers, and are not to be recommended where an alternative is available. There is little point in dwelling on the specific configurations of various rotational instruments; the basic concept is that fluid is sheared in a zone near the rotating part, and that du/dy is determined from angular velocity, while τ is found from simultaneous measurements of torque. Some designs, such as the cone-and-plate viscometer, attempt to obtain a constant value of du/dy throughout the sheared zone, but for other designs, such as those having a bob which rotates within a beaker of fluid, du/dy is far from uniform. In general, some calibration formula is required to reduce the torque and speed readings into data for plotting the rheogram. These calibration formulas themselves depend on the rheological nature of the fluid, and are thus subject to considerable uncertainty.

On the other hand, if tube viscometers are used they provide direct geometric similarity with the prototype pipeline, so that scale-up techniques can be applied without the intermediary of a rheological model. Conditions within the tube viscometer are geometrically similar to those in the pipeline, assuring the similarity of the stress distributions. The measured quantities of mean velocity and pressure gradient have the same meaning in both the viscometer and the pipeline, and thus are amenable to the direct scale-up techniques described below in Sections 3.4 and 3.5. In a tube viscometer the measured pressure gradient dp/dx gives a direct evaluation of the wall shear stress τ_o (from $\pi D\tau_o = [\pi D^2/4]dp/dx$). If the usual form of rheogram is to be constructed, the shear rate du/dy must also be evaluated at the wall. This shear rate can be obtained from the mean velocity V_m (discharge divided by tube area), but several steps are required. Only laminar flow is considered at present, and it is helpful to look first at the simplest case – that of a Newtonian fluid.

As shown in Chapter 2, for flow in a tube (Newtonian or non-Newtonian) the shear stress varies linearly from zero at the tube axis to τ_o at the wall. With du/dy evaluated in terms of the known local shear stress, the velocity distribution can be obtained by integration. A second integration gives the discharge, and hence the mean velocity. When these integrations are carried out for the laminar flow of a Newtonian fluid it is found that the value of du/dy at the wall is given by $8V_m/D$ (*cf.* Eqs. 2.19 to 2.25). Thus for laminar Newtonian flow in a tube, the rheogram is equivalent to a plot of τ_o *versus* $8V_m/D$, which will be a straight line passing through the origin and having slope equal to the viscosity μ.

For non-Newtonian fluids the value of du/dy at the wall is no longer equal to $8V_m/D$, but it is proportional to that quantity, as determined by Rabinowitsch (1929) and Mooney (1931). In applying the Rabinowitsch-Mooney technique, the first step is to plot τ_o *versus* $8V_m/D$ on logarithmic co-ordinates. A plot of this sort is shown on Fig. 3.4 for a phosphate-slimes slurry tested at the GIW Hydraulic Laboratory in a pipe of internal diameter 203 mm. The slope of this plot, denoted n', represents $d\ln(\tau_o)/d\ln(8V_m/D)$. For a non-Newtonian which follows the power-law model (to be discussed below) n' would be constant. For other types of behaviour n' varies along the rheogram, but can usually be taken as approximately constant over considerable portions of the curve of τ_o *versus* $8V_m/D$, as shown on Fig. 3.4. For each such portion of the curve n' is found and used to calculate $(3n'+1)/4n'$. As demonstrated by Rabinowitsch and Mooney, this ratio represents the proportionality constant between $8V_m/D$ and the strain rate at the wall (Metzner, 1961). The rheogram, i.e. τ versus du/dy, can thus be obtained from the tube data by using τ_o in place of τ and $\left(\frac{3n'+1}{4n'}\right)\frac{8V_m}{D}$ in place of $\frac{du}{dy}$. For the special case of a Newtonian fluid both n′ and the ratio (3n′ + 1)/4n′ are unity.

This technique has been applied to the data plotted on Fig. 3.4 to produce the rheogram shown on Fig. 3.5. From Fig. 3.4, The slope n′ of the logarithmic plot was found to be 0.110, and hence the scaling ratio (3n′ + 1)/4n′ is 3.02. (In this case, a single line on logarithmic co-ordinates fits all the test points, so only one value of the scaling ratio is required. If a piecewise fit had been needed, each portion would require its own scaling ratio.) Thus, in the present case, du/dy for each point is obtained by multiplying $8V_m/D$ by 3.02, resulting in the rheogram plotted on Fig. 3.5. On the Cartesian axes of this figure, all points but the lowest can be represented by a straight line, with an intercept of 52.7 Pa and a slope of 0.020 Pa·s.

The Rabinowitsch-Mooney transformation is equally applicable in the opposite direction. If data are available in the form of the normal rheogram, a logarithmic plot of τ versus du/dy will also give the slope n′ for each portion of the curve. As before, τ_o can be equated to τ; now n′ is used to obtain $8V_m/D$ by multiplying du/dy by 4n′/(3n′+1).

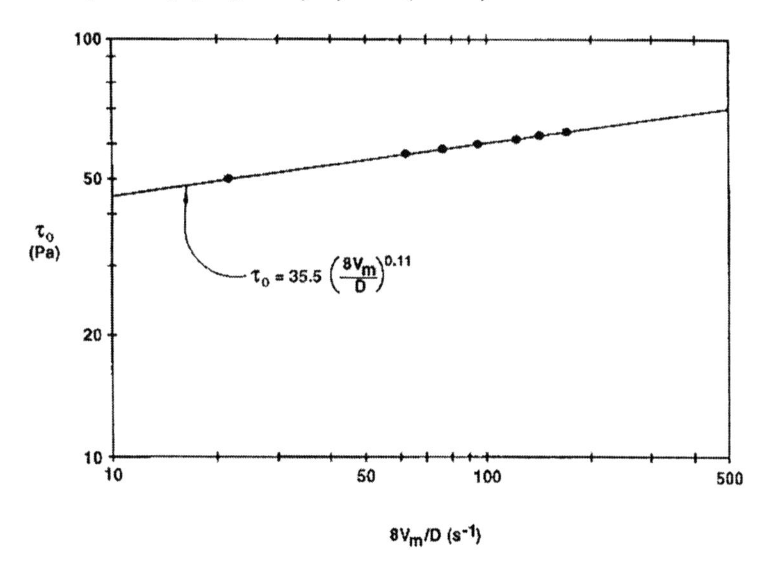

Figure 3.4 Logarithmic plot of τ_o *versus* $8V_m/D$

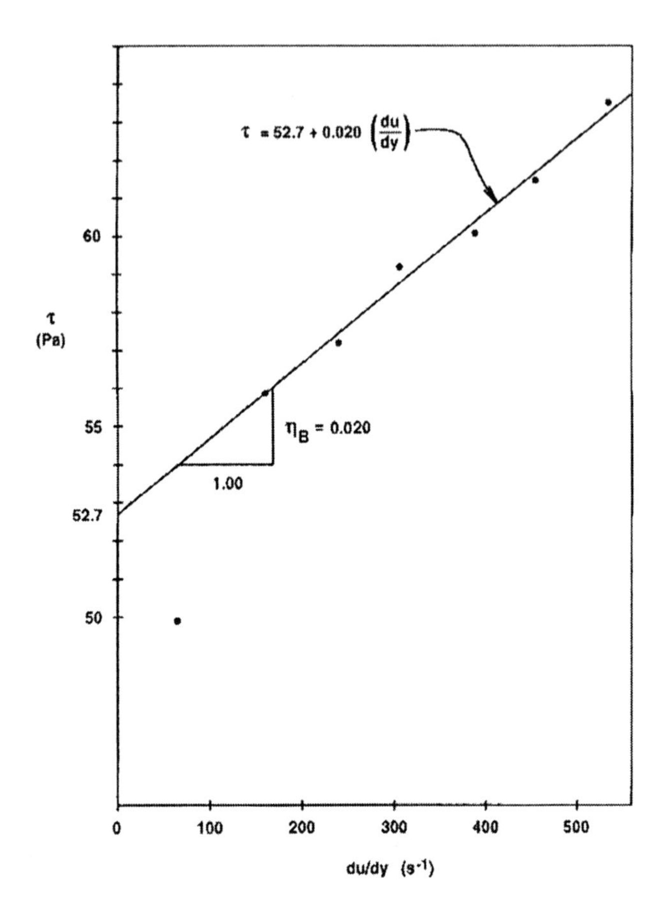

Figure 3.5. Rheogram obtained from data of Fig. 3.4.

It has just been shown how rheological data can be transformed from one axis system to another. As will be demonstrated in Sections 3.4 and 3.5, the system employing $8V_m/D$ is most convenient in scaling up to larger pipe sizes. On the other hand, descriptions of flow configurations require relationships between local shear stress and deformation rate, so that this is the more fundamental basis and is employed in mathematical representations of rheological behaviour. Strictly speaking, both stresses and deformation rates are tensors, and full rheological modelling requires tensor relationships. However, this level of complexity is not necessary for pipeline design. The discussion here is therefore presented at the engineering level, in terms of relationships between τ and du/dy.

Only time-independent rheological models are considered here. However, a full description of rheological behaviour also includes variation with time, because many homogeneous slurries show shear-thinning or shear-thickening behaviour. Nevertheless, once a slurry has passed through a pump, it has usually been so thoroughly sheared as to reach its asymptotic rheological behaviour and, for purposes of pipeline design, time-dependent effects can be ignored. There are, however, some practical cases where time-dependent behaviour should be considered. A notable example is suction design for a pump handling a very concentrated slurry such as thickener or hydrocyclone underflow.

The simplest rheological model is the Newtonian fluid discussed previously, represented by equation (3.1) with a single rheological parameter, μ. The next step upward in complexity is represented by two-parameter models. True fluids have rheograms which pass through the origin; *i.e.* any imposed shear stress causes flow. However, the rheogram may not be a straight line, and the power law model may then be employed. It has two parameters, k and n, and is expressed by the equation,

$$\tau = k\left(\frac{du}{dy}\right)^n \tag{3.2}$$

For slurry flows the usual behaviour, called pseudoplastic, corresponds to $n < 1$.

For a plastic material, the rheogram does not pass through the origin, and no strain rate occurs until the shear stress exceeds some yield value, τ_y. A simple two-parameter model called the Bingham plastic is often employed to quantify this behaviour. Its rheogram is linear, passing through the Bingham yield stress τ_B at $du/dy = 0$ and having a slope η_B that is called 'plastic viscosity', or 'tangent viscosity', thus:

$$\tau = \tau_B + \eta_B\, du/dy \tag{3.3}$$

If a material shows non-linear behaviour at low shear rates, a better fit may be obtained with the Casson model, given by the equation:

$$\sqrt{\tau} = \sqrt{\tau_c} + \sqrt{(\eta_c du/dy)} \tag{3.4}$$

where τ_c is the Casson yield stress and η_c is the Casson viscosity.

Introducing additional parameters might appear to improve the fit of the model to the rheological data; but in practice it is very difficult to obtain an adequate calibration of a three-parameter model, and thus the additional complexity is rarely worthwhile. One three-parameter model that may be mentioned is the yield-power-law or Herschel-Bulkley model, with the equation:

$$\tau = \tau_y + k\left(\frac{du}{dy}\right)^n \tag{3.5}$$

As can readily be seen, this model simplifies to the power-law model if $\tau_y = 0$, and to the Bingham model if $n = 1$ (and if both these conditions apply, it simplifies to the Newtonian model).

In use, the rheological model that has been selected is fitted to the data points of the rheogram, thus determining the specific values of parameters such as τ_y, k and n. However, these parameters are not of direct interest for the pipeline flow of non-Newtonian fluids. The designer wants to know whether the flow in the prototype pipeline will be laminar or turbulent, and to be able to predict pressure drop in terms of mean velocity. For laminar flow the velocity distribution, and hence the relation of pressure-drop to mean velocity, can be determined by integration of the rheological-model equation, but this method is indirect and is often tedious. It is easier and more accurate to use a tube viscometer and to scale up the results directly to prototype size, as described below in Section 3.4. Even if a tube viscometer had not been used, the rheogram can be transformed onto the axes of τ_o *versus* $8V_m/D$ by the inverse of the Rabinowitsch-Mooney transformation described above. The latter graph can then be used directly for scale-up.

If a tube viscometer or a test loop is used, it may be possible to extend the experimental range into the turbulent flow region. This is the recommended technique, since the turbulent-flow data can be scaled directly to prototype size, as shown in Section 3.5. The transition from one flow type to the other can then be found from the intercept of the scaled-up relations for laminar and turbulent flow, as demonstrated by example calculations given later in this chapter.

3.3 Turbulent Flow of Non-Newtonians

In order to understand and analyse non-Newtonian turbulent flow, some sort of flow model is required. One method is to begin with Eq. 2.28, which gives the mean velocity of Newtonian turbulent flow in a smooth-walled pipe as $V_m = 2.5\,U_* \ell n(\rho D U_* / \mu)$ where the shear velocity U_* is evaluated at the pipe wall (*i.e.* $\sqrt{\tau_o/\rho}$). For tests of turbulent flow of a non-Newtonian, all quantities except μ will be available for each data point . Therefore the equation can be solved for the viscosity in each case; the viscosity thus determined is called the equivalent turbulent-flow viscosity, and given the symbol μ_{eq}. The equation is then written:

$$V_m = 2.5\ U_* \ell n\,(\rho\ D\ U_* /\ \mu_{eq}) \tag{3.6}$$

At this point, two approaches are available to the pipeline designer. On the one hand, μ_{eq} can be taken as a basic quantity, to be obtained by testing in turbulent flow conditions. This approach leads to the scale-up technique that will be discussed below and applied in Sections 3.5 and 3.9. On the other hand, if turbulent-flow tests have not been made, then an indirect approach must be taken. This can be considered as equivalent to developing a model for predicting μ_{eq} from the rheogram, *i.e.* on the basis of laminar behaviour.

The modelling of the turbulent flow of non-Newtonians must begin from Newtonian turbulent flow, which itself is a complex phenomenon that is not completely understood. However, all that is usually required is a reasonable approximation to the velocity distribution. For this purpose the 'engineering' model outlined in Chapter 2 should be adequate. In the simplest version of this model, based on a smooth boundary, the effect of fluid viscosity is confined to a thin sub-layer which extends from the wall to a distance δ given by

$$\delta \approx 11.6\ \mu / \rho U_* \tag{3.7}$$

where μ and ρ are the viscosity and density, respectively, of the Newtonian fluid and U_* is the shear velocity at the wall, i.e. (*i.e.* $\sqrt{\tau_o/\rho}$). In the main flow it is assumed that all momentum transfer takes place by turbulent mixing, which is an inertial process rather than a viscous one. This gives rise to the logarithmic velocity profile in the turbulent region, (*cf.* Eq. 2.26 and the associated text), as opposed to the linear velocity increase in the sub-layer, where

$$u = \tau_o\, y / \mu \qquad (y < \delta) \tag{3.8}$$

As the velocity gradient in the viscous sub-layer is much steeper than that in the logarithmic zone, the sub-layer has an influence on mean velocity which is disproportionate to its small thickness (see Fig. 2.9). As seen in Section 2.3, it is the sub-layer that determines the influence of viscosity (*via* the Reynolds number) on the friction factor for Newtonian turbulent flow.

If a non-Newtonian fluid is to be substituted for the Newtonian one, certain reasonable prognostications can be made. In the logarithmic zone the momentum transport is inertial in nature; as a result, the velocity gradient here is not affected by Newtonian viscosity, and it would be expected that the same should apply for non-Newtonian rheological properties. (In the part of the flow nearest the pipe axis, some change in the velocity profile would result from non-zero τ_y, but the effect on mean velocity is very small.) Within the sub-layer, which occupies such a small portion of the flow that variations in shear stress and velocity gradient are negligible, laminar conditions are approached. Here a linear velocity increase equivalent to Eq. 3.8 should also apply for a non-Newtonian fluid.

Two questions arise at this point. The first concerns the value of μ that applies in Eqs. 3.7 and 3.8, and the second refers to the coefficient that specifies the thickness of the viscous sub-layer (for Newtonian flows this is 11.6, as given in Eq. 3.7, but the value may be larger for non-Newtonians, as discussed below). The first question can be resolved by referring to Fig. 3.3. The 'secant' slope on the figure, μ, is equivalent to the Newtonian viscosity in that it is the ratio of τ to du/dy. It is not constant like the Newtonian viscosity, but can be expressed in terms of τ. Thus Eq. 3.8 should still apply for non-Newtonians provided μ is understood in the sense of Fig. 3.3 and evaluated at $\tau = \tau_o$.

The remaining question, concerning thickening of the viscous sub-layer, can be related to a conceptual model proposed by Lumley (1973, 1978) to explain the phenomenon of drag reduction in aqueous flows, obtained by the addition of small quantities of certain long-chain molecules. These substances act to increase the size of the smallest, dissipative, turbulent eddies. Figure 3.6, which illustrates Lumley's model, graphs the distance from the wall, y, as the ordinate and representative eddy sizes as the abscissa. As indicated on the figure, the size of the large eddies (the macro scale of turbulence) is directly proportional to the distance from the wall, while the size of the smallest eddies (the dissipative, or Kolmogorov, scale) does not change much with this distance. At large values of y the inertial macro-eddies are much bigger than the dissipative micro-eddies, and between them is a whole range of turbulent eddy sizes, shown shaded on Fig. 3.6. As the wall is approached, the range of possible eddy sizes shrinks until the size of the largest and smallest eddies are equal, indicating the elimination of turbulence. At this point y equals δ, the thickness (in a statistical sense) of the viscous sub-layer.

The dashed line on Fig. 3.6 shows that increasing the size of the dissipative micro-eddies leads to an increase in the thickness of the viscous sublayer. It is known that the micro-eddy size is increased by long-chain molecules, and non-Newtonian rheological properties can produce a similar effect, as shown by Wang & Larsen (1994). If other quantities are unaffected the thickened sublayer will, in turn, produce a higher mean velocity for the same wall shear stress, giving a lower friction factor.

A predictive model for non-Newtonians has been developed on this basis (Wilson and Thomas, 1985; Thomas and Wilson, 1987). It was noted that the interaction of the eddies in turbulent flow causes them to increase in axial length, often very rapidly. This 'vortex stretching' process produces abrupt increases in strain rates and in the viscous dissipation of the smallest eddies. As the energy available for dissipation is fixed by the 'turbulent energy cascade', the result is an increase in the micro-eddy size. The model uses the ratio of the integrals under the non-Newtonian and Newtonian rheograms – denoted by α, and defined above in connection with Fig. 3.3 -- to estimate the size increase of the micro-eddies. From the relationship shown on Fig. 3.6, it follows that the thickness of the sub-layer should also be multiplied by a factor equal to α. As shown by Wilson and Thomas (1985), the value of the mean velocity V_m which results from this thickened viscous sub-layer is given by

$$\frac{V_m}{U_*} = \frac{V_N}{U_*} + 11.6\,(\alpha - 1) - 2.5\,\ell n\,\alpha \qquad (3.9)$$

where V_N is the mean velocity for the equivalent smooth-wall flow of a Newtonian fluid with viscosity μ, and the deviation from the logarithmic profile near the pipe axis has been ignored. Manipulation of Eq. 3.9, together with Eq. 3.5 gives the following expression for the equivalent turbulent-flow viscosity

$$\mu_{eq} = \mu\,\alpha\,e^{-4.64(\alpha-1)} \qquad (3.10)$$

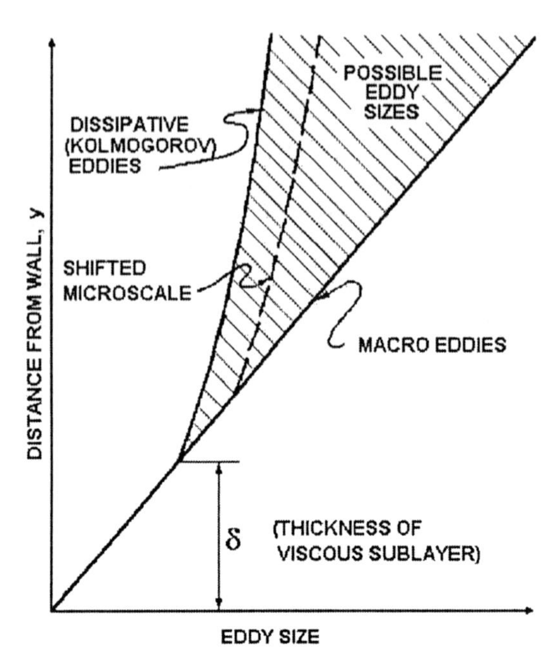

Figure 3.6. Eddy scales in turbulent flow (from Wilson & Thomas, 1985)

It should be noted that this relation is not based on any specific rheological model, and if necessary a series of values of the secant viscosity μ and the area ratio α could be abstracted directly from the rheogram. For the power-law and Bingham models and for the yield-power-law, expressions for α have been worked out (Thomas and Wilson, 1987), and used to obtain μ_{eq}.

Example 3.1

For a power-law fluid with n = 0.15 (a typical value) find the ratio of η_t/μ and μ_{eq}/μ. The tangent viscosity is obtained by differentiating Eq. 3.2, *i.e.* $\eta_t = d\tau/d(du/dy)$ giving

$$\eta_t = n\tau(du/dy) = n\mu$$

Thus, for n = 0.15, $\eta_t = 0.15\ \mu$.

To determine μ_{eq}, it is first necessary to find α, for substitution into Eq. 3.10. Integration of Eq. 3.2, followed by division by the triangular area $0.5\tau(du/dy)$ gives

$$\alpha = 2/(n+1)$$

It follows from Eq. 3.10 that

$$\mu_{eq} = \mu \bullet \left(\frac{2}{n+1}\right) exp\left[-4.64\left(\frac{1-n}{1+n}\right)\right]$$

Thus, for n = 0.15

$$\mu_{eq} = 0.056\ \mu$$

It is worth emphasizing that the power-law model with sub-layer thickening produces simple proportionalities linking μ_{eq}, η_t and μ, with the proportionality constants depending only on the power n. It should also be noted that μ_{eq} is significantly less than the tangent velocity η_t (which in turn is much smaller than the secant viscosity μ). For this model μ_{eq} decreases with increasing shear rate (and hence with increasing shear stress and U*), but real fluids may differ in this respect. For example, data obtained by Thomas (1981) showed μ_{eq} increasing with increasing U*, although both μ and η_t displayed the opposite trend.

This completes Example 3.1, and we can next consider other methods that have been used, in effect, to evaluate μ_{eq}. A common method for predicting μ_{eq} has been to equate it to the 'tangent' viscosity of the rheogram, *i.e.* η_t as shown on Fig. 3.3. This approach is usually applied to Bingham materials (Hedström, 1952) where η_t is a constant, η_B (which is less than the secant viscosity μ). However, no physical mechanism has been proposed to account for this use of the tangent viscosity, which generally does not accord with the experimental data (these normally yield values of μ_{eq} less than η_B, see Wilson and Thomas, 1985).

Another model, proposed by Dodge and Metzner (1959) for power-law fluids, contains two parameters that they had evaluated empirically. Although their technique does not provide a direct formulation of μ_{eq}, calculations show that it gives equivalent-viscosity values which are smaller than η_t and decrease with increasing U_* (similar to the results for μ_{eq} based on power-law fluids, noted above).

Rheological modelling will be mentioned again in connection with the settling of coarse particles in non-Newtonian media, which is covered in Chapter 7. The next topic to be dealt with here concerns techniques for scale-up of non-Newtonian data.

3.4 Scale-up of Laminar Flow

The laminar flow of any given fluid can be scaled from one pipe size to another by use of the Rabinowitsch-Mooney technique. The only stipulations are that there is no 'slip' of the fluid at the wall (see Metzner, 1961) and that the flow remains laminar in both pipes (turbulent flow will be dealt with separately below).

As mentioned in Section 3.2, the basis of the Rabinowitsch-Mooney technique is the plot of wall shear stress τ_o *versus* $8V_m/D$. Rabinowitsch (1929) and Mooney (1931) proved that for all steady uniform laminar flows in a pipe τ_o and $8V_m/D$ are linked in a functional relationship. In other words, for a given material the values of both τ_o and $8V_m/D$ can be determined from experiments in a single pipe. This experimentally-determined plot or function linking the two variables can then be applied to all laminar flows of the material in question, whatever the pipe diameter.

The application of the method is based on the logarithmic plot of τ_o *versus* $8V_m/D$ which was discussed previously. An example is Fig. 3.4, based on data obtained from a pipe of internal diameter 203 mm. For some other pipe, of diameter D_2, the pressure gradient dp/dx can be obtained from the plotted values of τ_o by multiplying them by $(4/D_2)$. Similarly the values of mean velocity in the pipe of diameter D_2 equal the values of $8V_m/D$ multiplied by $(D_2/8)$. Since the factors by which each of the co-ordinates of Fig. 3.4 are multiplied depend only on D_2, a simple re-scaling of the axes of this figure gives a plot of dp/dx versus V_m for a pipe of diameter D_2. For the logarithmic co-ordinates of this figure, this type of scaling amounts to a pair of linear translations, which do not affect the shape of the plot. In this regard it should be noted that a Newtonian fluid shows a slope of unity on a logarithmic plot of pressure gradient versus mean velocity, whereas the slope of the non- Newtonian logarithmic plot which appears on Fig. 3.4 is much flatter.

An alternate way of dealing with scale-up is to transform the data from the experimental pipe (diameter D_1) to the prototype (diameter D_2) on a point-by-point basis. This process is similar to scaling head and discharge data for a centrifugal pump by using the affinity laws (see Chapter 9). For the pipe the analogous quantities are the frictional gradient i_m (m of water/m, see Eq. 2.44) and the mean velocity V_m. With the subscripts 1 and 2 denoting experimental and prototype conditions, respectively, the scaling relations (affinity laws) for laminar flows in pipes are:

$$(j_m)_2 = (j_m)_1 \left(\frac{D_1}{D_2} \right) \tag{3.11}$$

and

$$(V_m)_2 = (V_m)_1 \left(\frac{D_2}{D_1} \right) \tag{3.12}$$

3.5 Scale-up of Turbulent Flow

As mentioned in the previous section, if it is foreseen that flow in the prototype pipeline may be turbulent, it is important to extend the small-scale experimental tests into the turbulent-flow region, and then to scale these test results up to prototype size. For turbulent scaling, as for the laminar version, the quantity which remains unchanged is the wall shear stress τ_o. As in the laminar case, the wall shear stress fully determines the stress distribution within the pipe. For example, with a material having a yield stress τ_y a specified value of τ_o is sufficient to give the fraction of the pipe area where $\tau < \tau_y$, a factor which may affect the mean velocity (Wilson and Thomas, 1985; Thomas and Wilson, 1987). Likewise, a given value of τ_o, and hence of U_*, determines conditions within the viscous sub-layer and hence establishes both μ and the area ratio α (used in the previous section). The result is that τ_o (or U_*) is sufficient to determine the equivalent turbulent-flow viscosity μ_{eq} . For each value of τ_o two scaling laws apply to turbulent pipe flow. The first, for pressure gradient, is the same as that established above for laminar flow, i.e.

$$(j_m)_2 = (j_m)_1 \left(\frac{D_1}{D_2} \right) \tag{3.11}$$

The second scaling law, for mean velocity, is given by

$$(V_m)_2 = (V_m)_1 + 2.5 U_* \ell n (D_2 / D_1) \tag{3.13}$$

As with Eq. 3.11 and Eq. 3.12 for laminar flow, the combined use of Eq. 3.11 and Eq. 3.13 allow scale-up of data for turbulent non-Newtonian flow from one pipe size to another. The velocity scaling relation (Eq. 3.13) was originally derived on the basis of a hydraulically smooth boundary (Wilson, 1986), for which Eq. 3.6 applies. However, Eq. 3.13 is not limited to smooth boundaries. A more general friction relation that takes account of the boundary roughness size ε is the Colebrook-White equation (Eq. 2.33), which can be written

$$V_m = -2.43\, U_* \ln\left[\frac{\varepsilon}{3.7D} + \frac{0.89\,\mu}{\rho_f U_* D}\right] \tag{3.14}$$

or

$$V_m - 2.43\, U_* \ln D = -2.43\, U_* \ln\left[\frac{\varepsilon}{3.7} + \frac{0.89\,\mu}{\rho_f U_*}\right] \tag{3.15}$$

For a non-Newtonian μ can be replaced by μ_{eq}, but this depends only on U_*, as noted previously. Thus the right-hand side of Eq. 3.15 depends on U_*, but not on D, and hence is invariant for constant τ_o and roughness size. Ignoring the slight differences in the evaluation of the reciprocal of von Karman's constant (2.43 versus 2.5, see the comments following Eq. 2.33) the scaling law derived from Eq. 3.15 is found to be the same as Eq. 3.13. Incidentally, it is seen that Eq. 3.13 is also valid for cases where μ_{eq} (or μ) is constant, *i.e.* for Newtonian fluids. Moreover, it has been demonstrated (Wilson, 1989) that this scale-up technique for turbulent flows is also applicable to dilute polymer solutions exhibiting drag reduction.

In short, the affinity laws represented by Eq. 3.11 and Eq. 3.13 give a generally applicable method of scaling turbulent flows which complements the Rabinowitsch-Mooney scaling technique for laminar flows (Eqs. 3.11 and 3.12).

Figure 3.8 (from Wilson, 1986) shows an example of scale-up of turbulent flow, using data for a 7.5 percent kaolin slurry tested by Thomas (1981) in pipes of internal diameter 18.9 mm and 105 mm. The figure plots pressure gradient *versus* mean velocity for the data from both pipe sizes, and also the scaled-up points from the smaller pipe, transformed to the larger pipe diameter by Eq. 3.11 and Eq. 3.13. Although the points do not plot as straight lines on the logarithmic co-ordinates of Fig. 3.7, (not unexpected for non-Newtonian flows), the points scaled up from the smaller pipe are in excellent agreement with the observed points from the larger pipe.

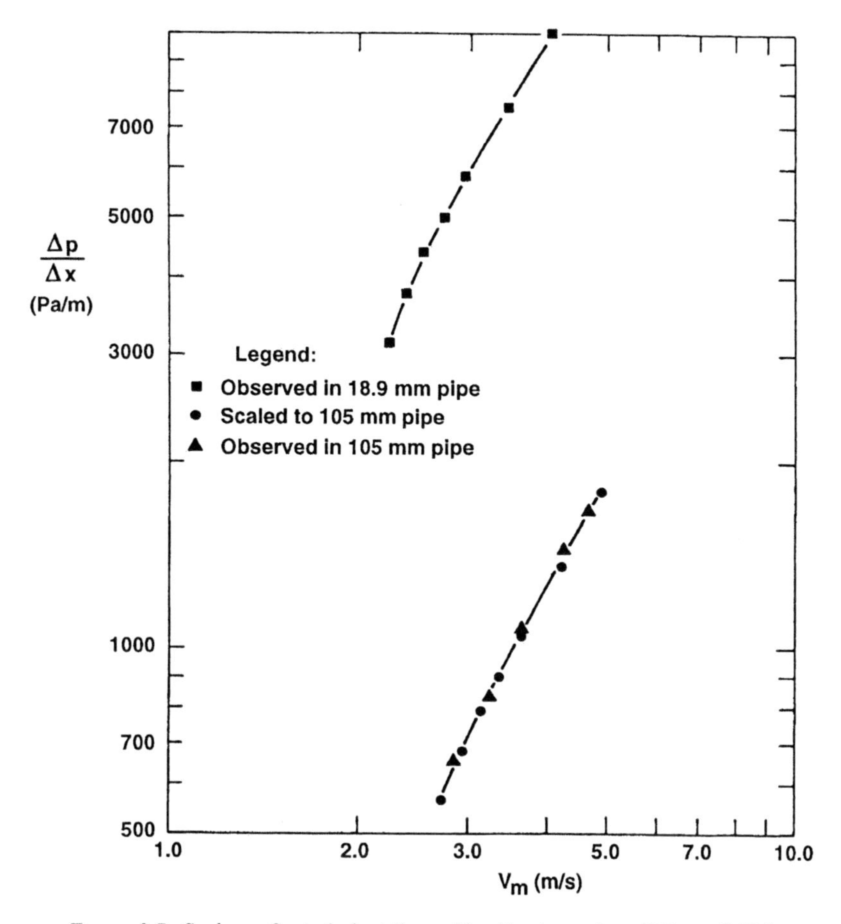

Figure 3.7. Scale-up for turbulent flow of kaolin slurry, from Wilson (1986)

The velocity scaling law (Eq. 3.13) can be written in an alternative form using the fluid friction factor f_f, i.e.

$$(V_m)_2 = (V_m)_1 \left[1 + 2.5\sqrt{\frac{f_f}{8}}\, \ell n\left(\frac{D_2}{D_1}\right) \right] \tag{3.16}$$

In practice, the final term within the brackets of Eq. 3.16 is usually rather small, and hence the ratio $(V_m)_2/(V_m)_1$ is generally not greatly different from unity in turbulent-flow scale-up. For example, for the case illustrated by Fig. 3.7, the ratio $(V_m)_2/(V_m)_1$ is not constant (because of variation in f_f) but clusters around a mean value of about 1.23. In comparison, for laminar flows the velocity ratio equals the diameter ratio (Eq. 3.12), which is much greater for these pipes, having a value of 5.35.

Up to this point laminar and turbulent scale-up techniques have been considered separately, but it should be noted that for non-Newtonian flows there is a special interest in conditions near the laminar-turbulent transition. There are two reasons for this interest, one economic and the other based on reliability of operation. The question of operability is explored in Chapter 13. The economic reason is associated with the flat slopes of logarithmic plots of τ_o *versus* $8V_m/D$ which, as mentioned previously, are typically found for laminar non-Newtonian flows. In this situation slight increases in pressure gradient produce much larger increases in mean velocity, and hence tend to be attractive economically. The converse applies after the transition to turbulent flow, which is accompanied by an abrupt increase in the slope of the log-log plot; as a result the favoured operating point for many non-Newtonians is at velocities close to the transition to turbulence.

So long as the slurry is truly non-settling, operation on the laminar side of the transition will be reliable, provided that solids concentration and size are not variable (see Chapter 13). If the slurry contains larger particles which might be subject to settling (see Chapter 7), many designers would prefer to operate in the turbulent-flow range, just above transition. There are many cases where this reliability condition may be over-conservative (Thomas, 1978; Duckworth *et al.*, 1982), but in all events the laminar-turbulent transition is of considerable interest.

This transition can be displayed conveniently on the logarithmic plot of τ_o *versus* $8V_m/D$. An example is shown on Fig. 3.8, using data obtained at the GIW Hydraulic Laboratory for a clay-water slurry in a pipe of 0.077 m internal diameter. The transition is shown by the slope change on the figure, which takes place at $8V_m/D$ of about 420 s^{-1}. On the laminar side of this transition the slope is virtually constant at $n' = 0.12$, and on the turbulent side the slope rises rapidly to approach a value near 2.0. These data are now to be used to predict conditions in a 'prototype' pipeline twice as large (D_2 = 0.154 m). As discussed previously, for the axis system used on Fig. 3.8, laminar flows in all pipe sizes are superposed. Thus the laminar plot for the larger pipe will coincide with that of the smaller one, and on these axes only the turbulent points require scaling. In this scaling τ_o remains unchanged, and a new value of $8V_m/D$ is required. Turbulent-flow scaling by Eq. 3.13 gives this transformed value, say $8(V_m)_2/D_2$ as

$$\frac{8(V_m)_2}{D_2} = \frac{8}{D_2}\left[(V_m)_1 + 2.5\,U_*\,\ell n\left(\frac{D_2}{D_1}\right)\right] \tag{3.17}$$

where the shear velocity U_* is taken from the value of τ_o which applies to the data point in question. The transformed points for turbulent flow are shown on Fig. 3.8; they can then be replotted onto the i_m *versus* V_m graph for the larger pipe in the same way as the laminar-flow points.

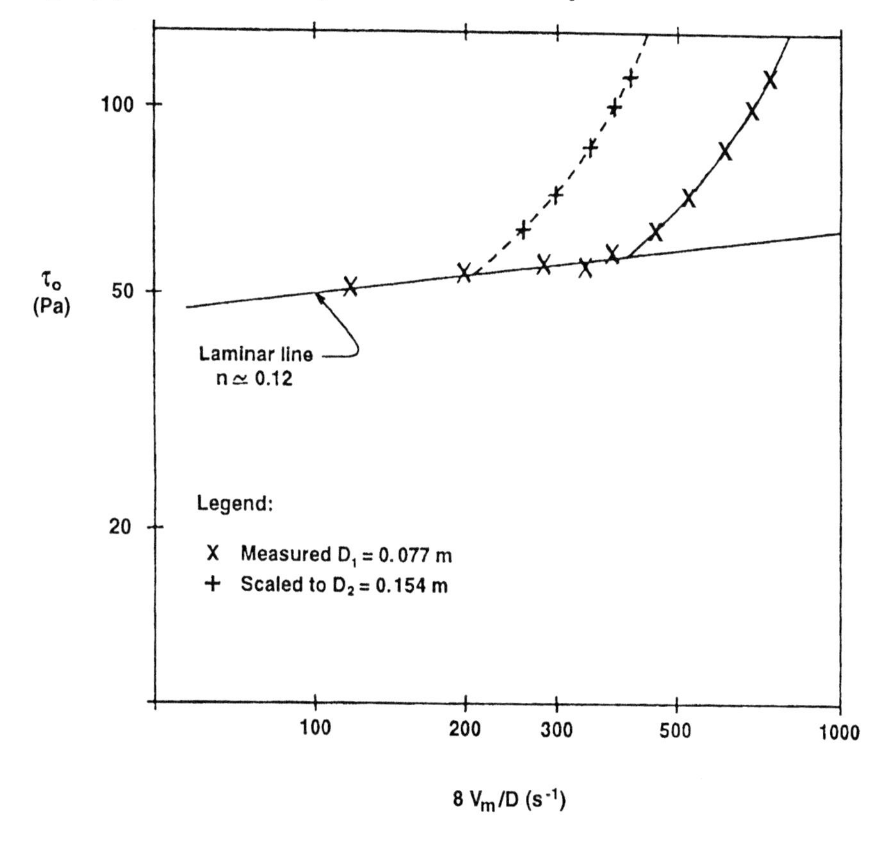

Figure 3.8. Turbulent-flow scale-up using τ_o and $8V_m/D$

It can be seen from the figure that there is a gap between the transformed turbulent points and the laminar line. This missing segment must be interpolated in order to obtain the transition point, which is shown on Fig.

3.8 at $8(V_m)_2/D_2 \approx 210\ s^{-1}$, giving $(V_m)_2 \approx 2.0$ m/s. In this instance the transition velocity for the larger pipe does not differ from that found in the smaller pipe, which also occurred at about 2.0 m/s. Other example calculations will be given in the case study of scale-up presented in Section 3.8.

3.6 Effects of Solids Concentration

The previous sections have implicitly assumed slurries with a fixed concentration of solids, but it is now time to return to the concepts raised in Section 3.1 regarding the effects of varied solids concentration. The case of laminar flow will be considered first, and for this purpose it is convenient to employ the Rabinowitsch-Mooney axis system -- a log-log plot of wall shear stress τ_o *versus* $8V_m/D$. (The y-axis could also be $\log(\tau_o/\rho_m)$, where ρ_m is mixture density.) In this system, a straight line segment represents a power-law rheological model (Eq. 3.2) with the value of n equal to the slope n' of the line. If the concentration of solids is increased, the result will be a new line which lies above the original one (*cf.* Fig. 3.1). It is found that the lines for various concentrations are often essentially parallel within sizeable regions of a log-log plot such as Fig. 3.2 (and likewise on the log-log Rabinowitsch-Mooney plot, implying a constant value of n', and n). The separation between the lines usually increases sharply with increasing solids concentration. This variation can often be expressed by a power law, based either on the volumetric concentration C_v or on the quantity $(S_m - 1)$, which is proportional to C_v. The latter basis gives an equation of the type

$$\tau_o = G(S_m - 1)^m \left(\frac{8V_m}{D}\right)^n \tag{3.18}$$

where G, m and n are numerical coefficients selected to fit the experimental data. An alternative form of this equation uses τ_o/ρ_m for the left hand side, with somewhat different numerical values for G and m.

A considerable corpus of non-Newtonian data obtained at the GIW Hydraulic Laboratory has been analysed by Dr. M. R. Carstens on the basis of Eq. 3.18. An example of the effect of $(S_m - 1)$ is shown on Fig. 3.9. The material is a 'red mud' associated with the production of aluminium, and comparable to the material used for Fig. 3.2. It was found that n is approximately 0.09 in this case. Thus the ordinate of Fig. 3.9, equal to $G(S_m - 1)^m$, can be evaluated as $\tau_o/(8V_m/D)^{0.09}$. The abscissa is $(S_m - 1)$, and both axes are plotted logarithmically. The slope of the fit line on this figure, *i.e.*

the value of m, is about 3.5. Such a value, which is not unusual for non-Newtonian slurries, implies that minor changes in the relative density S_m have a grossly disproportionate effect on the ordinate of Fig. 3.9. Assuming no change in pipe diameter and V_m, the shear stress τ_o, and hence the pressure gradient, will be directly proportional to this ordinate. For example, a change of S_m from 1.2 to 1.3 (*i.e.* a change of mixture density from 1200 to 1300 kg/m^3, which is less than 10%) produces a four-fold increase in pressure gradient. It is worth stressing that large changes of this sort are normal, rather than exceptional, for homogeneous non-Newtonian slurries.

Another example, showing a similar effect of (S_m - 1), deals with phosphate slimes tested in a pipe loop of internal diameter 0.101 m. A limited range of slurry density (1090 to 1115 kg/m^3) produced large changes in pressure gradient. When Eq. 3.18 was applied to the laminar data for this material, it was found that n was virtually constant at 0.20, and that m had a value of about 3. This value of m indicates a doubling of pressure gradient when the mixture density changes from 1090 to 1115 kg/m^3. As with the red-mud plot of Fig 3.2, the laminar points have nearly the same slope on a log-log plot, with a wide separation between different concentrations, but there is comparatively little separation of the turbulent points.

Before proceeding, it is appropriate to re-iterate that the behaviour of non-Newtonians varies greatly between one material and another. The red mud referred to above was pumpable up to a maximum mixture density of approximately 1500 kg/m^3. However, for phosphate slimes mentioned in connection with Figs. 3.3, 3.4, and in the previous paragraph, slurries with mixture density beyond about 1170 kg/m^3 were not pumpable. For example, the slurry of Figs. 3.3 and 3.4 has density 1130 kg/m^2, and requires a wall shear stress of more than 50 Pa to set it in motion.

3.7 Transition from Laminar to Turbulent Flow

As stressed earlier in this chapter, the appropriate operating point for homogeneous pipeline flow is usually near the laminar-turbulent transition, and thus it is important to be able to predict transition conditions. In Section 3.5 it was proposed that the best approach is to test the slurry in a tube viscometer, making sure that the test range includes both laminar and turbulent flows. The laminar and turbulent points are then scaled separately to the prototype pipe size, and projected to obtain the transition point (see Section 3.5, and Case Study 3.1).

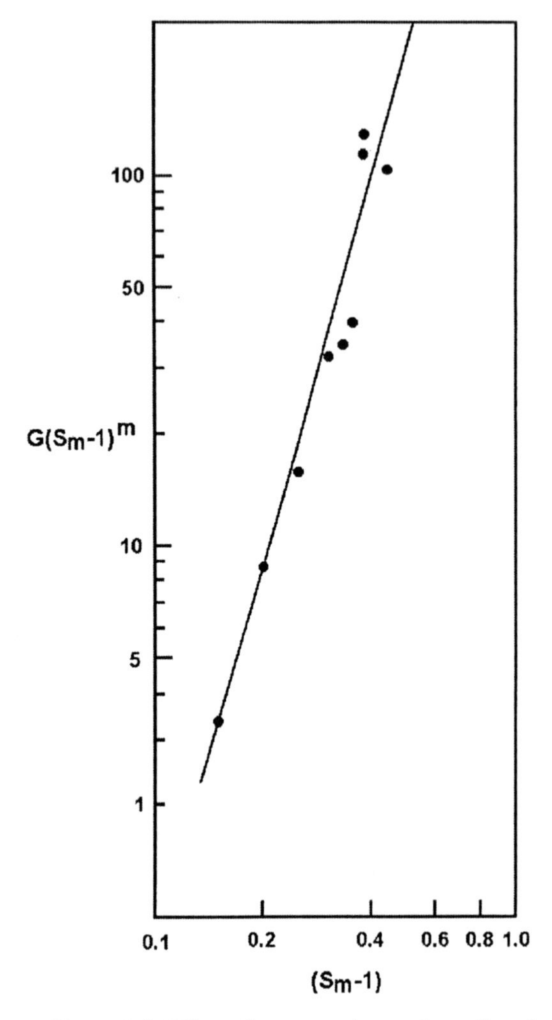

Figure 3.9. Effect of concentration on flow of a red mud

In other instances, turbulent flow data may not be available for the slurry at the solids concentration being considered. In such a case, experimental results like those plotted on Fig. 3.2 can be very useful in extrapolating or interpolating from other concentrations. To put this process on a sound basis we will now analyse the locus of the laminar-turbulent intercepts, where both laminar and turbulent frictional relationships apply.

Both power-law and Bingham-type slurries will be considered. As shown in Section 3.6, the laminar formulation of Eq. 3.18, with constant n, is equivalent to a power-law rheological model. It was also seen, in Example

3.1, that a model of this type produces simple proportionalities linking μ_{eq}, η_t and μ (with the proportionality coefficients depending only on n). For the power-law model it is has been shown (Wilson, 1996) that the friction factor at the laminar-turbulent intercept depends only on the value of n, with constant n producing a constant value of f. For the commonly encountered range $0.10 \leq n \leq 0.20$, this function, say f(n) is closely approximated by the straight line

$$f(n) = 0.0095 + 0.0430\, n \tag{3.19}$$

For example, with n = 0.15 (the typical value investigated in Example 3.1) the value of f is found to be 0.016. The transition locus can then be calculated using this value of f.

If at all possible, for each material the transition should be evaluated from at least one set of experimental runs extending into the turbulent flow zone (rather than by using the value of n estimated from laminar data together with Eq. 3.19). The transition velocity V_T can then be scaled to other conditions by noting that, for constant n, it is proportional to $[k/\rho]$ raised to the power 1/(2-n) and inversely proportional to the pipe diameter D to the power n/(2-n).

The transition velocity, V_T, can also be estimated based on fitting a Bingham rheological model to the laminar data. An early expression for the transition of a Bingham material from laminar to turbulent flow was proposed by D.G. Thomas (1963) in the form:

$$\frac{V_T D \rho}{\eta_B} = 2100\left[1 + \frac{\tau_B D}{6 \eta_B V_T}\right] \tag{3.20}$$

This equation can be re-formulated in terms of the Hedström number ($He = \rho D^2 \tau_B / \eta_B$) and the ratio $V_T / \sqrt{(\tau_B / \rho)}$ (a dimensionless measure of transition). It is found that $V_T / \sqrt{(\tau_B / \rho)}$ diminishes monotonically with He, dropping from a value near 30 for $He \approx 10^4$ to 22 for $He \approx 10^5$ and then diminishing slowly to slightly below 19 at $He = 10^8$.

Experience at the GIW Hydraulic Laboratory has shown that Eq. 3.20 under-predicts V_T. The result is the following empirical expression

$$V_T \geq 20\sqrt{(\tau_B / \rho)} \tag{3.21}$$

An example using the techniques discussed above is given in the case study presented in the next section.

3.8 Case Study

Case Study 3.1 - Non-Newtonian Slurry of Phosphate Slimes

A phosphate-slimes clay-water slurry is to be pumped over a horizontal distance of 700 m, using a pipe of internal diameter 305 mm. The slurry will be taken from a pond in which the mixture density is 1130 kg/m^3. Tests have been carried out with this material at the GIW Hydraulic Laboratory, using a pipe of internal diameter 203 mm. Test data for j_m and V_m appear in Table 3.1 together with values of $8V_m/D$ and τ_o (the wall shear stress equals $\rho g j_m D/4$). Figure 3.10 shows the values of j_m plotted *versus* V_m.

Table 3.1. Data for phosphate slimes slurry in 203 mm pipe

Run	V_m (m/s)	$8V_m/D$ (s^{-1})	j_m	τ_o (Pa)
1	0.53	21.0	0.089	49.9
2	1.52	60.1	0.100	56.2
3	2.00	78.8	0.102	57.2
4	2.59	102.1	0.105	59.1
5	3.24	127.9	0.108	60.1
6	3.81	150.3	0.110	61.5
7	4.43	174.7	0.113	63.4
8	5.12	202.0	0.119	67.0
9	5.64	222.6	0.130	73.2

The preliminary design of the pumping system calls for a discharge Q_m of 0.30 m^3/s in the 0.305 m pipe. However, this value is not yet definite, and Q_m values of 0.20 m^3/s and 0.40 m^3/s are also to be considered.

(a) Find the values of i_m for the three values of Q_m noted above. On dividing by the pipe area of 0.0731 m^2, the required velocities are found to be 2.74 m/s, 4.11 m/s and 5.48 m/s. Also calculate the required head and pressure at the pump for each case.

The first step is to scale the data points from the 0.203 m pipe to the 0.305 m one. For the laminar-flow points the appropriate scaling laws are given by Eq. 3.11 and 3.12, which show that i_m scales inversely with the diameter ratio while V_m scales directly with this ratio. Points from test runs 1 to 7 have been scaled on this basis and are shown on Fig. 3.10. For runs 8 and 9 the flow is turbulent. The hydraulic gradient can still be scaled by Eq. 3.11, but V_m must now be scaled by Eq. 3.13. This requires evaluation of U_*, i.e. $\sqrt{(\tau_0 / \rho)}$. For example, run 8 with τ_o = 67.0 Pa has U_* = 0.243 m/s, for which Eq. 3.13 gives a scaled-up value of V_m equal to 5.37 m/s. The laminar-turbulent transition point is obtained by projecting the turbulent line back to intercept the laminar line. For the test data in 0.203 m pipe this intercept occurred to the right of point 7, at a velocity of about 4.9 m/s. For the larger pipe the intercept lies between points 4 and 5, with a slightly reduced velocity, about 4.7 m/s.

Figure 3.10 shows that conditions will be laminar for the two lower flows to be investigated, and turbulent for the highest flow. The values of j_m can be taken directly from the figure, and are listed in Table 3.2. As i_m is expressed in m of mixture per m of pipe, $\Delta p/\Delta x$ equals $\rho_m g j_m$ or 11085 j_m, and the required pressure at the pump is the product of this quantity and the pipe length.

Table 3.2. Gradients and Pressures

Q_m (m^3/s)	0.20	0.30	0.40
V_m (m/s)	2.74	4.11	5.48
j_m	0.0673	0.0704	0.0810
$\Delta p/\Delta x$ (Pa/m)	746	780	898
Pump pressure (kPa)	522	546	628
Pump head (m water)	53.2	55.6	64.0
Pump head (m slurry)	47.1	49.2	56.6

(b) Find the power which must be supplied to the motor driving the pump, assuming motor efficiency of 95% and pump efficiency of 75%. Also find the specific energy consumption, both per tonne of mixture and per tonne of delivered solids on a dry-weight basis, assuming $S_s = 2.65$.

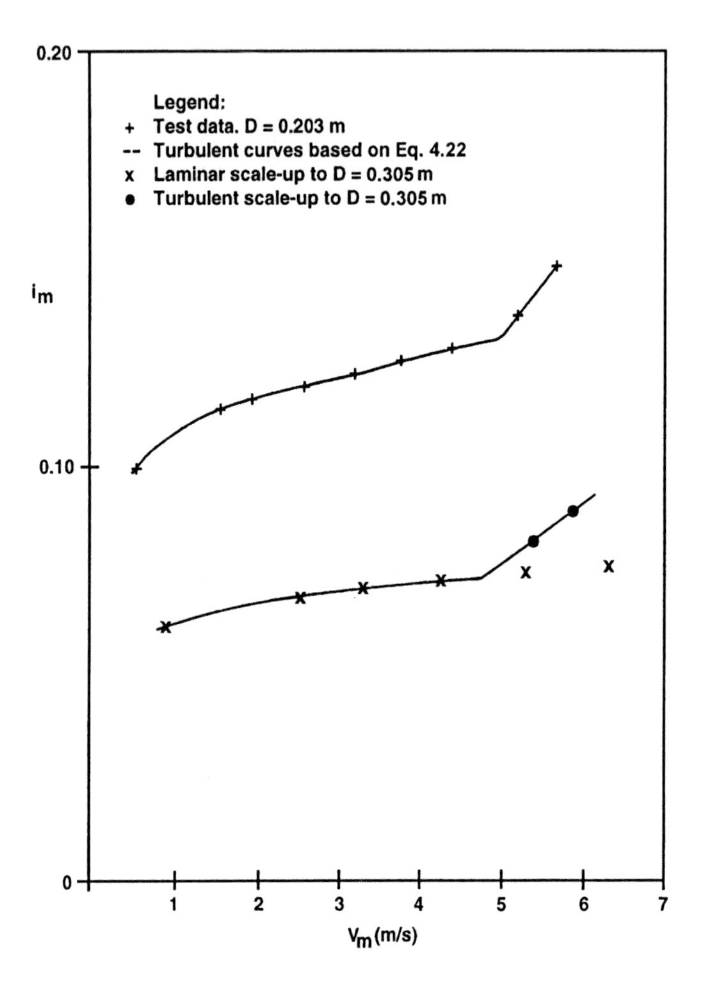

Figure 3.10. Test data and scale-up for Case Study 3.1

The output power from the pump, in kW, equals $g\rho_w QH/1000$ or 9.81 QH with H in m of water, and the power required by the motor will equal the input power divided by product of the efficiencies, giving 13.77 QH. For example at Q_m of 0.30 m^3/s H is 55.6 m of water, and the power required by the motor is 230 kW. To find the specific energy consumption in this case, we also require the transport in tonnes per hour. On the basis of mixture at 1.13 tonnes/m^3, this is 0.3 (1.13)(3600) or 1220 tonnes/h, so that the tonne-km/hour is 854.3, and the specific energy consumption is 230/854.3 or 0.269 kWh/tonne-km. On a dry-solids basis the useful throughput tonnage is based on $Q_mS_sC_v$. The relative density of the mixture, *i.e.* 1.13 equals $1 + (S_s - 1)C_v$, from which, with $S_s = 2.65$, C_v is equal to 0.0788, giving a dry-solids throughput of 225.5 tonnes/h for SEC = 1.458 kWh/tonne-km. The same result could be obtained by using the formula that SEC = 2.73 i_m/S_sC_v, giving 1.040 kWh/tonne-km out the basis of pump output. For kW input to the pump motor, this figure must be divided by (0.95)(0.75), giving 1.458 kWh/tonne-km, as before. The various results, for all three values of Q_m, are given in Table 3.3.

Table 3.3. Power and Energy Consumption

Q_m (m^3/s)	0.20	0.30	0.40
Power required (kW)	147	230	353
SEC for mixture (kWh/tonne-km)	0.257	0.269	0.310
SEC for dry solids (kWh/tonne-km)	1.394	1.458	1.678

As the solids concentration is fixed, the specific energy consumption varies only with j_m, in a direct proportionality. The flat curve for laminar flow, plotted on Fig. 3.11 shows a very small rise in j_m as Q_m goes from 0.20 to 0.30 m^3/s, and the rise in SEC for this step is less than 5%. On the other hand, the equal increase of Q_m from 0.30 to 0.40 m^3/s requires an increment in SEC of more than 15%, as a result of the shift from laminar to turbulent flow.

(c) Suppose now that testing had stopped after run 6, so that no turbulent-data flow points were available. What effect would this have on the results found above?

In this case the transition to turbulent flow must be predicted from the parameters of the laminar-flow rheogram. The first step is to prepare a logarithmic plot of τ_o *versus* $8V_m/D$ and find the slope n′. As it happens, the data from the laminar test runs for this material were used in plotting Fig. 3.4, from which it can be seen that all points lie close to a line with n′ = 0.110. The ratio (3n′ + 1)/4n′ is thus 3.02, which is multiplied by the values of $8V_m/D$, to give du/dy, as shown on Fig. 3.5. It was seen that all points on that figure except the lowest one can be considered as obeying a Bingham formulation of the rheogram, with τ_y = 52.7 Pa and η_B = 0.020 Pa·s.

If the power law model is considered appropriate, the locus of laminar-turbulent intercepts will correspond to a constant f evaluated (from Eq. 3.19) as 0.0142. On this basis the intercept velocity V_T should be almost independent of pipe diameter, with a predicted value of 5.53 m/s in the prototype pipe (D = 0.305 m). (For the test pipe, with D = 0.203 m, the predicted V_T is slightly larger, 5.66 m/s). Comparison with the points plotted on Fig. 3.10 show that in this case the intercept-locus found using the power-law model underpredicts friction loss in this area and hence is not appropriate for the present example.

An alternative approach is to use a Bingham-plastic model. As seen above, Fig. 3.5 showed the Bingham parameters of this material to be τ_B = 52.7 Pa and η_B = 0.020 Pa.s. Application of Eq. 3.21 gives $V_T \geq 4.3$ m/s. It is also found that the Hedström number is $6.1(10^6)$ for the laboratory pipe and $1.4(10^7)$ for the prototype one. A study presently underway by K.C. Wilson and A.D. Thomas suggests that, for a for a Bingham plastic with sub-layer thickening and this range of Hedström number, $V_T/\sqrt{(\tau_B/\rho)}$ is about 25. This value gives $V_T \approx 5.4$ m/s. This result is very similar to that obtained above for the power-law model, overestimating the experimental determination of 4.9 m/s (for which $V_T/\sqrt{(\tau_B/\rho)}$ is near 23). In short, predictions for the laminar-turbulent intercept can give helpful indications, but testing in the turbulent flow regime is desirable if it is foreseen that the prototype conditions may be turbulent.

In the present instance, the flow at $Q_m = 0.20$ m^3/s is clearly laminar, and so is that at $Q_m = 0.30$ m^3/s. Thus neither of these points will be influenced by the omission of the turbulent flow data. However, for $Q_m = 0.40$ (V_m = 5.48 m/s), the value of j_m could be altered significantly. To be conservative, we will assume the transition value of 4.3 m/s as estimated from Eq. 3.21. On the basis, j_m is predicted to be 0.115 m slurry/m pipe, *i.e.* some 42% in excess of the value obtained previously on the basis of scale-up of the turbulent-flow data (and 56% above the result of extrapolating the laminar flow line to this velocity). Compared to the change of about 5% found previously for a velocity change of 1.37 m/s in the laminar-flow region, it is now seen that the same change accounts for a 63% increase in j_m, and hence in specific energy consumption. Nothing could give a clearer indication of the differing effects of laminar and turbulent flow of non-Newtonian slurries.

REFERENCES

Dodge, D.W. & Metzner, A.B. (1959). Turbulent flows of non-Newtonian systems. *AIChEJ.,* Vol. 5, No. 2, pp. 189-204.

Duckworth, R.A., Pullum, L. & Lockyear, C.F. (1982). The hydraulic transport of coarse coal at high concentration. *Proc. 4th Intern'l Symposium on Freight Pipelines,* Atlantic City, NJ, USA.

Hedström, B.O.A. (1952). Flow of plastics materials in pipes. *Ind. and Eng. Chem.*, Vol 44, No. 3, pp. 651-656.

Lumley, J.L. (1973). Drag reduction in turbulent flow by polymer additives, *J. Poly Sci., Macromol, Rev.* 7, A. Peterlin (Ed.), Interscience, New York, pp. 263-290 (1973).

Lumley, J.L. (1978). Two-phase flow and non-Newtonian flow. In *Turbulence,* P. Bradshaw (Ed.), Topics in Applied Physics, Vol. 12, Springer-Verlag, Berlin (1978), Chap. 7.

Metzner, A.B. (1961). Flow of non-Newtonian fluids. *Section 7 of Handbook of Fluid Dynamics,* Ed. V.L. Streeter, McGraw Hill, New York.

Mooney, M. (1931). Explicit formulas for slip and fluidity. *J. Rheol.* Vol. 2, p. 210 ff.

Rabinowitsch, B. (1929). Ueber die Viscosität und Elastizität von Solen. *Zeitschrift physik. Chem.*, **A** 145, p. 1 ff.

Thomas, A.D. (1978). Coarse particles in a heavy medium -- turbulent pressure drop reduction and deposition under laminar flow. *Proc. Hydrotransport 5*, BHRA Fluid Engineering, Cranfield, UK.

Thomas, A.D. (1981). Slurry pipeline rheology. *Proc. 2nd Nat'l Conf. on Rheology*, Sydney, Australia.

Thomas, A.D. & Wilson, K.C. (1987). New analysis of non-Newtonian turbulent flow--yield-power-law fluids. *Canad. J. Chem. Engrg.*, Vol. 65, pp. 335-338.

Thomas, D.G. (1963). Non-Newtonian suspensions. Part 1, physical properties and laminar transport characteristics. *Ind. and Eng. Chem.* Vol. 55, No. 11, pp. 18-29.

Wang, Z. & Larsen, P. (1994). Turbulent structure of water and clay suspensions with bed load. *J. Hydraulic Engrg.*, ASCE, Vol. 120, No. 5, pp. 577-600.

Wilson, K.C. (1986). Modelling the effects of non-Newtonian and time-dependent slurry behaviour. *Proc. Hydrotransport 10*, BHRA Fluid Engineering, Cranfield, UK, pp. 283-289.

Wilson, K.C. (1989). Two mechanisms for drag reduction. *Drag Reduction in Fluid Flows, Techniques for Friction Control,* Ed. H.R.J. Sellin and R.T. Moses, pp. 1-8. Ellis Horwood Ltd., Chichester, UK.

Wilson, K.C. (1999). Transitional and turbulent flows of Bingham plastics. *Min. Pro. Ext. Met.Rev.* Vol. 20, pp 225-237. Overseas Publishers Association N.V.

Wilson, K.C. (1996). Laminar-turbulent transition locus for power-law non-Newtonians. *Proc. Hydrotransport 13,* BHR Group, Cranfield, UK, pp 61-74.

Wilson, K.C. &Thomas, A.D. (1985). A new analysis of the turbulent flow of non-Newtonian fluids. *Canad. J. Chem. Engrg.*, Vol 63, pp 539-546.

Chapter 4

PRINCIPLES OF PARTICULATE SLURRY FLOW

4.1 Introduction

The old fable of the blind men and the elephant, (Saxe, 1872), was introduced in Chapter 1. Each blind man had touched a different portion of the animal, and they subsequently found themselves in greater disagreement than before. Slurry flow, like elephants, can be visualised in many ways, with results that sometimes bring to mind the acrimonious deliberations of the blind men. In order to maintain a holistic approach, it is necessary first to examine various aspects of flow and then see how these come together to form the 'whole elephant'. In general terms, slurries are mixtures of solid particles in a carrier fluid. On the one hand, the carrier fluid may be water, or some other Newtonian liquid: on the other, it may be a non-Newtonian liquid, or a mixture of water and small particles which may behave like a non-Newtonian. It was shown in Chapter 3 that homogeneous non-Newtonian flows can be either laminar or turbulent, with operation often taking place near the laminar-turbulent transition.

Slurries of larger particles are generally not homogeneous. The force of gravity will promote particle settling, and particles that have settled will be supported by granular contact. There are also several forces which inhibit settling. Some, i.e. Brownian motion and surface forces, are significant only for small particles. For larger particles, hydraulic lift forces may be significant -- these are produced by strong velocity gradients and by particle rotation. For turbulent flow, the normal condition for settling slurries, turbulent diffusion tends to even out differences in solids concentration.

4.2 Homogeneous and Pseudo-Homogeneous Flow

A slurry of small particles (say less than 40 μm or 270 mesh, one conventional division between silt and sand sizes) in turbulent motion tends to behave in a *homogeneous* fashion. The particles are distributed throughout the flow, and there is little change of solids concentration with height. When the particle grading is broader, the small particles continue to behave in this fashion, and are known as the homogeneous load. The fraction of the total mass solids which is less than 40 μm is known as the carrier-fluid fraction, denoted by X_f.

An aqueous slurry of somewhat larger particles (say 100 μm, which is in the fine-sand range) exhibits somewhat different behaviour. If the mean velocity of flow is several metres per second, particles will be distributed throughout body of the flow by turbulent diffusion, but there will be a measurable decrease of concentration with height. This type of flow is described as *pseudo-homogeneous* and shares with truly homogeneous flow the property that the pressure gradient increases with throughput velocity in a fluid-like fashion. An increase of this type can be expressed, at least to a reasonable approximation, by the statement that the pressure drop for turbulent flow of a pseudo-homogeneous mixture is proportional to that obtained for an equal discharge of carrier fluid alone.

Before this statement can be put into mathematical form, it is necessary to consider various ways of expressing the pressure drop due to friction. The change of pressure per unit length of pipe, $\Delta p/\Delta x$, is not commonly employed in practice, in part because it includes the effect of differences in elevation as well as friction losses. For liquids, especially water, the hydraulic gradient, i, is normally used. This quantity is based on the hydraulic grade line, which, as shown on Fig. 2.6 is defined by the height to which fluid would rise if a series of vertical sight tubes were tapped into the pipeline. The hydraulic gradient represents the slope of the hydraulic grade line; i.e. the drop in level per unit length of pipe. As noted in Section 2.4, for water (density ρ_w) flowing in a horizontal pipe, i is related to the pressure gradient as follows

$$i = \frac{1}{\rho_w g}\left(-\frac{\Delta p}{\Delta x}\right) \tag{4.1}$$

As water forms the carrier liquid in the majority of slurry flows, it is convenient to use it for a standard for comparison of frictional pressure losses. Thus, throughout this book the usual expression for frictional losses is in terms of i (height of water per unit length of pipe). The value of i for clear water flow is written i_w, and that for a mixture (i.e. frictional loss in height of water per unit length) is written i_m. The statement made above for pseudo-homogeneous flows amounts to a direct proportionality between i_m and i_w.

In non-slurry applications, if a liquid other than water is conveyed the hydraulic gradient concept is applied by substituting the density of this liquid for that of water in Eq. 4.1, so that the hydraulic gradient is expressed in height of flowing liquid per unit length. In considering, say, a pseudo-homogeneous slurry of fine sand, it may be of interest to compare its behaviour to that of a liquid with density equal to that of the flowing mixture ρ_m. Substituting ρ_m for ρ_w in Eq. 4.1 gives the mixture-height hydraulic gradient (i.e. measured in height of mixture rather than of water). For clarity, a different symbol must be used, and j is employed for this purpose. As noted in Section 2.4, for a mixture flowing in a horizontal pipe:

$$j = \frac{1}{\rho_m g}\left(-\frac{\Delta p}{\Delta x}\right) \tag{4.2}$$

This definition of j has already been employed for non-settling slurries in Chapter 3. Although j is expressed in height of mixture per unit length of pipe, actual measurements based on columns of mixture in vertical tubes are not used, and j is obtained from pressure measurements. It can be seen from Eq. 4.1 and 4.2 that the ratio of i to j equals the relative density of the mixture S_m (i.e. ρ_m/ρ_w).

The equivalent-fluid model of slurry flow assumes that the solids have little effect on friction factor, and that the mixture acts as a liquid as far as the relative-density effect is concerned. The resulting hydraulic gradient for homogeneous mixture flow, i_{mh}, is equivalent to the product of S_m, and i_w.

Although the equivalent-fluid model has been employed in the past by numerous workers, being called 'la loi classique' by Durand (1951a, 1951b), it is not generally supported by the experimental evidence. For example, sand-water experiments by Carstens & Addie (1981) show that for some pseudo-homogeneous flows i_{mh} does not exceed i_w at all. An appropriate general equation for the hydraulic gradient, capable of representing either type of behaviour, is:

$$i_{mh} = [1 + A'(S_m - 1)]\, i_w \tag{4.3}$$

Setting the coefficient A' equal to unity gives the relative-density effect of an equivalent-fluid model, whereas $A' = 0$ gives the behaviour observed by Carstens & Addie. Intermediate types can be represented by values of A' between zero and unity. In subsequent chapters, Eq. 4.3 will be referred to as the 'homogeneous flow' equation, with the specific case of $A' = 1.0$ called the 'equivalent-fluid' model.

4.3 Flow of Settling Slurries

When particle diameters are increased from the small-sand to the medium-sand range (say 400 μm), the flow will shift to *heterogeneous* behaviour, with smaller values of V_m producing greater non-uniformity of solids concentration, and increases in hydraulic gradient beyond that given by i_{mh}.

In heterogeneous flow, lines of constant delivered concentration show minimum points on plots of i_m against V_m, as illustrated by Fig 4.1. It appears that Blatch (1906) was the first to produce experimental evidence of this behaviour. She compared experimental results from pipes of 1 inch and 3 inch nominal diameter with results from dredge discharge pipes of approximately 32 inch diameter, and found that the velocity for minimum i_m increased with pipe diameter (the 30-fold increase in diameter changed the velocity by a factor of 2.6). Blatch did not provide a mathematical formulation of her results, though it is apparent that the observed minimum in i_m implies that the quantity (i_m-i_w), i.e. the 'solids effect', must decrease with increasing velocity.

Functional relations for behaviour of this type were proposed in the 1950's. Two major groups of workers were involved; one in France, led by Durand, the other in England, led by Newitt. The group in France (Durand & Condolios, 1954; Gibert, 1960) expressed the excess gradient in heterogeneous flow by means of the dimensionless ratio Φ defined by:

$$\Phi = \frac{i_m - i_w}{C_{vd} i_w} \tag{4.4}$$

This form invites comparison with the 'homogeneous flow' relationship of Eq. 4.3. S_m represents the relative density of the delivered mixture, and $(S_m\text{-}1)$ equals $(S_s\text{-}1)C_{vd}$. Hence Φ can be considered as formally equivalent to $A'(S_s\text{-}1)$. The latter quantity can be constant for homogeneous flow, but in heterogeneous flow Φ is velocity-dependent. To express this dependence, Durand's group proposed that Φ could be related to a second dimensionless variable, denoted Ψ, which includes the flow-related quantity $V_m^2/[gD(S_s\text{-}1)]$ and the particle drag coefficient C_D, i.e.

$$\Psi = \frac{V_m^2}{gD(S_s - 1)} (C_D)^{1/2} \tag{4.5}$$

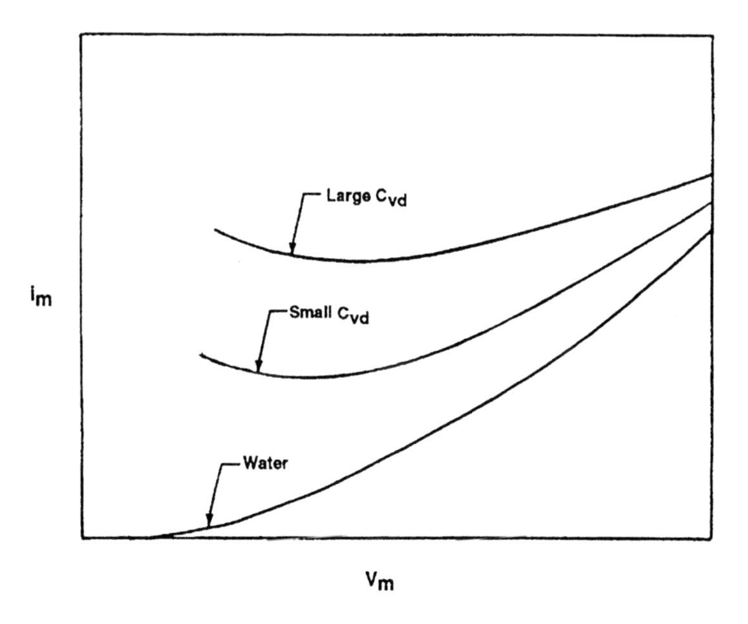

Fig.4.1. Constant-concentration lines for heterogeneous slurry flow

The selection of the two components, and their multiplication to obtain the quantity Ψ, was not based on theory, but rather on the necessarily limited corpus of empirical results available in the 1950's. As shown in Chapter 2, the drag coefficient approaches a constant value for large heavy particles; so that for such particles (e.g. sand larger than 2 mm) Ψ no longer depends on particle size or fall velocity.

Durand's group sought to establish an empirical link between their two dimensionless variables by a logarithmic Φ - Ψ plot of their data. The general trend of their points suggested a slope of about -1.5, and they proposed that the proportionality of Φ to $\Psi^{-1.5}$ represented a law of general applicability. In fact, relationships of this nature, based simply on data correlation, present difficulties which are well-documented in the literature (Clemens, 1874). The Durand equation was used extensively in the early 1960's for design calculations and presentation of experimental results, but despite its usefulness for some applications, its general validity became more and more doubtful when scrutinized in the light of the developing body of experimental data.

Next, it is appropriate to consider the work of Newitt *et al.* (1955), who divided particulate slurry flows into three types. At one extreme is homogeneous or pseudohomogeneous flow, typical of the fine sand mentioned previously. Newitt's group employed the equivalent-fluid model for this type of flow. The other extreme involves small velocities or large particles, with the particles travelling by saltation, or as a sliding bed. This type of flow is described as stratified (Wallis, 1969, Shook *et al.,* 1973), while the term 'heterogeneous flow' is now restricted to conditions intermediate between pseudo-homogeneous and fully stratified. For stratified flow Newitt's group utilized the concept of mechanical sliding friction between moving solids and the pipe. Assuming an essentially two-dimensional fully-stratified flow with negligible difference between delivered and *in situ* concentration, they suggested that i_m - i_w should be directly proportional to (S_m - 1) or

$$\frac{i_m - i_w}{S_m - 1} = B \tag{4.6}$$

From their data, Newitt *et al.* evaluated B as about 0.8, and proposed that Eq. 4.6 applies provided $V_m \leq 17\, v_t$, where v_t is the terminal settling velocity of a solid particle.

It can be seen that the above equation for fully-stratified flow reflects a domain in which variations of particle size or fall velocity do not affect the pressure gradient. Durand's formulation gives a similar independence for large heavy particles for which C_D is virtually constant, but despite this similarity between the two formulations, there are also profound differences.

As mentioned earlier, the Durand equation gained considerable popularity as a design tool throughout the early 1960's, but its validity began to be questioned as the corpus of experimental data grew. For example, Zandi & Govatos (1967) had assembled a data bank of some 1400 experimental points which had been obtained by various workers. Their plot of these points on Durand's Φ and Ψ coordinates showed a scatter that is truly horrendous. A pointed criticism was raised by Babcock (1971), who reported the results of a large number of closely-controlled experiments carried out at the Colorado School of Mines. He plotted his results on Durand's co-ordinates and found that the situation was simply appalling, with no way of putting a meaningful line through the galaxy of points. Babcock noted that a great many of his runs could be correlated to Newitt's equation for stratified flow (Eq. 4.6). He further remarked that the use of the drag coefficient in Durand's abscissa cannot be correct since, with stratified flow, particles with a wide range of C_D all obey Eq. 4.6, but the results are widely scattered on the Durand co-ordinates. Babcock's work clearly demonstrated the merit of Newitt's proposal to deal separately with the three zones of flow: fully stratified flow with no suspension, heterogeneous flow with partial stratification and partial suspension, and homogeneous (or pseudo-homogeneous) flow.

Although Babcock employed Durand's Φ - Ψ axes to plot his results, a different co-ordinate system can give a clearer picture. In the new system the ordinate is the relative solids effect $(i_m\text{-}i_w)/(S_m\text{-}1)$, and the abscissa is the ratio of mean velocity to particle fall velocity, V_m/v_t. A log-log plot of Babcock's results on these co-ordinates is shown on Fig. 4.2. As numerical values for all individual data points were not included in Babcock's report, averages (or in three instances, scatter bands) have been used for each velocity range. The data now fall within a single cycle of the ordinate, and the greater part of the data, shown on Fig. 4.2, also falls within a single cycle of the abscissa. The exceptions are the points for steel shot (S_s = 7.49,d = 2.9 mm, $(i_m - i_w)/(S_m - 1) = 0.55$, $V_m/v_t = 2.4$), which plot on a leftward continuation of the trend of the other coarse particles, and those for the finest sand (S_s = 2.65, d = 0.17 mm.) which lie to the right of Figure 4.2, extending to $V_m/v_t \approx 285$. This fine sand shows behaviour indicative of pseudo-homogeneous flow, which is not surprising in view of the data of Carstens and Addie (1981) which show this type of flow for velocities in excess of 200 v_t.

It can be seen that Fig. 4.2 brings the data for $V_m < 30\ v_t$ together in a single band (which Φ - Ψ co-ordinates failed to do). It also pulls together the sets of rapidly dropping heterogeneous data for particles of two sizes (which were widely separated on the Φ - Ψ axes). Nevertheless, the overall support which this figure gives to Newitt's three-fold division of flow behaviour does not extend to the detailed formulations for fully stratified and heterogeneous flow given by Newitt *et al.* (1955). The fully-stratified flow data on the left of Fig. 4.2 are not strictly constant as proposed by Newitt's group (Eq. 4.6), but slope downward slightly with increasing velocity, as is required to produce the observed minima in plots of i_m versus V_m.

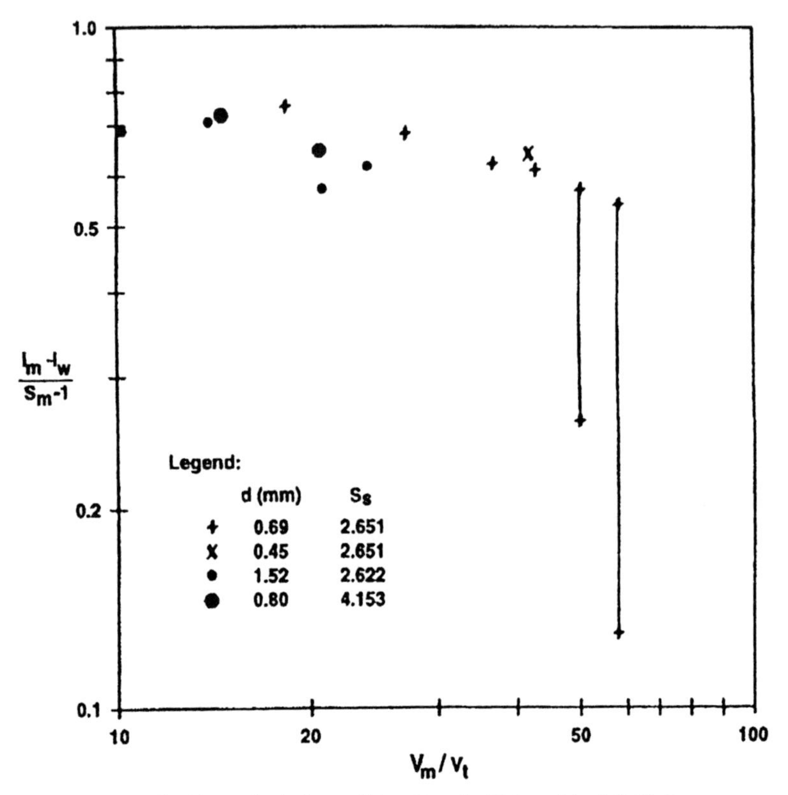

Fig. 4.2. Plots of relative solids effect for Babcock's (1971) data

The data for partially-stratified (heterogeneous) flow show a much more rapid descent with V_m/v_t, but the scatter is considerable and the rate of descent seems more abrupt than would be indicated by the logarithmic slope of -1 proposed by Newitt's group. (This slope coincides with that implied by Durand's proposed dependence of Φ on $\Psi^{-1.5}$.) Thus plots like Fig. 4.2, which are very useful in distinguishing the different flow types, cannot be considered as an end point in themselves, but indicate the need for further work to gain improved understanding of both fully-stratified and heterogeneous flows.

4.4 Physical Mechanisms Influencing Type of Flow

Up to this point we have described the phenomena observed in slurry flow, rather than dealing with the underlying physical mechanisms. The next step is to consider some these mechanisms, specifically the ones which are important for fully-stratified and partially-stratified flow of granular solids in a Newtonian carrier fluid.

When the flow is fully stratified, the particles are concentrated in the lower portion of the pipe and will come in contact with each other and with the pipe wall. This contact can be continuous, as occurs with a stationary or sliding bed of solids, or it can be sporadic, as occurs when the particles travel in a jumping motion known as saltation. The case of continuous contact is well known in the mechanics of granular soils and other materials composed of large collections of cohesionless particles. The method of analysis is based on concepts dating back to the classic work of Coulomb (1773), and considers the normal stress or pressure to be composed of the hydrostatic pressure p, carried by the interstitial fluid, and the intergranular normal stress σ_s, which is based on the forces carried by intergranular contacts. Together with the intergranular pressure σ_s, there is an intergranular shear stress τ_s. The basic Coulombic relation between them is that for non-cohesive grains τ_s cannot exceed $\sigma_s \tan\varphi$, where φ is the friction angle. If the granular mass is not stressed to its limit, τ_s will be less than $\sigma_s \tan\varphi$, whereas a value of τ_s equal to $\sigma_s \tan\varphi$ implies a stress state associated with an internal failure, i.e. incipient motion. If the motion takes place between the mass and a rigid boundary, the equivalent relation for τ_s is not based on $\sigma_s \tan\varphi$, but on $\mu_s \sigma_s$, where μ_s is the coefficient of mechanical sliding friction between the granular mass and the boundary. (In more advanced treatments the stresses σ_s and τ_s are replaced by a tensor with six independent elements, but this degree of sophistication is not required here.)

The first step in applying granular mechanics to particles that are already in motion was made by Du Boys (1879). In dealing with motion of particles on the bed of the river Rhône, he noted that the shear stress applied to the bed by the flow could move the whole upper portion of the bed, possibly comprising several layers each of thickness equal to a particle diameter. Du Boys proposed that the normal granular stress σ_s acting on each layer was equal to the sum of the submerged weight of the grains in all the higher layers. At the bottom of the moving layers the product $\sigma_s \tan\varphi$ just equals the applied shear stress τ, and for higher layers the resisting capacity of the granular mass, $\sigma_s \tan\varphi$, will be less than the applied shear stress. The shear stress excess (τ - $\sigma_s \tan\varphi$) sets up a shear strain rate, leading to a velocity profile within the moving layers. Unfortunately, Du Boys was unable to determine a valid expression for this velocity profile, as the required understanding of turbulent flow had not yet been achieved . Thereafter, analysis of the bed load, as the particles rolling and saltating near the bed of river and canals came to be called, was not actively pursued for some time.

When the new understanding of the mechanisms of turbulent flow was applied to particle motion, it was to the particles found throughout the flow in many natural channels. These particles, which are often smaller than the bed-load particles, are suspended by fluid turbulence, and thus are known as suspended load. The analyses of Schmidt (1932) and Rouse (1937) give a method for calculating the variation of suspended-load concentration with height, indicating that this variation is dictated by the value of the parameter v_t/U_*. Here v_t is the settling velocity of a particle and U_* is the shear velocity, given by $\sqrt{(\tau_o/\rho_f)}$, where τ_o is the shear stress at the boundary of the channel and ρ_f is the fluid density. The theory was developed for small concentrations of solids, and thus the evaluation of U_* ignores any effect of the particles on shear stress. Likewise, no account is taken of the difference between the settling velocity of a single particle and the hindered settling velocity (see Chapter 2).

The variation of relative concentration with height was later verified for pipe flow (with low concentrations of fine sand) by Hsu *et al.* (1980), but the difficulty remains that the actual concentration must be known at some reference level before a concentration profile can be determined. It appeared that the reference concentration (or, equivalently, the average concentration in the flow) could be obtained by linking the lower end of the suspended-load profile to the motion of the bed-load, but it was not easy to see how this link could be analysed.

Further insights into bed-load motion, and its relation to suspended load, were developed by Bagnold (1956), based on his observations of moving desert sand which he combined with his duties as an officer in North Africa both before and during World War II.

Bagnold's treatment of bed-load marked a significant advance over that of Du Boys and the researchers who had followed him. Bagnold continued to employ the concepts of granular pressure and shear stress, noting that the lower non-sheared portion of a granular bed must have a value of $\sigma_s \tan\varphi$ in excess of the applied shear stress τ_o. Shearing motion begins at the point where the normal intergranular pressure σ_s equals $\tau_o/\tan\varphi$, and this intergranular pressure must be produced by the submerged weight of the moving bed-load, transferred downward by intergranular contacts (if this intergranular pressure were absent a stable stationary bed could not be maintained). In considering conditions within the sheared layer where the bed-load is moving, Bagnold, like Du Boys, utilised the Coulombic equality, written here as

$$\tau_s = \sigma_s \tan \phi' \tag{4.7}$$

where ϕ' is the dynamic friction angle (Bagnold recognized that ϕ' may differ somewhat from the static friction angle φ). The remaining shear stress τ_o - τ_s is carried by the fluid, and hence should be associated with the velocity gradient.

Bagnold noted that the intergranular normal stress at a given height within the sheared layer equals $(\rho_s - \rho_f)g$ times the integral of the *bed load* concentration above that height, but he also saw that the bed load need not correspond to all the solids. Bagnold considered the total solid concentration in any unit volume to consist of elements of both bed load and suspended load, with the submerged weight of the bed load transmitted downward through the intergranular normal stress and that of the suspended load "transmitted not to the bed grains but between them as an excess static fluid pressure" (Bagnold 1956, p. 250).

In another publication Bagnold (1955) noted that, for conditions where suspended load is absent, the concentration of the remaining (bed-load) particles diminishes with height up to a 'ceiling' above which particles are not usually found. He stated that "the ceiling at which zero concentration occurs, gets higher as the shear stress is increased" and suggested that the concentration profile for bed-load must have a characteristic shape to account for these observations. Later work by Shook and Daniel (1965) used γ-rays to obtain concentration profiles for many flows where only bed-load was involved. These profiles have a characteristic shape with a 'ceiling', giving support to Bagnold's proposal.

Bagnold had also carried out experiments to measure the stress components associated with intergranular contact, σ_s and τ_s (Bagnold, 1954). These experiments used a rotating-drum apparatus with an annular space between drums which was filled with mixtures of various concentrations of neutral-density wax spheres. The results gave a valuable experimental demonstration of the validity of the Coulombic equality (Eq. 4.7). Bagnold also employed them in suggesting empirical relations to link the granular stress components to the shear rate, solids concentration and particle size. Unfortunately, his experiments were confined to one particle size and one size of annulus. Thus his data base was insufficient to demonstrate or disprove the validity of his proposed correlations, and this caused some subsequent controversy [see Wilson, 2004].

Bagnold's concept of the two mechanisms of particle support – fluid action and intergranular contact – provides the basis for understanding the heterogeneous-flow results presented earlier in this chapter. The presence of particles suspended by the fluid increases the fluid pressure at the bottom of the pipe, as noted above, but this pressure has no direct influence on fluid shear, and thus should not affect the pressure gradient. This concept is in accord with the data of Carstens & Addie (1981) which showed slurry pressure gradients equal to those for water in cases with mean velocity greater than 200 times the particle fall velocity. The results of more recent experimental work will be presented in Chapter 6.

The situation is different for particles whose submerged weight is carried by granular contact. This weight produces normal granular stresses against the bottom of the pipe, and shear stresses proportional to the normal stresses must be provided in order to set the particles in motion. Detailed analysis of these stresses will be presented in Chapter 5, but a simplified linearization indicates that an extra driving force is needed, roughly proportional to the submerged weight of the contact-load (bed-load) particles. Provided the flow is fully stratified, i.e. all particles travel as contact load, this result is in essential accord with the formula of Newitt *et al.* (Eq. 4.6).

4.5 Stationary Deposits of Solids

The previous sections of this chapter have been concerned with the pressure gradients associated with slurry flow, but there are other points which must also be considered by pipeline designers and operators in selecting the desirable operating range of the system. A major one concerns the conditions for which a stationary bed of solids is found in the line. In general, a stationary bed is not desirable, but it is precisely for this reason that it is important to know (and hence to avoid) the conditions for which bedding occurs. This point was first studied by Durand and his co-workers (Gibert, 1960) who defined the 'vitesse limite de dépôt' as the value of mean velocity which marks the upper limit of the region in which a stationary deposit can be found. This velocity will be denoted here as V_{sm}, in accord with the nomenclature to be used when this matter is dealt with in detail in Chapter 5. Durand's group set up the ratio $V_{sm} / \sqrt{(2g(S_s - 1)D)}$, which they found experimentally to be of the order of unity, and to be affected by particle size. Graphs of the Durand deposition-velocity ratio constituted a significant practical advance at the time , but a check against the results of more recent research shows that they are only roughly correct. In many cases, beds begin to move at values considerably smaller than those predicted. It would appear that the Durand relation gives at best an upper limit or envelope which already contains a significant safety margin.

During the 1970's work on the limit of the region of stationary deposition shifted from a data-correlation approach to one based on analysing intergranular stresses to arrive at a force balance (Wilson, 1970, Wilson *et al.*, 1973). Results will be presented in Chapter 5.

4.6 Heterogeneous Flow

There remains the problem of dealing with heterogeneous flows, which have behaviour intermediate between the pseudo-homogeneous and fully-stratified cases. Heterogeneous or partially-stratified behaviour is typically observed within the range of V_m/v_t between 20 and 200, as indicated on Fig. 4.2. It should be noted, however, that the axes of this figure should be scrutinized for possible improvement. The ordinate represents a normalised form of the solids effect, or excess energy gradient, and here it may be desirable to replace the clear-water friction gradient i_w by an experimentally-measured friction gradient for homogeneous flow, a quantity which can be evaluated from vertical-flow data. If rising and descending vertical pipes are included in the pilot-plant loop, vertical and horizontal friction gradients can be observed simultaneously (Wilson, 1973). The mean vertical friction gradient $\bar{i}_v$ can be obtained by averaging the pressure gradient for rising and descending pipes, thus eliminating the effect of the weight of solids in the vertical columns (see Chapter 8). For two-phase flow, pressure gradients can often be measured with greater precision than discharges, and the random error in the quantity $i_m - \bar{i}_v$ may be significantly less than that in $i_m - i_w$.

In addition to this scatter-reducing aspect, it is believed that expressing the excess gradient as $(i_m - \bar{i}_v)$ is preferable from a mechanistic standpoint, since it gives a direct indication of the additional energy required to overcome the effect of gravitational attraction at right angles to the direction of flow. It is well known that the vertical friction gradient is affected by solids concentration and the properties of the particles, and thus deviates from the clear water friction gradient in either a positive or a negative sense. If one imagines two-phase flow taking place in a weightless environment, it seems most improbable that the friction gradients for flow with all types and sizes of solids would be identical to that for the liquid alone. On the contrary, it is much more likely that the friction gradients in the gravity-free condition would show the same type of variation with solids concentration and other parameters that have been observed for vertical flow (see Chapter

8). In other words, the friction gradient in a weightless environment would not be far from the mean friction gradient of ascending and descending flow in a normal environment.

The effect of adding a gravitational field at right angles to pipe axis is to exert a submerged-weight force on each particle, which drives it toward the lower boundary of the pipe. In the absence of turbulent diffusion and fluid lift the sum of these submerged-weight forces must be transferred to the boundary, and the cross-pipe gravitational component will produce stratification (Wallis, 1969), setting up the normal stresses between the solid grains and the pipe wall which were discussed in Section 4.3. In partially stratified (heterogeneous) flow the submerged weight of the particles is not entirely transferred to the pipe wall in this fashion, but the fact that i_m exceeds $\bar{i}_v$ for this sort of flow indicates some significant degree of particle-boundary interaction. It follows that in heterogeneous flow, as in fully-stratified flow, the particles act to transfer momentum from the moving fluid to the boundary.

Although it has been found that the use of $\bar{i}_v$ in place of i_w is not required for sand-water slurries in large pipes, this technique can prove very valuable for small pipes sizes (Wilson, 1973) and for slurries with non-Newtonian carrier fluids (see Chapter 7).

Following the discussion of the best form of the ordinate of plots such as Fig. 4.2, it is also appropriate to consider the best form of the abscissa. Changes in the ordinate indicate differences in the stratification ratio or fractional stratification (Shook *et al.,* 1973), which represents the fraction of solids supported by granular contact (see Chapter 6). Hence, the abscissa should represent the dominant variable influencing the stratification ratio. It was suggested previously, on empirical grounds, that the velocity ratio V_m/v_t might be suitable for this purpose. However, the mechanisms of particle support are not simple, and it is now known that a more complex form of the abscissa is required.

As noted earlier, support of particles by turbulent diffusion depends on the ratio of particle fall velocity to the velocity of the turbulent eddies in the fluid. It is usual to approximate the vertical fluctuating velocity of a typical eddy by the shear velocity U_*, which can be written $(\tau_o/\rho_f)^{1/2}$ or $V_m\,(f/8)^{1/2}$. In this formulation it is assumed that the solids concentration is too low to affect the wall shear stress or the friction factor f, both of which are evaluated as if for a flow of liquid only. Nevertheless, with large concentrations of solids in the system the pressure gradient, and hence the observed value of the friction factor, may be much greater than that found for an equal discharge of clear liquid. This increase in pressure gradient indicates that there is an increase in the total shear stress over the pipe boundary. However, the submerged weight of all particles not in suspension must be transferred to the boundary in the form of normal stress; and as noted in Section 4.4 when discussing the work of Bagnold (1956), a normal stress of this type is accompanied by a shear stress between the moving particles and the pipe wall. It would appear that the observed increase in pressure gradient in the presence of contact-load solids can be accounted for by this contact friction between the particles and the wall. Thus it is probable that the turbulent fluctuating velocities would not be significantly increased, which indicates that the friction factor derived from the elevated pressure gradient found with solids in the line would not be appropriate for use in the relationship under consideration. A lower friction factor should be employed, and the reasoning outlined above suggests that it can be equated, at least approximately, to the friction factor for an equal discharge of clear liquid, f_f (or, for an aqueous slurry, f_w), giving U_{*f} as $V_m(f_f/8)^{1/2}$.

This point was investigated experimentally in an earlier publication (Wilson, 1973), based on 0.7 mm sand in a 50 mm pipe, with results shown on Fig. 4.3. This figure employs the mean velocity V_m (proportional to U_{*f}) as abscissa, and displays remarkably little scatter. In the same publication it was shown that the use of U_* as abscissa results in much greater scatter, demonstrating the advantage of using V_m.

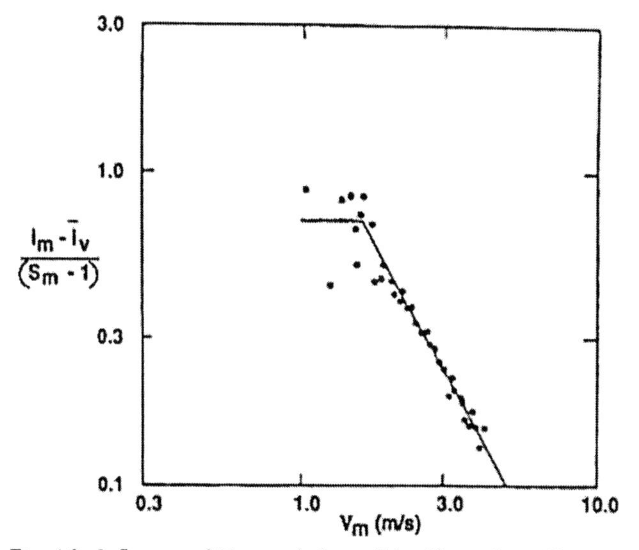

Fig. 4.3. Influence of V_m on relative solids effect, after Wilson (1973).

Although the lift force is connected to particle size, and fall velocity, the resulting stratification ratio may depend on more than simply the ratio of

V_m/v_t, which had been used on Fig. 4.3. Specifically, the correlation visible on Fig. 4.3 only applies if the particles are small with respect with the pipe size. Wilson & Watt (1974) proposed that the lack of suspension for relatively large particles is directly influenced by the ratio of particle size to typical eddy size, as small eddies cannot be expected to support large particles. This point will be considered further in Chapter 6.

4.7 Specific Energy Consumption

In order to compare the merits of different transport systems, it is necessary to have a measure of the energy required to move a given quantity of product over a given distance. This measure is the specific energy consumption, which in industrial usage is often expressed in inconsistent units such as kilowatt-hours per tonne-kilometre, horsepower-hours per ton-mile or British thermal units per ton mile. For slurry pipelines all of these expressions are directly proportional to a simple dimensionless ratio, denoted below by SEC.

In slurry transport, the solids are normally the 'payload' while the conveying liquid is merely the 'vehicle'. The specific energy consumption is therefore to be related to the solids transported rather than to the mixture. However, the power which the pump must supply is used to drive the slurry as a whole, and is given by $\rho_w g Q_m H$ where H is the head in height of water and Q_m is the volumetric flow rate of the mixture. The payload of solids is delivered over the line length, L, at a volumetric rate that is the product of the delivered solids concentration C_{vd} and the mixture flow rate. If the application is a hoisting system, with the line directed vertically upwards, the rate at which energy is being added to the solids is $\rho_s g C_{vd} Q_m L$ or $\rho_w g S_s C_{vd} Q_m L$, and the efficiency of the system can be obtained by dividing by the power input, giving $S_s C_{vd} L/H$. In this case the specific energy consumption is the inverse of the efficiency i.e.

$$SEC = \frac{H}{S_s\, C_{vd}\, L} \tag{4.8}$$

where, as before, H is the head supplied by the pump in height of water.

In the more usual case of an essentially horizontal pipeline, potential energy will not be added to the solids, and thus from the physicist's viewpoint no work will be done. However, in industrial practice transporting solids along the line represents useful work in the economic sense. We can no longer speak of efficiency in the physicist's terminology, but the appropriate economic measure of specific energy consumption is still given by Eq. 4.8. With H/L now equal to the friction gradient, i_m, the expression is written

$$SEC = \frac{i_m}{S_s\, C_{vd}} \tag{4.9}$$

To obtain a value in kWh/tonne-km this ratio is multiplied by 2.73. (For horsepower-hour/ton-mile the factor is 5.33.) If the power supplied to the pump is required, the expression must be divided by pump efficiency, and the efficiency of the drive train or motor can be taken into account in the same way.

The lower the SEC, the more energy-effective the pipeline is as a means of transport. Since S_s is fixed by the nature of the solids, the quotient i_m/C_{vd} is the basic variable to be considered. There is no need here to follow in detail the method used by Gibert (1960) and other authors for minimising i_m/C_{vd} analytically. Not surprisingly, it is found that the basic requirements are a large value of C_{vd}, together with operation at the minimum in the i_m curve for this concentration. The existence of such a minimum was mentioned in connection with Fig. 3.1, and further information on the effects of very high C_{vd} will be given in Chapter 6. As will be shown in Chapter 13, other factors shift the operating point somewhat to the right of the minimum point , but if energy consumption or installed pump power is a critical element in the cost of a slurry system, there is an unquestionable incentive to use large values of C_{vd}. Consequently, great importance is given in subsequent chapters to stable operation at relatively high concentrations of solids.

REFERENCES

Babcock, H.A. (1971). Heterogeneous flow of heterogeneous solids. *Advances in Solid-Liquid Flow in Pipes and its Application* (Ed. I. Zandi), Pergamon Press, pp 125-148.

Bagnold, R.A. (1954). Experiments on a gravity-free dispersion of large solid spheres in a Newtonian fluid under shear. *Proc. Roy. Soc.*, London, Scr. A, Vol. 225, pp. 49-63.

Bagnold, R.A. (1955). Some flume experiments on large grains but little denser than the transporting fluid, and their implications. *Proc. Instn. of Civil Engineers,* Part III, pp 174-205.

Bagnold, R.A. (1956). The flow of cohesionless grains in fluids. *Phil. Trans. Roy. Soc.*, London, Ser. A, Vol 249, pp. 235-297.

Blatch, N.S. (1906). Discussion of "Works for the purification of the water supply of Washington D.C." (Hazen, A. and Hardy, E.D.) *Trans Amer. Soc. Civil Engrs.* Vol. 57, pp. 400-409.

Carstens, M.R. & Addie, G.R. (1981). A sand-water slurry experiment. *Jour. Hydr. Div.* ASCE, Vol. 107, No. HY4, p. 501-507.

Clemens, S.L. (Mark Twain) (1874). *Life on the Mississippi,* Chapter 17, H.O. Houghton & Co., republished Harper and Row, New York.

Coulomb, C.A. (1773). Essai sur une application des règles de maximis et minimis à quelques problèmes de statique, relatifs à l'architecture. *Memoires de mathématique et de physique presentés à l'Academie Royal des Sciences, VII,* Paris, pp. 343-382.

Du Boys, P. (1879). Étude du régime du Rhône et de l'action exercée par les eaux sur un let a fond de graviers indéfiniment affouillable. *Annales des Ponts et Chausées,* Vol 18, No. 49 pt 2, pp. 141-195.

Durand, R. (1951a). Transport hydraulique des matériaux solides en conduite, études expérimentales pour les cendres de la central Arrighi. *Houille Blanche,* Vol. 6, No. 3, pp. 384-393.

Durand, R. (1951b). Transport hydraulique des graviers et galets en conduite. *Houille Blanche,* Vol. 6, No. B, pp. 609-619.

Durand, R. & Condolios, E. (1954). The hydraulic transport of coal. *Proc. Colloq. on Hydraulic Transport of Coal,* National Coal Board, London, U.K.

Gibert, R. (1960). Transport hydraulique et réfoulement des mixtures en conduites. *Annales des Ponts et Chausées,* Mai-Juin 1960, Juil-Août, 1960

Hsu, S.T., Van der Beken, A., Landweber, L., & Kennedy, J.F. (1980). Sediment suspension in turbulent pipeline flow. *Jour. Hydr. Div.* ASCE Vol 106. No. HY11, pp. 1783-1792.

Newitt, D.M., Richardson, J.F., Abbot, M., & Turtle, R.B. (1955). Hydraulic conveying of solids in horizontal pipes. *Trans. Inst. of Chem Engrs.,* Vol. 33, London, U.K.

Reynolds, O. (1883). An experimental investigation of the circumstances which determine whether the motion of water shall be direct or sinuous, and the laws of resistance in parallel channels. *Phil. Trans Roy. Soc.,* London, Vol 174, pp. 935ff.

Rouse, H. (1937). Modern conceptions of the mechanics of fluid turbulence. *Trans. Amer. Soc. Civil Engrs.* Vol. 102, pp. 436-505.

Saxe, J.G. (1872). The blind men and the elephant. in *Fables and Legends of Many Countries Rendered in Rhyme.* Republished in *The Home Book of Verse* (Ed. B.E. Stevenson) Holt, Reinhart and Winston, New York, 9th Ed. (1953) Vol. 1 pp. 1877-1879.

Schmidt, W. (1932). Der Massenaustausch in freier Luft und werwandte Erscheinungen. *Die Wasserwirtschaft,* No. 5-6 (Mitteilung 10, Inst. für Wasserbau, Tech. Hochschule, Berlin).

Shook, C.A. (1985). Experiments with concentrated suspensions of slurries with densities slightly greater than the carrier fluid. *Canad. Jour. Chem. Engrg.* Vol. 63, pp. 861-869.

Shook, C.A. & Daniel, S.M. (1965). Flow of suspensions of solids in pipelines part I, flow with a stable stationary deposit. *Canad. Jour. of Chem. Engrg.* April 1965. pp. 56-61.

Shook, C.A., Haas, D.B., Husband, W.H.W., Schriek, W. & Smith, L. (1973). Some experimental studies of the effect of particle and fluid properties upon the pressure drop for slurry flow. *Proc. Hydrotransport 2,* BHRA Fluid Engineering, Cranfield, UK, pp. D2-13-D2-22.

Shook, C.A., Gillies, R., Haas, D.B., Husband, W.H.W., & Small, M. (1982). Flow of coarse and fine sand slurries in pipelines. *Jour. Pipelines,* Elsevier, Vol 3, pp. 13-21.

Wallis, G.B. (1969). *One-dimensional Two-phase Flow.* McGraw-Hill, New York.

Wilson, K.C. (1970). Slip point of beds in solid-liquid pipeline flow. *Proc. Amer. Soc. Civil Engrs., 96* No. HY1, pp. 1-12.

Wilson, K.C. (1973). A formula for the velocity required to initiate particle suspension in pipeline flow. *Proc. Hydrotransport 2*, BHRA Fluid Engineering, Cranfield, UK, pp. E2-23 - E2-36.

Wilson, K.C. (2004). Rotating-drum experiments for particle-laden flows: a new view, *Granular Matter*, Vol. 6 No. 2-3, Springer-Verlag, pp 97-101.

Wilson, K.C., Streat, M. & Bantin, R.A. (1973). Slip-model correlation for dense two-phase flow. *Proc. Hydrotransport 2*, BHRA Fluid Engineering, Cranfield, UK, pp. B1-1 - B1-10.

Zandi, I. & Govatos, G. (1967). Heterogeneous flow of solids in pipelines. *Proc. Amer. Soc. Civil Engrs., 93.* No. HY7, pp. 145-159.

Chapter 5

MOTION AND DEPOSITION OF SETTLING SOLIDS

5.1 Introduction

Chapter 4 presented the concept developed by Bagnold (1956) that there are two mechanisms of particle support -- fluid suspension and intergranular contact. It was seen that intergranular contact can be continuous, as is the case for stationary or moving beds, or it can be sporadic, as for particles saltating along the upper surface of a bed. In the terminology of river engineering, saltating and rolling particles are called bed load (as opposed to the suspended load carried by the fluid), but in pipeline usage it is more convenient to treat this bed load together with the grains in continuous contact (e.g. in a sliding bed) and call the combination 'contact load'. The present chapter deals with the analysis of contact load. This includes flows of coarse particles which travel entirely as contact load, and also the contact-load component of heterogeneous flows. The suspended-load component will be discussed in Chapter 6.

As mentioned, contact load comprises stationary beds, sliding beds and individual rolling and saltating particles. It tends to have a significant adverse effect on pressure gradient, and for the stationary-bed case problems of flow stability can also occur. Thus flows with a stationary deposit are generally avoided in practice. However, as mentioned in Chapter 4, designers and operators must have a method of predicting stationary-deposit conditions, and hence a major requirement of contact-load analysis is the prediction of the limit of stationary deposition. This limit occurs when the forces tending to move the deposit just equal those resisting motion, since if the driving forces exceed the resisting forces the stationary bed will either break up or travel *en bloc* as a sliding bed. Thus a force-balance analysis is required to determine the limit of stationary deposition. The first step is to develop expressions for the forces resisting and promoting motion.

5.2 Resisting and Driving Forces

The driving force, tending to set the bed in motion, arises from the action of the pressure gradient on the portion of the cross-section occupied by the bed, plus the effect of shear stress on the top surface of the bed. The resisting force, acting to hold the bed in place, is the mechanical friction between the pipe wall and the particles comprising the bed. Once the bed is in motion, there is an additional fluid-friction component of the resisting force.

Figure 5.1 shows a pipe of diameter D with a bed of solids occupying the lower portion. On the figure, α represents an angular co-ordinate defining the position of any point on the pipe wall, and β gives the position of the upper surface of the bed. The normal stress between the solid grains and the pipe wall is denoted σ_s. The total normal force F_N is obtained by integrating the normal stress over the boundary between the bed and the wall, giving

$$F_N = D \int_o^\beta \sigma_s \, d\alpha \tag{5.1}$$

At the point of incipient bed slip the total frictional force between the solid grains and the pipe boundary will equal $\mu_s F_N$, where μ_s is the coefficient of mechanical friction between the particles and the wall material.

It might be thought that the total normal force could be equated to the submerged weight of the solids which occupy a unit length of pipe, but as indicated in earlier publications (Wilson, 1970; Wilson *et al.*, 1973) this is not so. A summation of forces in the vertical direction shows that only the vertical component of σ_s can act to support the bed weight, and thus the submerged bed weight F_W will be inherently less that the expression for F_N found above, which includes both vertical and horizontal components.

In order to proceed further, it is necessary to introduce a relationship for the variation of σ_s with depth. Although a rigorous solution to this problem in the stress analysis of a granular medium may not be attainable, it is reasonable to make the simple assumption (Wilson, 1970) that the rate of increase of granular pressure with depth is a constant which equals the submerged unit weight of solids, i.e.

$$-\frac{d\sigma_s}{dz} = \rho_w g(S_s - S_f)C_{vb} \tag{5.2}$$

Here z is height and g is gravitational acceleration, ρ_w is the density of water, and S_s and S_f are the relative densities of the solids and the fluid, respectively. C_{vb} is the volumetric fraction of solids in the bed, and the product $\rho_w g(S_s - S_f)C_{vb}$ represents the submerged weight of the solids which are contained in a unit volume of the bed. Equation 5.2 has the same form as the hydrostatic pressure law for liquids, and it is well known that the pressure distributions of this type satisfy the required condition that the vertical components of pressure balance the weight.

If all the contact-load grains were located within the bed, σ_s would be zero at the bed surface, which would mark the origin of z. However, in reality some of the contact-load particles are dispersed above the bed in the bed-load layer of rolling and saltating particles. The submerged weight of these moving bed-load particles sets up a normal intergranular stress at the bed surface (equal to $\tau/\tan\varphi'$ where τ is shear stress and φ' is the angle of dynamic internal friction of the particles). This normal stress can be included in the analysis (Wilson, 1970), but for present purposes it is more convenient to define an equivalent interface which represents the surface of the bed which would occur if the rolling and saltating particles were allowed to settle. The angle β on Fig. 5.1 is now considered as defining this equivalent interface, and Eq. 5.2 can be applied with the origin of z at this level. The resulting normal granular stress at any point on the pipe wall, σ_s, is found to be

$$\sigma_s = \rho_w g(S_s - S_f)C_{vb}\frac{D}{2}(\cos\alpha - \cos\beta) \tag{5.3}$$

F_N can then be obtained by performing the integration indicated by Eq.5.1, yielding,

$$F_N = \rho_w g(S_s - S_f)C_{vb}\frac{D^2}{2}(\sin\beta - \beta\cos\beta) \tag{5.4}$$

The submerged weight of the bed F_W is given by

$$F_W = \rho_w g(S_s - S_f) C_{vb} \frac{D^2}{4} (\beta - \sin\beta \cos\beta) \tag{5.5}$$

and on taking the ratio of these quantities one obtains:

$$\frac{F_N}{F_W} = \frac{2(\sin\beta - \beta\cos\beta)}{(\beta - \sin\beta\cos\beta)} \tag{5.6}$$

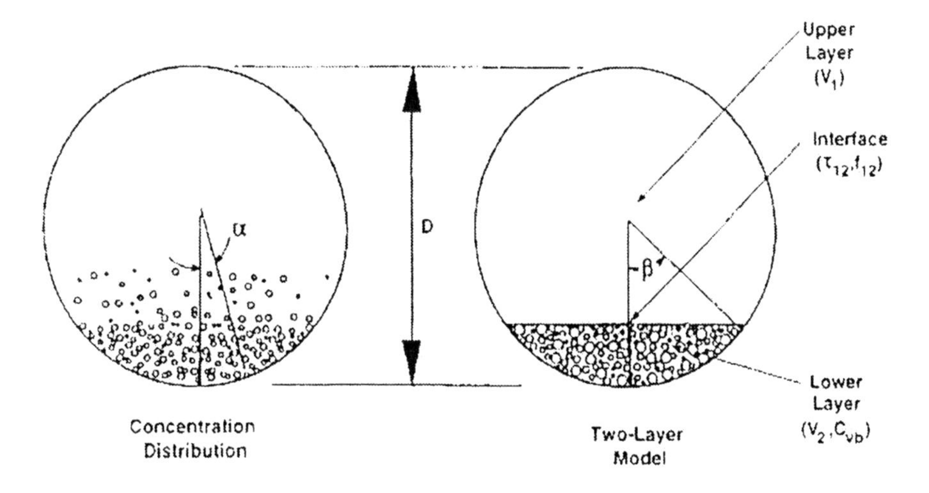

Figure 5.1. Definition sketch for two-layer model of stratified flow

The values calculated from Eq. 5.6 show that for β less than 60° the total normal force is only slightly greater than the submerged bed weight, while at larger values of β there is a marked difference. With a deposit which completely fills the pipe (β = π) the force F_N is twice F_W.

As noted previously, the force which tends to hold a stationary bed in place cannot exceed the product $\mu_s F_N$, where μ_s is the coefficient of mechanical sliding friction. Opposed to this resisting force are two forces which tend to set the bed in motion: the downstream-directed force set up within the bed and the force resulting from interfacial shear stress, which is imposed on the top of the bed by the flow above it.

The force within the bed is associated with the seepage flow set up by the pressure gradient, $\Delta p/\Delta x$, where x is measured horizontally in the axial direction. (Only horizontal pipes will be considered here; inclined transport will be dealt with in Chapter 8.) The intergranular stress σ_s does not vary with x, but the seepage flow creates drag forces on the particles which, in total, produce a force per unit length equivalent to $\Delta p/\Delta x$ times the cross-sectional area occupied by the bed.

The shear stress at the top of the bed can be expressed in terms of the velocity difference between the velocity U_1 in the waterway above the bed and the velocity of the bed U_2. (For a stationary bed U_2 is zero, of course, but the general expression is used so that it is also applicable to sliding beds). With the interfacial shear stress denoted as τ_{12} and the associated friction factor as f_{12}, the expression is written

$$\tau_{12} = 0.125\,\rho_f\, f_{12}(U_1 - U_2)^2 \qquad (5.7)$$

The evaluation of the interfacial friction factor f_{12} will be discussed below, but it is fortunate that there are two configurations for which the interfacial friction is not important, so that the other portions of the model can be verified without requiring a value of f_{12}.

The first configuration is a bed of solids in an inclinable tube, with no discharge. For beds of various values of β, the tube is tilted in order to measure the inclination angle at which the bed begins to move. In this case there is no shear stress on the top of the bed, and the drag forces on the individual particles in the bed are replaced by the axial component of gravitational attraction. Verification using this method was first reported by Wilson (1970), and the tilting tube apparatus has been used since for evaluating the mechanical friction coefficient μ_s.

Flows with very high delivered solids concentration, approaching the loose-poured value C_{vb}, provide the second configuration for verifying the portions of the force balance model not associated with interfacial shear stress This dense-phase flow, also know as plug flow, has been the subject of extensive experimental investigations at Imperial College, London, and the results of these experiments were compared directly with the force-balance model. For the plug-flow configuration β approaches π, giving a negligible interfacial force, and F_N equal to twice the submerged weight of the particles. The force balance for these conditions predicts that the hydraulic gradient (m water/m pipe) required to set up the plug in motion, i_{pg}, is given by

$$i_{pg} = 2\,\mu_s (S_s - S_f) C_{vb} \tag{5.8}$$

The experimental verification used particles of S_s = 2.5 and 4.5, and showed excellent agreement with the theory (Wilson *et al.* 1973).

The analysis was later extended to cases where the plug was in motion, and hence an additional term was required for the velocity-associated component of hydraulic gradient arising from the fluid shear between the moving dense-phase core and the pipe wall. The friction factor associated with this fluid resistance was found (Wilson & Brown, 1982) to be a function of a Reynolds number written $V_m d/v$ where V_m is mean velocity, d is particle diameter and v is the kinematic viscosity of the fluid.

Following the verification of the other elements of the force balance model, it was necessary to consider how to evaluate the interfacial friction factor f_{12}. If the bed is composed of coarse particles, then it can be expected that the interface at the top of the bed is similar to a hydraulically-rough boundary, for which the friction factor depends on the ratio of the particle size to the hydraulic radius of the waterway above the bed, equivalent to a variation with the ratio of particle size to pipe size, d/D. It would appear that the relationship should be similar in nature to that normally found for granular roughness, but with higher values of friction factor because the saltating grains of a mobile bed are more effective than fixed grains in transferring shear stress. On the basis of data for the limit of stationary deposition (Wilson, 1975) and the results of the experiments of Fowkes and Wanchek (1969) with large particles, the interfacial friction factor was initially estimated to be roughly twice that given by the Nikuradse formula for fixed-grain roughness (Wilson, 1976).

As the particle size is decreased, the effect of d/D is less important in determining the interfacial friction factor. During the 1970's and early 1980's, the rough-boundary law was simply extrapolated to smaller particle sizes, but recent work on this type of friction has produced a modified analysis of f_{12} (Wilson, 1988). The results of this analysis will be outlined in the final portion of Section 5.3.

5.3 Velocity at the Limit of Stationary Deposition

The force-balance model described in the previous section is rather complex for manual calculation, and thus was incorporated into a computer program which can be used to evaluate the limit of stationary deposition and the friction loss for fully-stratified coarse-particle flow. The deposition limit will be discussed in this section and in Section 5.4, and the modelling of coarse-particle flow at velocities above the deposition limit in Section 5.5.

The results of the force-balance model for the limit of stationary deposition showed that the throughput velocity V_m at this limit is concentration-dependent, having small values at low concentration, rising to a maximum (denoted V_{sm}) at some intermediate concentration (which depends on pipe size and particle size and density) and then dropping off again as the delivered solids concentration approaches the loose-poured value for plug flow, C_{vb}. This behaviour is shown on Fig. 5.2. It should be noted that the velocities used here are obtained simply by dividing the mixture flow rate (Q_m) by the pipe area ($\pi D^2/4$).

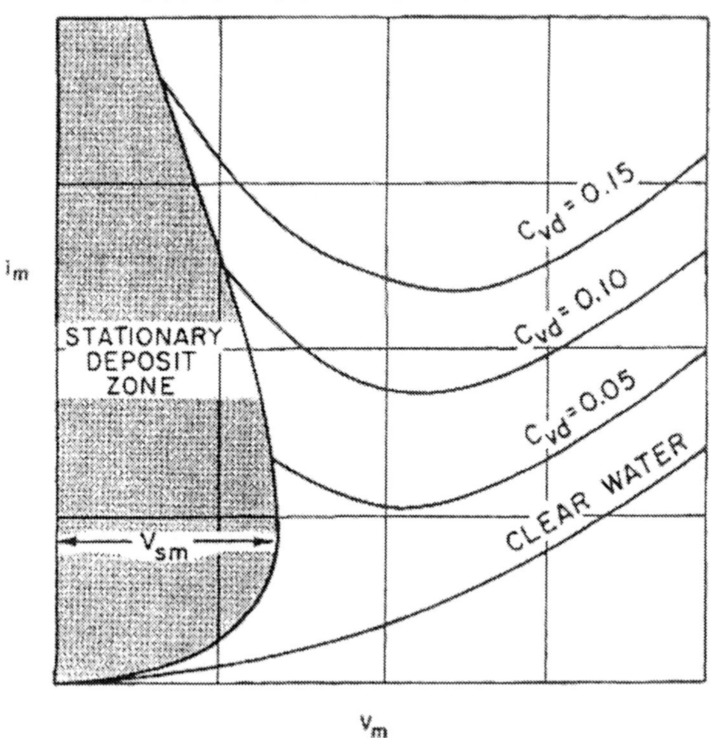

Figure 5.2. Definition sketch for limit of stationary deposit zone

The computer output, though manageable for detailed studies, soon becomes unwieldy for a designer concerned with many alternative proposals. Moreover, the conservative designer may be content to know only the maximum velocity at the limit of deposition, V_{sm}, since maintaining the operating velocity above this value ensures that deposition will not occur. The value of V_{sm} depends on internal pipe diameter, particle diameter and relative density, and the effect of these variables is expressed concisely by a nomographic chart which was developed at Queen's University (Wilson & Judge, 1978; Wilson, 1979) with the help of Professor F.M. Wood's expertise in nomography (Wood, 1935). This chart, reproduced here as Figure 5.3, is recommended as a practical design aid.

It should be noted, by way of explanation of the chart, that the left-hand panel deals with sand-weight materials (S_s = 2.65). The internal pipe diameter appears on the left vertical axis, with V_{sm} on the central vertical axis. The particle diameter is plotted on a curve known, on the basis of its shape, as the 'demi McDonald'. This shape is associated with behaviour noted in the previous section, *i.e.* for large particles, but not for small ones, the shear stress at the interface between the upper and lower layers increases with increasing particle size. It follows that for coarse-particle transport the velocity at the limit of deposition (*i.e.* the velocity beyond which no bed can remain stationary) will decrease with increasing particle size. This finding shows how the contact-load mode of transport can display behaviour quite different from that of the heterogeneous mode. Although engineers with experience only in fine-solids pipelining might find this aspect of coarse-particle behaviour surprising, is in good accord with the experimental evidence.

To demonstrate the particle size effect consider particles of S_s = 2.65 in a pipe 300 mm in diameter. This diameter is located on the vertical scale on the left-hand side of the chart, and connected by straight-edge to any desired particle size on the curved scale. V_{sm} is then obtained by projecting to the central vertical scale. For instance, a particle size of 0.6 mm gives V_{sm} of almost 4 m/s, which is the largest value found for this pipe diameter and solids density. For a larger particle of, say, 5 mm the deposition-limit velocity is diminished to about 2.7 m/s.

When operating with centrifugal pumps it may be difficult to take advantage of the decrease of V_{sm} with increasing particle size. The question is one of obtaining a stable intercept of pump and pipeline characteristics, and will be discussed in detail in Chapter 13. For applications where control of particle size is limited, the conservative designer may wish simply to assume particles of 'Murphian' size, i.e. those which give the largest value of V_{sm} for the pipe under consideration. In this case it should be noted that the values of V_{sm} obtained from Fig. 5.3 tend to be conservatively high, especially for large pipe diameters. As shown in a Case Study 5.1 (Section 5.8), these values can sometimes be used as operating velocities.

A particularly useful feature of nomographic presentation of results is that it gives an immediate indication of the sensitivity of the output to variations in the input. Thus on Figure 5.3 it is seen that the value of V_{sm} for sand-weight solids in a 0.30 m pipe is virtually unaffected by a variation of the particle diameter between 0.4 mm and 1.0 mm. However, the right limb of the particle-diameter curve shows greatly increased sensitivity, so that a change in d from 0.15 mm to 0.20 mm alters V_{sm} by more than 25 percent.

It is found that a change of the relative density does not influence the form of the relation shown on Fig. 5.3, but does require its 'recalibration.' This is accomplished graphically by using the inclined relative-density axis on the right-hand side of the figure. To illustrate the use of the right-hand panel of Fig. 5.3, consider again the pipe 0.30 m in diameter, now carrying particles with diameter 1.0 mm and relative density 1.50. With 1.0 mm entered on the particle diameter scale, a straight line joining it to the pipe diameter can be projected to the central axis of the figure, showing that V_{sm} is 3.5 m/s for the equivalent sand size. In order to correct this for the relative density of 1.50, the latter value is entered on the sloping axis on the right-hand side of the figure and joined by straightedge to the point just found on the central axis (3.5 m/s). The projection of this line to the vertical axis on the right of the figure gives the adjusted value of the maximum deposition velocity, approximately 1.9 m/s.

Particles which differ in density from sands may also have different values of other properties, including the solids fraction in the deposit and the mechanical friction coefficient between the particles and the pipe. The values of these quantities which were employed in the computer program, and hence are reflected in the nomographic chart, apply to sands but not necessarily to other materials. Therefore the values of V_{sm} determined from Fig. 5.3 for materials other than sands must be treated as somewhat less accurate than values for sand. However, both the solids fraction in the bed and the particle-pipe mechanical friction coefficient occur in the computer program only as multiples of the submerged relative density of the particles, and it is found that any change in these quantities merely gives rise to a multiplicative factor which must be applied to the values of V_{sm} obtained from the figure. This permits calibration of the output of the nomographic charts from a few pilot-plant tests with the material of interest.

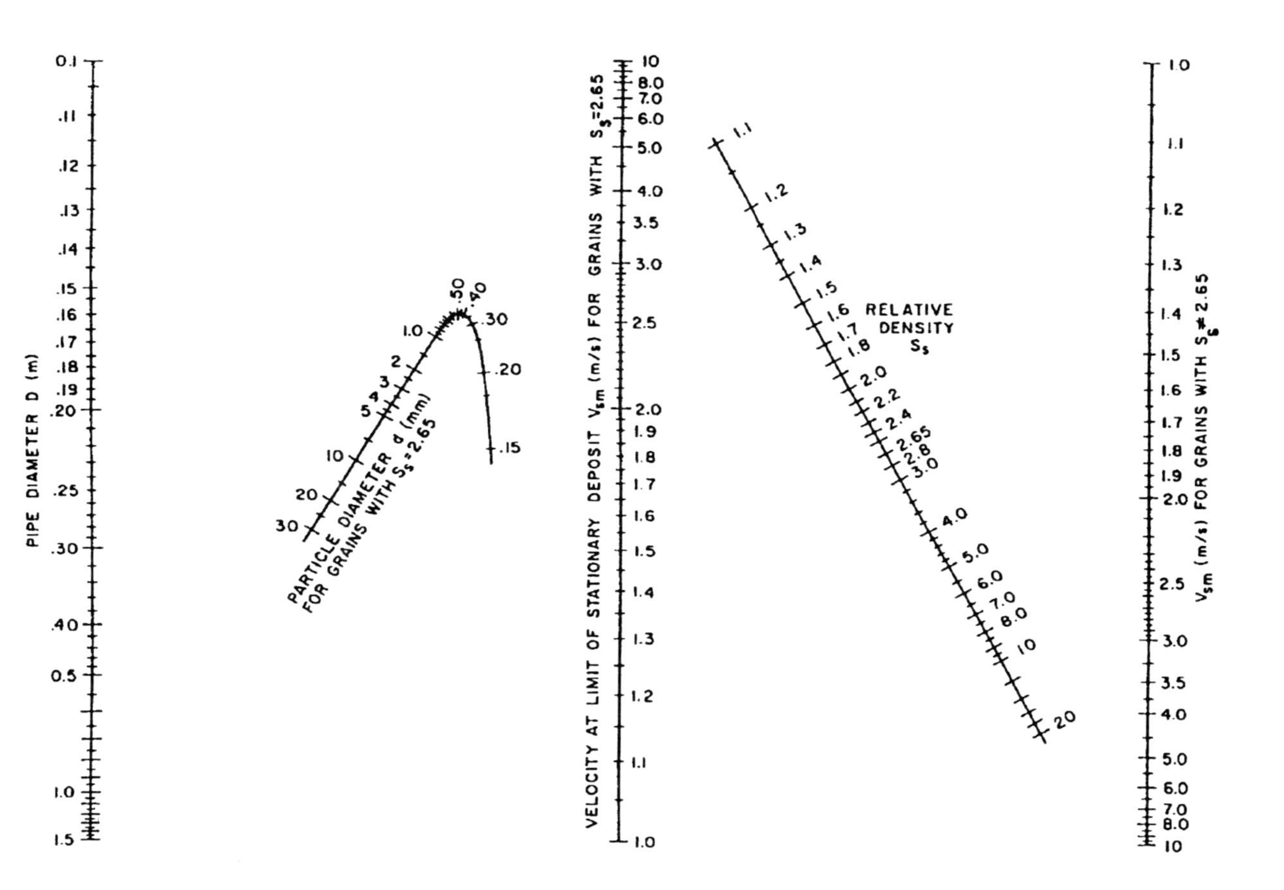

Figure 5.3. Nomographic chart for maximum velocity at limit of stationary deposition, from Wilson (1979)

The particle diameter scale of Fig. 5.3 has not been extended below 0.15 mm since smaller particles tend to be influenced by mechanisms not included in the mathematical model on which the nomogram is based. As shown by Thomas (1979) the viscous sublayer can have a significant effect, and all particles which are small enough to be completely embedded in this sublayer will behave in a fashion which is no longer dependent on particle diameter. In this limiting case certain simplifications can be made in the mathematical model. Thomas found that these lead to a simple expression for the shear velocity at the limit of stationary deposition, and on employing a power law approximation for the friction factor, he then obtained a corresponding expression for V_{sm}

$$V_{sm} = 9.0[g\nu(S_s - S_f)]^{0.37}(D/\nu)^{0.11} \tag{5.9}$$

This equation gives the minimum value of deposition velocity for small particles, assuming turbulent flow. In Eq. 5.9 ν represents the kinematic viscosity of the carrier fluid, and the coefficient of 9.0 applies in any consistent system of units.

For particles larger than the viscous sublayer but smaller than 150 μm, the following equation obtained by Wilson & Judge (1976) is recommended. Here C_D is the particle drag coefficient (see Section 2.5).

$$\frac{V_s}{\sqrt{2gD(S_s - S_f)}} = 2.0 + 0.3\,Log_{10}\left[\frac{d}{DC_D}\right] \tag{5.10}$$

A recent publication by Sanders et al. (2004) presented experimental results that support the use of both Eq. 5.9 and Eq. 5.10, and give criteria for the conditions of applicability of each equation.

As mentioned previously, a new analysis of interfacial friction factor has been developed for cases where the interfacial friction sets up a sheared layer several grain diameters in thickness (Wilson, 1988, Wilson & Nnadi, 1990). For large pipes, and particles near the Murphian size, the new analysis shows that the particle size no longer influences V_{sm} directly. In such cases the value of V_{sm} is less than that predicted by Fig. 5.3, a point which is in accord with experimental evidence. It is then necessary to obtain the value of V_{sm} from the new analysis of interfacial friction, for comparison with the value from the nomographic chart.

When the parameters C_b, μ_s and $\tan\varphi'$ are given values typical for sands, the new analysis of interfacial friction shows that the dimensionless maximum deposition velocity (i.e. the Durand deposition variable) depends only on the fluid friction factor for the portion of the pipe wall above the deposit. A power-law approximation was found to be an appropriate fit function for this effect, giving

$$\frac{(V_{sm})_{\max}}{(2gD(S_s - S_f))^{1/2}} = \left(\frac{0.018}{f_f}\right)^{0.13} \tag{5.11}$$

where f_f is the friction factor for fluid alone. If the value of V_{sm} found from the nomographic chart exceeds the value of $(V_{sm})_{max}$ from Eq. 5.10 the latter value should be used for V_{sm}.

5.4 Effect of Solids Concentration on Deposit Velocity

Although the use of the upper-limit deposit velocity V_{sm} ensures that operation with a stationary bed will be avoided, occasionally there are cases where operating at throughput velocity less than V_{sm} will be economically attractive and yet deposition-free. Specifically, this occurs when the solids concentration at the desired operating point is markedly different from that for V_{sm}, so that the deposit velocity, V_s , at the desired concentration is significantly less than V_{sm}. To deal with such cases the technique was generalised to obtain values of V_s for any delivered volumetric solids concentration (Wilson, 1986). Since this technique is seldom required, it will not be detailed here, and the interested reader is referred to the original publication. A computational algorithm incorporating this technique, together with the nomographic chart and the associated equations presented in the previous section, can be found in SLYSEL, a proprietary product of GIW Industries Inc.

5.5 Fully Stratified Coarse-Particle Transport

Fully-stratified flow occurs where almost all of the particles travel as contact load (*i.e.* fluid suspension is ineffective). The ratio of particle diameter to pipe diameter is of major importance in determining the presence of this flow type, which does not normally occur for d/D ratios less than 0.015. Fully-stratified flow is less likely if the particles are broadly graded, especially if there is a significant homogeneous fraction (*i.e.* a significant fraction of particles smaller than 40 μm, see Section 6.5 and Section 6.7). Calculations made for narrow-graded slurries with water as a carrier fluid indicate fully-stratified behaviour for values of d/D above 0.018. For d/D between 0.015 and 0.018 both types of behaviour can occur, with fully stratified flow more likely for larger values of S_s and smaller values of V_m. In uncertain cases such as this it is best to carry out analyses using both the method given here and that for heterogeneous flow given in Chapter 6 (see Case Study 6.2). The intercept of the two methods establishes the upper velocity of fully-stratified flow for the pipe-particle combination being analysed.

The uncertainty in determining the modes of particle support in various instances of slurry flow, mentioned above, had led to significant difficulties in interpreting research results. This problem came to the fore during the 1970's, when it was represented by another denizen of the slurry bestiary: Kazanskij's Black Cat, who might possibly be found among the more obscure research results (see Proceedings of Hydrotransport 5, Vol. 2, p vii). An area relatively free from this beastly uncertainty is coarse-particle flow for cases where virtually all the solids are larger than, say 0.018 D. As noted above, for this condition fluid suspension of particles is not effective and the flow is fully stratified. The analysis presented below is confined to this type of flow. Although it is less energy-efficient than heterogeneous flow, this transport mode can be economically attractive for pipelines of moderate length as it obviates problems of head-end processing and subsequent separation problems at the tail end of the line.

In the limit, as the concentration approaches that of loose-packed particles, C_{vb}, the type of transport is known as plug flow (*i.e.* coarse-particle dense-phase flow). This type of flow was discussed in Section 5.2, where the friction gradient required to set plug flow in motion was defined by Eq. 5.8. For the more usual cases of coarse-particle flow, with the delivered volumetric concentration of particles C_{vd} substantially less than the loose-poured value C_{vb}, the difference between mixture pressure gradient and that of fluid alone (the excess pressure gradient or solids effect) will be less than that for plug flow. It can be expected that the relative excess pressure gradient, obtained by dividing by the plug-flow gradient, will depend, in part, on the relative concentration C_r, *i.e.* C_{vd}/C_{vb}. The simplest relation would be a linear one, equivalent to the early proposal of Newitt *et al.* (1955), but in reality the situation is much more complicated. Even at high velocities (*i.e.* those greatly in excess of V_{sm}) the relationship with relative concentration is non-linear, and at lower velocities the relative excess pressure gradient is also velocity-dependent. As shown by detailed computer simulations (Wilson, 1976, 1988a), this velocity effect arises from the variation of the lag ratio (related to the ratio of *in situ*/delivered concentration, See Eq. 2.43) which depends chiefly on the ratio V_r between throughput mixture velocity and the velocity at the limit of deposition, i.e. V_m/V_{sm}. Thus the limit of deposition is important not only in its own right, but also as a parameter which influences the solids effect at larger velocities which fall within the normal operating range.

The behaviour of fully-stratified slurries in this operating range has been obtained from computer simulation with the two-layer mathematical model mentioned previously, but despite its value as a research tool, this model is not convenient for the designer. A major difficulty is that the throughput velocity and delivered concentration are found only at the end of the computations, while the required input includes quantities of little engineering interest, such as the fraction of the pipe occupied by moving solids. To obviate this difficulty, and to increase the efficiency of the algorithm as a design tool, certain simplifying fit functions have been used in place of the complete force-balance relations. In addition to delineating the locus of the deposition limit (Wilson, 1986), these fit functions describe both the high-velocity asymptote and behaviour at velocities nearer the deposition limit.

The high-velocity asymptote for the relative solids effect is based on the excess-gradient curves found from the detailed modelling of coarse particle transport (Wilson, 1988a). For any given delivered solids concentration, the difference between the mixture pressure gradient and that for an equal flow of fluid alone depends not only on the relative concentration C_r but also on the velocity ratio V_r. The nature of this dependence can be seen from Fig. 5.4. On this figure the excess pressure gradient, or solids effect, has been expressed in relative terms by dividing by the gradient required to move a dense particulate plug filling the pipe (given by Eq. 5.8). Thus the relative excess pressure gradient, denoted ζ, is given by

$$\zeta = \frac{(i_m - i_f)}{i_{pg}} \tag{5.12}$$

Figure 5.4 (from Wilson, 1988a) plots ζ versus V_r, with C_r as parameter. At large values of V_r the curves flatten, approaching asymptotic values which represent the relative excess gradient evaluated at velocities high enough to eliminate any hold-up of solids.

In dealing with fully-stratified flows it is often desirable to employ a simpler method than that outlined above, even if some loss of accuracy may be entailed. The simplest method is that of Newitt *et al.* (1955). As shown earlier (Chapter 4, Eq. 4.6), this method considers the ratio $(i_m - i_w)/(S_m - 1)$ to be a constant, equal to about 0.8. In fact, this term is not a constant. For a specific solid-liquid mixture with a given delivered concentration of solids, the quantity (i_m - i_w) *i.e.* (i_m - i_f), is proportional to the relative excess pressure gradient ζ. As shown graphically on Fig. 5.4 ζ decreases with increasing V_r (*i.e.* V_m/V_{sm}). Wilson & Addie (1995) found that the curves can be approximated to reasonable accuracy by the expression

$$\frac{i_m - i_w}{(S_m - 1)} = \left[\frac{V_m}{0.55\,V_{sm}}\right]^{-0.25} \tag{5.13}$$

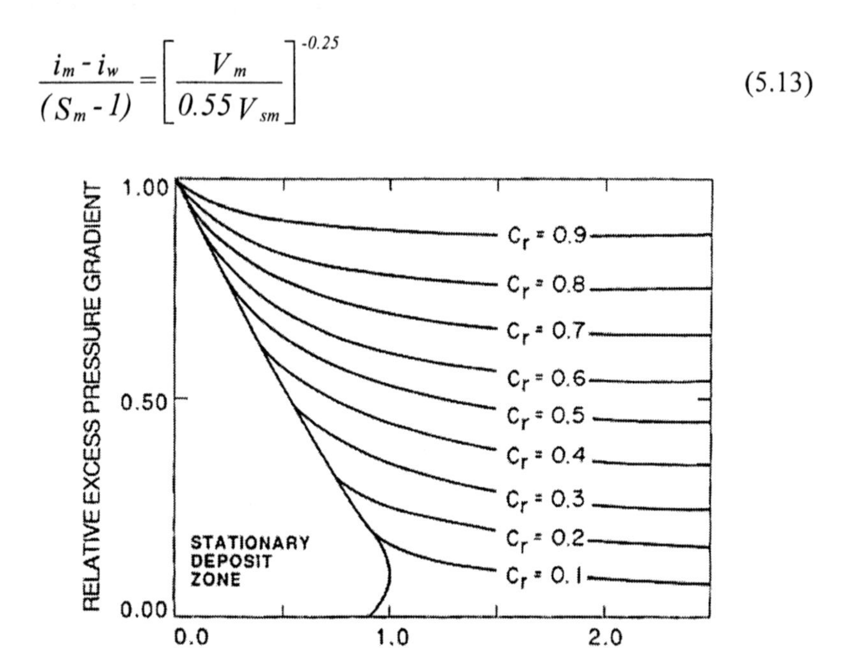

Figure 5.4. Curves of relative-excess pressure gradient, from Wilson (1988a)

The ratio on the left hand side of this equation is the same as that used by Newitt *et al.* (1955), and the right hand side shows the decrease of this ratio with increasing V_m/V_{sm}. The coefficient of 0.55 was evaluated on the implicit assumption that the coefficient of sliding friction μ_s equals 0.40. This value is appropriate for sand or gravel sliding in a steel pipe. If a better evaluation of μ_s is available for a specific case, then the right-hand side of Eq. 5.13 should be multiplied by this value of μ_s and divided by 0.40. This technique is illustrated below, in Case Study 5.2.

As fully-stratified flow has higher energy consumption than other flow types, Eq. 5.18 can be thought of as representing the upper limit of excess pressure gradient (*i.e.* solids effect). Heterogeneous slurry flows, which are covered in the following chapter, will generally display smaller values of the solids effect.

Another type of stratified flow that may be encountered is flow above a stationary bed of solids. This type of operation is generally uneconomic, and thus is seldom a deliberate design choice, but it is sometimes found in existing pipelines. Flow over a stationary bed has been studied experimentally, and analysed by a computer model (Nnadi & Wilson, 1992; Pugh, 1995). For most cases of this type of flow encountered in pipelines a rough approximation to the computer output may be sufficient. In such instances, the following equation for estimating i_m may be useful. It is based on particles near the 'Murphian' size, which tend to be disproportionately represented in stationary deposits

$$i_m = 0.32\,(S_s - 1)^{1.05}\, C_{vd}^{0.6} \left(\frac{V_m}{(2gD)^{1/2}} \right)^{-0.1} \tag{5.14}$$

In some cases it will be necessary to incorporate the effects of pipe slope, which will be dealt with in Chapter 8. Before considering sloping flows, however, it will be necessary to investigate fluid suspension of particles, and show how suspended load combines with contact load in heterogeneous slurry flow. These are the subjects of Chapter 6.

5.6 Case Studies

Case Study 5.1 - Preliminary Pipe Sizing

A pipe is to be designed to convey solids at 4000 tonnes/hour (on a dry-weight basis). The material is described only as 'coarse sand', and it is estimated that the largest volumetric solids concentration which can be fed to the pumps is C_{vd} = 0.24. Preliminary estimates are required of suitable combinations of pipe size, C_{vd} and V_m. For convenience it will be assumed that pipes are available with inside diameter D in increments of 50 mm.

First, the tonnes per hour is converted to m^3/s of solids, Q_s, by dividing by the estimated sand density of 2.65 tonnes/m^3 and by 3600 s/hour, giving Q_s = 0.419 m^3/s. The mixture discharge Q_m equals Q_s/C_{vd} and hence the minimum Q_m, corresponding to the maximum C_{vd} of 0.24, equals 1.747 m^3/s.

The next step is to select a pipe size which gives a reasonable velocity for this value of Q_m. For example, a pipe of D = 0.60 m has a cross-sectional area of 0.283 m^2, and hence the velocity in this pipe must be at least 1.747/0.283 or 6.17 m/s in order to satisfy the maximum - C_{vd} condition.

This velocity seems to be in the right 'ball park', but must be compared with the deposition velocity V_{sm}. The particle size distribution of the coarse sand may not be known, but it is safe to assume that it contains grains of the 'Murphian' size, which is about 0.7 mm for this pipe diameter. For this particle and pipe size combination, the nomographic chart (Fig. 5.3) gives $V_{sm} \approx 5.8$ m/s, but it should be recalled that the value of V_{sm} estimated from the chart tends to be high. This point can be checked by means of Eq. 5.11, using a reasonable value of f_w, say 0.012. The result is a revised value of V_{sm} of only 4.65 m/s, so that the minimum operating velocity of 6.17 m/s provides a generous margin against deposition.

Other pipe sizes should also be investigated. The next smallest pipe (D = 0.55 m), is found to have a minimum operating velocity of 7.35 m/s, based on $C_{vd} = 0.24$. This can be compared to V_{sm} values of 5.4 m/s from Fig. 5.3 and 5.45 m/s from Eq. 5.11. Using the latter number as the upper limit for deposition, it is seen that the minimum operating velocity would be 65 percent in excess of the deposition limit, producing stable operation at the cost of high frictional losses. This pipe size could merit further investigation of the sort illustrated in case studies in subsequent chapters, but it is obvious that no smaller pipe would be suitable, and the 0.55 m pipe appears at present to be considerably less attractive than the 0.60 m size.

Turning to a larger pipe, with D = 0.65 m, it is found that for $C_{vd} = 0.24$ the minimum velocity is 5.26 m/s, while the V_{sm} values are 6.1 m/s from Fig. 5.3 and 5.84 m/s from Eq. 5.11. The minimum operating velocity of 5.26 m/s is less than 10 percent in excess of V_{sm} by Eq. 5.11, which is not a sufficient margin. A margin of 30 percent should be adequate, giving $V_m \approx 6.30$ m/s. Increasing the velocity to this value requires a lower solids concentration, i.e. 0.20 instead of 0.24. This combination (D = 0.65 m, V_m = 6.3 m/s, C_{vd} = 0.20) merits further investigation, but any larger pipe size (which would involve higher operating velocity and lower delivered concentration), would probably not be suitable.

It is worth noting that this case study shows that a preliminary selection of appropriate pipe sizes can be made on the basis of criteria of continuity and deposition velocity, even though very little is known about the solids in the slurry. In the present instance, it is found that to transport 4000 tonnes of coarse sand per hour, only pipes of internal diameter 0.60 m and 0.65 m appear to be attractive.

Case Study 5.2 - Transporting Clay Balls

The dredging industry is normally concerned with pumping sand slurries for which heterogeneous flow is expected (this type of flow will be dealt with in Chapter 6), but on occasion very coarse particles must be transported. Cases of well-defined coarse-particle transport arise when dredging cohesive clays. The cutter head of the dredge shaves slices off the surface of the clay, and after passing through the pump these emerge as fist-size lumps, with a typical dimension of roughly 100 mm. The clay balls are then transported through the discharge pipe to a disposal area.

The following data, for a typical application of this sort, were provided by Ing. K. Oudmaijer of the Netherlands (personal communication). The clay had density 1790 kg/m^3 and the diameter of the balls could be taken as 100 mm. Ing. Oudmaijer measured μ_s by wiring a group of balls together, placing them (under water) on a steel plate and inclining the plate until slip occurred. He obtained a value of 0.31. The carrier fluid was seawater with density 1020 kg/m^3. The pipe was steel, with internal diameter 0.70 m, and the pressure drop was measured over a horizontal reach of pipe 1850 m in length. The measured pressure gradient was 580 Pa/m at V_m = 4.6 m/s and C_{vd} = 0.0714.

The friction factor of the pipe is not known; for commercial pipe at the Reynolds number of this application a value of 0.012 might be expected. However, it should be recalled that the actual friction losses in dredge lines are raised considerably because of dents in the pipe and misaligned joints between lengths. Thus a higher friction factor must be employed, and a value of 0.020 is suggested.

The pressure gradient is to be calculated and compared with the observed value. The first step is to find the plug flow gradient i_{pg}, using Eq. 5.8, with μ_s = 0.31, S_s = 1.79 and S_f = 1.02. The value of the loose-packed concentration, C_{vb}, was not measured, and a typical value of 0.6 will be employed, giving

$$i_{pg} = 2(0.31)(1.79 - 1.02)(0.6) = 0.286$$

The relative concentration C_r equals C_{vd}/C_{vb}, i.e. 0.0714/0.60 or 0.119.

In order to use Fig. 5.4, it will be necessary to know the relative velocity V_r, which is based on the velocity at the deposition limit V_{sm}. The latter quantity can be estimated from the nomographic chart displayed on Fig. 5.3. It will be recalled that with this chart the first step is to use the left-hand panel to find V_{sm} for S_s = 2.65, correcting later for different density. The pipe diameter of 0.70 m is readily located on the vertical axis at the left of the chart, but it is seen that the particle diameter of 100 mm lies off the plotted portion of the 'demi McDonald'. However, the large-particle branch of this curve is virtually straight, and the numbers follow an essentially logarithmic scale on which the distance from 30 mm to 100 mm will be the same as that from 9 mm to 30 mm. The line starting at D = 0.70 mm and passing through this extrapolated point intersects the central vertical axis at a value of about 2.4 m/s. As the relative density is not 2.65, the right hand panel of the chart is also needed. For this purpose, the relative density is taken as 1790/1020 or about 1.75. A line starting from 2.4 m/s on the central axis is projected through this value of 1.75 on the inclined axis to give the corrected V_{sm} on the right-hand vertical axis, about 1.55 m/s.

The relative velocity V_r is $V_m/1.55$, and for the operating velocity V_m = 4.6 m/s, V_r is approximately 3.0.

Extrapolating from Fig. 5.4, with $C_r \approx 0.12$, gives a value of ζ of about 0.1. When multiplied by i_{pg} (found previously to be 0.286) this gives the solids effect (i_m - i_f) as 0.029 metres of water per metre of pipe. The fluid friction gradient i_f is calculated as $S_f\, f\, V_m^2/(2gD)$, or 0.031, from which i_m is predicted to be 0.060. The equivalent pressure gradient is 590 Pa/m, which very close to the observed value of 580 Pa/m.

Alternatively, the simplified form of Eq. 5.13 can be used. For the present case V_{sm} is 1.55 m/s, as found above, and 0.55 V_{sm} equals 0.853 m/s. As μ_s has been evaluated at 0.31, the right hand side of Eq. 5.13 is multiplied by the ratio 0.31/0.40 or 0.775, and the solids effect is given by

$$i_m - i_f = (S_m - 1)(0.775)(V_m / 0.853)^{-0.25}$$

For the specific case referred to above, (S_m - 1) equals (1.75-1)(0.0714), or 0.0536, and V_m = 4.6 m/s, giving

$$i_m - i_f = 0.027$$

This number is a reasonably good approximation of the solids effect of 0.029 found above using Fig. 5.4. The resulting pressure gradient is now predicted to be 570 Pa/m, which is again very close to the observed value.

REFERENCES

Bagnold, R.A. (1956). The flow of cohesionless grains in fluids. *Phil. Trans. Roy. Soc.*, London, UK. Ser. A., Vol. 249, pp. 235-297.

Fowkes, R.S. & Wancheck, G.A. (1969). Materials handling research: hydraulic transportation of coarse solids. *Report 7283*, U.S. Department of the Interior, Bureau of Mines, USA.

Newitt, D.M., Richardson, J.F., Abbot, M., & Turtle, R.B. (1955). Hydraulic conveying of solids in horizontal pipes. *Trans. Inst. of Chem Engrs.*, Vol. 33, London, U.K.

Nnadi, F.N. & Wilson, K.C. (1992). Motion of contact load at high shear stress. *J. Hydr. Engrg.*, ASCE 118 (12), pp. 1670-1684.

Pugh, F.J. (1995). *Bed-load Velocity and Concentration Profiles in High Shear Stress Flows.* Ph.D. Thesis, Queen's University, Canada.

Sanders, R.S., Sun, R., Gullies, R.G., McKibbem, M.J., Litzenberger, C. & Shook, C.A. (2004). Deposition velocities for particles of intermediate size in turbulent flow, *Proc. Hydrotransport 16*, BHR Group, Cranfield, UK, pp 429-442.

Thomas, A.D. (1979). The role of laminar/turbulent transition in determining the critical deposit velocity and the operating pressure gradient for long distance slurry pipelines, *Proc. Hydrotransport 6*, BHRA Fluid Engineering, Cranfield, UK, pp 13-26.

Wilson, K.C. (1970). Slip point of beds in solid-liquid pipeline flow. *Proc. Am. Soc. Civil Engrs., 96*, No. HY1, pp. 1-12.

Wilson, K.C. (1975). Co-ordinates for the limit of deposition in pipeline flow. *Proc. Hydrotransport 3*, BHRA Fluid Engineering, Cranfield, UK, pp. E1-1-13.

Wilson, K.C. (1976). A unified physically-based analysis of solid-liquid pipeline flow. *Proc. Hydrotransport 4*, BHRA Fluid Engineering, Cranfield, UK, pp A1-1-A1-16.

Wilson, K.C. (1979). Deposition-limit nomograms for particles of various densities in pipeline flow, *Proc. Hydrotransport 6*, BHRA Fluid Engineering, Cranfield, UK, pp 1-12.

Wilson, K.C. (1986). Effect of solids concentration on deposit velocity. *Jour. Pipelines* Elsevier, Vol 5, pp. 251-257.

Wilson (1988a). Algorithm for coarse-particle transport in horizontal and inclined pipes. *Proc. Intern'l. Symp. on Hydraulic Transp. of Coal and Other Minerals (ISHT.88)* CSIR and Indian Institute of Metals, Bhubaneswar, India, pp. 103-126.

Wilson, K.C. (1988b). Evaluation of interfacial friction for pipeline transport models. *Proc. Hydrotransport 11.*, BHRA Fluid Engineering, Cranfield, UK, pp 107-116.

Wilson, K.C. & Addie, G.R. (1995). Coarse-particle pipeline transport: effect of particle degradation on friction. *Proc. 8th International Freight Pipeline Society Symposium*, pp. 151-156.

Wilson, K.C. & Brown, N.P. (1982). Analysis of fluid friction in dense-phase pipeline flow. *Canad. Jour. Chem. Engrg.*, Vol 60, No. 1, pp. 83-86.

Wilson, K.C. & Judge, D.G. (1977). New techniques for the scale-up of pilot-plant results to coal slurry pipelines, *Proc. Int'l Symp. on Freight Pipelines*, Univ. of Pennsylvania, pp 1-29.

Wilson, K.C., & Judge, D.G. (1978). Analytically-based nomographic charts for sand-water flow, *Proc. Hydrotransport 5*, Solids in Pipes, BHRA Fluid Engineering, Cranfield, UK, pp A1-1-11.

Wilson, K.C.. & Nnadi, F.N. (1990). Behaviour of mobile beds at high shear stress. *Proc. 22nd Int'l Conference on Coastal Engineering*, Delft, Netherlands, Vol. 3, pp. 2536-2541.

Wilson, K.C., Streat, M. and Bantin, R.A.(1973). Slip-model correlation for dense two-phase flow. *Proc. Hydrotransport 2*, BHRA Fluid Engineering, Cranfield, UK, pp.B1-1- B1-10.

Wood, F.M. (1935). Standard nomographic forms for equations in three variables, Canadian Journal of Research, Vol. 12.

Chapter 6

HETEROGENEOUS SLURRY FLOW IN HORIZONTAL PIPES

6.1 Introduction

The three basic types of particulate slurry flow were described in Chapter 4. At one extreme (the anterior end of the elephant, as it were) are large, rapidly-settling particles. Their submerged weight cannot be carried by fluid support mechanisms, and must be transferred downwards by continuous or sporadic inter-granular contacts. When all particles behave in this way the flow is fully stratified. This behaviour was dealt with toward the end of Chapter 5. The other limiting case is represented by the behaviour found for fine-sand slurries, and known as pseudo-homogeneous flow. As noted in Chapter 4, for this type of flow the particles are carried by the fluid rather than by inter-granular contacts. This behaviour is represented by the homogeneous-flow equation (Eq. 4.3), and often approaches the equivalent-fluid case where A' = 1.0 in the equation.

The present chapter focusses on the intermediate case (known as heterogeneous flow) for which both inter-granular contact and fluid support mechanisms are significant. In Chapter 4 important early approaches to the heterogeneous flow problem were outlined. For example, Newitt's group (Newitt *et al.* 1955) proposed that heterogeneous flow could be characterised by an inverse relationship between excess pressure gradient and V_m/v_t, where v_t is the terminal settling velocity of a particle. However, subsequent research showed a decline in stratification ratio with V_m that is more precipitous than a simple inverse relationship (see Fig. 4.3 and the discussion in Section 4.3).

At this point it is helpful to examine recent experimental findings obtained by Whitlock at the GIW Hydraulic Laboratory with fine (85 μm) and medium (400 μm) sands in a pipe loop with internal diameter 100 mm. These results are displayed on Fig. 6.1, from Whitlock *et al.* (2004). The abscissa of Fig. 6.1 is the mean (throughput) velocity, i.e. $4Q/\pi D^2$, and the ordinate equals $(i_m = i_w)/(S_m - 1)$, also called the stratification ratio.

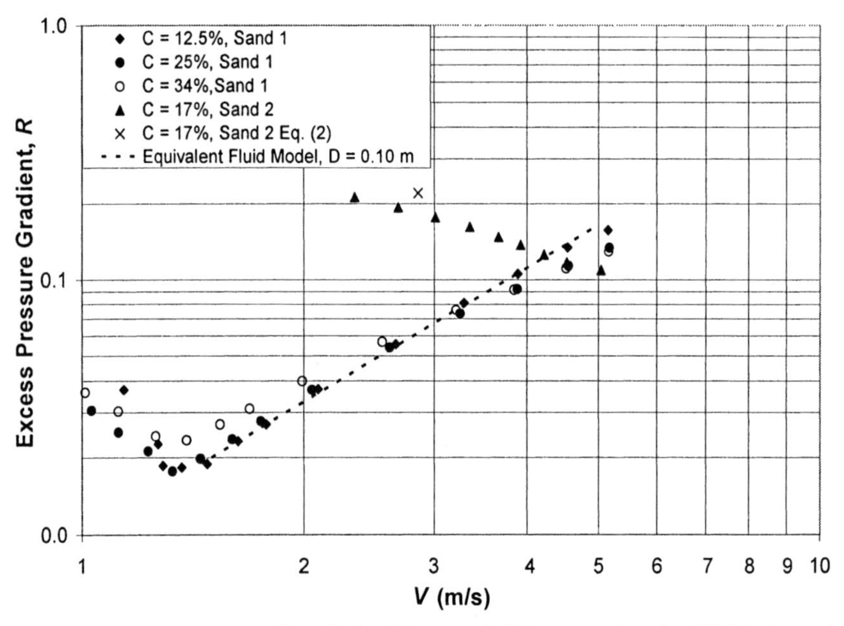

Figure 6.1. Excess pressure gradient R for 85 μm and 400 μm sands, after Whitlock et al. (2004)

The dashed line on the figure represents the 'equivalent-fluid' model is based on Eq. 4.3, with A' taken as unity. For this case $(i_m - i_w)$ equals $(S_m - 1)i_w$ so that $R = i_w = f_w V^2/2gD$. Note that data for the 85 μm sand closely follow the equivalent-fluid line for velocities above 1.5 m/s. The behaviour of the 400 μm sand is quite different, as here R decreases with V, following a straight line on the logarithmic plot of the figure. This is represents heterogeneous flow behaviour, to be described in detail below, in Section 6.3.

6.2 Concepts of Near-Wall Hydrodynamic Lift

It is particularly noteworthy that Fig. 6.1 shows, for velocities above 4 m/s, the stratification ratio (and hence the pressure gradient) for the slurry of 400 μm particles is less than that of the 85 μm ones. It is important to develop an explanation of this strongly counter-intuitive fact, and this has been the object of research at the GIW Hydraulic Laboratory. The early experimental work of Carstens & Addie (1981) was later followed by papers by Wilson et al. (2000) and Wilson & Sellgren (2002), and the most recent work is that of Whitlock *et al.* (2004), mentioned above. The paper by Wilson et al. (2000) noted that, within the turbulent portion of the flow, particle support is dominated by turbulent diffusion, introduced by Schmidt (1932) and Rouse (1937). Later Lumley (1978) pointed out the need for more sophisticated models of turbulence. Turbulent diffusion produces characteristic concentration profiles, observed by Hsu et al. (1980) and more recently in numerous publications associated with the Saskatchewan Research Council. The diffusion concept implies that any horizontal plane in the flow acts as a 'mirror' which ejects upward a flux of particles equal to those that move down through the plane. This condition is met in the fully-turbulent part of the flow, but causes difficulties near the lower part of the boundary. Here near the pipe wall (which can often be taken as hydraulically smooth) turbulence is ineffective in the viscous sub-layer and begins to be felt in the adjacent buffer layer. Thus the 'mirror' effect will be inhibited in these near-wall regions, and here the particles that fall out of the turbulent flow must be replaced by a flux of particles driven away from the wall by a force (the off-wall force) that can be considered to arise from hydrodynamic lift.

An indication of near-wall lift is given by concentration profiles first published in the early work of Shook & Daniel (1965) and Shook et al. (1968). They measured profiles of solids concentration versus height for horizontal flow in a rectangular closed channel. For sands of mean diameter above 150 μm and less than 1000 μm the concentration profiles showed maximum values somewhat above the lower boundary of the conduit, accompanied by a rapid decrease in concentration on closer approach to the boundary. A concentration profile of this sort for 180 μm sand is shown below on Fig. 6.9 (from Gillies & Shook, 1996). The observed concentration maximum indicates the presence of a significant off-wall lift force, which is most important for particles in a size range substantially larger than the thickness of the viscous sub-layer.

Like drag force, lift force can be either viscous or inertial in character (plus intermediate behaviour). Viscous lift, identified with the work of Saffman (1965, 1968), occurs in slow-moving viscous fluids such as

Poiseuille flow in blood vessels and capillaries, where it causes near-equal-density particles such as blood corpuscles to move away from the wall. In turbulent flows adjacent to hydraulically-smooth boundaries, only the viscous sub-layer approximates laminar flow. The boundary between the viscous sub-layer and the buffer layer is located at a distance from the boundary where $y^+ \approx 5$. The dimensionless variable y^+ is defined as yU^*/ν, where y is distance from the boundary, U^* is shear velocity at the wall (the square root of τ_w/ρ, where τ_w is shear stress at the wall and ρ is fluid density), and ν is kinematic viscosity of the fluid. For y^+ between 5 and 25, the flow combines viscous and inertial (turbulent) components in what is called the buffer layer; and for $y^+ \geq 25$ the velocity profile is logarithmic and the flow is fully turbulent. The middle of a particle cannot be closer to the wall than $d/2$, where d is the particle diameter. On this basis a dimensionless particle radius, r^+, can be defined as $dU^*/(2\nu)$. For the conditions with 400 μm sand, r^+ generally was found to be greater than 25, so the middle of the particles must be in the logarithmic layer (fully turbulent flow). The same applies for the backward-curving concentration profiles observed in Saskatoon.

Thus the observed near-wall lift force cannot be viscous lift, and it is logical to seek an inertial-based explanation. The first person to identify lift of this sort was Magnus (1852, 1853) who was concerned with musket balls and other military projectiles. He proposed that the lift force which caused projectiles to depart from their expected trajectories is proportional to the vector product of the velocity of the projectile through the air and its rotation. Since the lift does not occur in a vacuum, it is clear that surface friction must convert the particle rotation into a circulation in the surrounding fluid, which acts to produce the lift. Subsequently Lord Rayleigh (1877), made the same point concerning the more pacific application of Magnus's work to the flight of tennis balls.

For the cases of inertial lift mentioned above, the required velocity difference and circulation are provided by the initial linear and angular momentum of the projectile, but these are probably not significant for the particles of current interest. For particles travelling within the log-law zone, it is expected that the time-averaged fluid velocity, U, will not have a direct effect on lift force. However, as with viscous lift, the velocity gradient dU/dy is an important variable. It is also expected the second differential d^2U/dy^2 will be important, since for a velocity field with a second differential of zero, symmetry arguments show that no lift can occur. For a logarithmic velocity profile, $dU/dy = U^*/(\kappa y)$ and $d^2U/dy^2 = -U^*/(\kappa y^2)$, which suggests that U^* is the basic velocity variable to be used in the lift-force equation for a particle. On this basis, the equation for inertial lift in the log-law zone can be written:

$$F_L = C_L \cdot \left[\rho \cdot \frac{(U^*)^2}{2} \right] \cdot \left[\pi \cdot \frac{d^2}{4} \right] \tag{6.1}$$

Here F_L is the lift force on a particle, and the lift coefficient, C_L, is to be determined experimentally. It was found (Wilson & Sellgren, 2002) that for $r^+ \geq 25$, the lift coefficient C_L is approximately constant, with a value near 0.27.

For smaller values of r^+, the particle is in the buffer zone of the velocity profile, and in this zone the it would appear that the combination of velocity differentials which influence inertial lift diminish as the wall is approached (i.e. at lower values of r^+). The effect of r^+ on the lift coefficient (relative to that for $r^+ > 25$, and denoted by λ) is indicated schematically on Fig. 6.2.

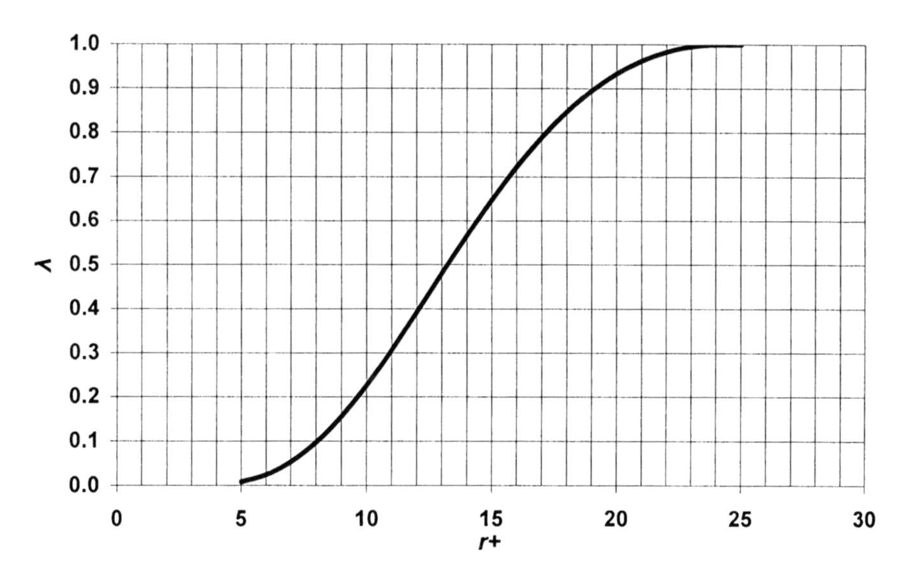

Figure 6.2. Lift-reduction ratio λ versus dimensionless particle radius r^+, from Whitlock et al. (2004)

The line in Fig. 6.2 is based on fitted curves derived from the dimensionless velocity profile, and is intended to illustrate the main features of the variation of inertial lift near the pipe boundary. For the log-law portion of the velocity profile, indicated by $r^+ > 25$, the ratio λ is 1, indicating a constant value of C_L (approximately 0.27, as noted above). The inertial lift associated with turbulence drops throughout the buffer layer, reaching a negligibly small value in the viscous sub-layer ($r^+ < 5$). Conversely, viscous lift, not shown here, would be maximum in the viscous sub-layer, and then decrease with increasing y^+ (and r^+).

It was noted earlier in this section that for the 400 μm particles the value of r^+ typically exceeded 25, indicating that the particles are effectively within the logarithmic part of the flow. On the other hand, the 85 μm particles have $r^+ < 10$, throughout the flow velocity range of Fig. 6.1, and thus will be engulfed in the near-wall layers. Additional experimental work has been done recently in the same pipe loop, with the intermediate particle size of 200 μm, and this showed essentially constant values of R (near 0.08) in the velocity range 2 to 4 m/s, followed by a rise to somewhat blow the equivalent-fluid line at higher velocities. In this case r^+ varies from 8 at 2 m/s to 20 at 5 m/s, indicating intermediate behaviour. It was found that if R were re-defined by replacing the water gradient i_w by the equivalent-fluid gradient, the expected pattern of R decreasing with increasing V would occur for the 200 μm sand slurry.

6.3 Stratification Ratio and Scale-up

In Chapters 4 and 5 it was implied that the transition marked by the intercept between fully-stratified and heterogeneous flow might be helpful in quantifying the heterogeneous regime. However, experience has shown that any formulation based on this threshold velocity is subject to considerable difficulties. Although the intercept can often be observed for small-pipe flow; for larger pipes the velocity at the deposition limit usually exceeds the expected threshold for turbulent suspension, precluding its experimental determination. Indeed, if the heterogeneous flow regime is considered as intermediate between fully-stratified flow at one extreme and pure carrier-fluid flow at the other, we would expect a sigmoidal (ogee) curve rather than an abrupt threshold.

As noted previously, contact-load solids can only be moved by overcoming mechanical friction forces between the particles and the pipe wall, which in turn leads to the large excess pressure gradients found for fully-stratified flow. For this type of flow, the relative excess gradient is almost velocity-independent, as shown on Fig. 4.2. For the other limiting case (at least for large pipes and no significant fraction of -40 μm solids), the friction gradient can be evaluated on the basis of clear-water flow at the same discharge (Carstens & Addie, 1981).

An appropriate sigmoidal curve for the transition region between the limiting cases is given by the integrated log-normal distribution, which is defined by two parameters: the mean and the standard deviation of the logarithms. It is noteworthy that the central portion of the sigmoid formed by the integrated log-normal function can be approximated, to reasonable accuracy, by a power-law function. Thus the sloping straight line passing through heterogeneous data on a logarithmic plot (such as that of the 400 μm sand data on Fig. 6.1) can best be regarded as a power-law approximation to the central portion of the sigmoidal distribution. The basic representation is then seen to be the ogee curve rather than the power law, and hence the major parameter defining the distribution should be based on a central value rather than some 'threshold'. The central-value parameter will be denoted here by V_{50}, signifying the value of V_m at which the stratification ratio is 0.50; i.e. half the mass of solids is supported by granular contact and half by fluid suspension.

From a practical standpoint it is necessary to match the 50-percent stratification point with a specific value of $(i_m-i_f)/(S_m-S_f)$. The mechanics of the situation suggests that this value should be approximately $0.5\mu_s$. Heterogeneous flow is more likely to involve particle impact against the pipe wall rather than continuous sliding motion, and although μ_s for sliding friction varies with the material of the particles and the boundary, for the impact mode there is much less variation, and μ_s can be assigned a value close to 0.44. Thus it is expected that V_{50} will correspond to a value of $(i_m-i_f)/(S_m-S_f)$ of about 0.22. Experimental data bearing on this point will be discussed in Section 6.5.

In establishing the factors that influence V_{50}, many of the points mentioned previously in connection with the earlier concept of a 'threshold' velocity can be used. Thus V_{50} should vary with $\sqrt{(8/f_f)}$ (see Chapter 4, Fig. 4.2 and 4.3). It should also depend on the diameter ratio d/D. The expression for the effect of this ratio suggested by Wilson & Watt (1974) has been modified in light of large-pipe data obtained more recently. The newer relation uses the hyperbolic cosine, and, when applied to V_{50}, is written cosh[60d/D].

In previous formulations, following Newitt *et al.* (1955) and others, the major effect of the particles had been expressed in the ratio V_m/v_t. On this basis it would be expected that V_{50} should vary directly with the terminal settling velocity v_t. However, the experimental evidence throws doubt on this postulate. Although V_{50} increases with settling velocity, it appears that the increase may not be a direct proportionality. The mechanisms involved in this relation will be considered below in Section 6.4. In the meantime, following earlier editions of this book, the observed behaviour will be taken into account by introducing a quantity, w, into the expression for V_{50}. This quantity has dimensions of velocity and a value which should depend only on particle and fluid properties. The resulting equation is:

$$V_{50} = w\sqrt{\frac{8}{f_f}}\cosh[60d/D] \tag{6.2}$$

The ratio V_m/V_{50} determines the stratification ratio R, which is generally taken as $(i_m - i_f)/(S_m - S_f)$. Thus, for a given ratio V_m/V_{50} it is assumed that the solids effect $(i_m - i_f)$ varies directly with solids concentration (and hence with $S_m - S_f$). As shown later in this chapter, this assumption has received extensive experimental verification for volumetric solids concentrations up to 25 %. On this basis we can write:

$$\frac{(i_m - i_f)}{(S_m - S_f)} = fn\left(\frac{V_m}{V_{50}}\right) \tag{6.3}$$

Here fn represents a functional relationship. As noted earlier, the form of this function must reflect the nature of heterogeneous flow as a transition between the fully-stratified and pseudo-homogeneous cases; and hence a sigmoidal (ogee) shape is expected. Moreover, the statistical nature of turbulence indicates that the fraction of particles carried by fluid suspension will vary in an essentially probabilistic fashion. This type of behaviour supports the proposal made above that the integrated log-normal distribution forms an appropriate basis for the stratification-ratio function.

The mathematical form of this distribution is rather awkward, and as mentioned previously, it is convenient to employ a simpler function as a working approximation. The approximating function selected here is the power law, which follows the central portion of the integrated log-normal distribution rather closely, and can be applied over a considerable range of V_m/V_{50}. This feature would appear to account for the essentially straight lines followed by plots of experimental data on logarithmic coordinates.

The equation for the approximating power law is readily obtained in terms of V_{50}, giving the relative solids effect as

$$\frac{(i_m - i_f)}{(S_m - S_f)} = 0.22\left(\frac{V_m}{V_{50}}\right)^{-M} \tag{6.4}$$

The coefficient 0.22 represents the value of the relative solids effect at the point where V_m equals V_{50}, as mentioned previously. The power M is typically near 1.7 for slurries with a narrow particle grading (Clift *et al.*, 1982). The variation of M with particle size distribution will be examined in Section 6.5.

The reference velocity V_{50} is specified by Eq. 6.2. It depends on the particle-associated parameter w and other quantities which are known. Expressions for estimating w from particle and fluid properties will be noted in Section 6.4; for the present it should be remarked that for given particles and carrier fluid w is fixed. Thus V_{50}, as well as M, can be evaluated directly by pipe-loop experimentation. Experiments are less costly for small-diameter pipe loops, and in many cases the quantity of slurry required for large-diameter testing may not be readily available. In such cases it is necessary to scale up experimental results from small pipes to the larger pipe diameter required for the prototype design. This scale-up will be dealt with now.

In scaling up small-pipe experiments it should be noted that, in the general case, scaling should be applied both to the experimentally-determined friction gradient and to the mean velocity at which it was observed. The scaling relationships are different for the two variables, forming a pair of scaling laws conceptually similar to the affinity laws used for scale-up of head and discharge for centrifugal pumps handling water (see Chapter 9). Laws of this type are based on maintaining dynamic similarity, which implies that certain quantities remain invariant as the size scale is changed. It was shown in Chapter 3 that for any fluid (Newtonian or non-Newtonian) the basic invariant is the wall shear stress (or the shear velocity U_*). For constant shear velocity, variations in pipe size cause changes in both pressure gradient and mean velocity V_m. For the turbulent flow which is of interest here, the effect of pipe size on V_m is rather small. Since U_* equals $\sqrt{\frac{f}{8}}V_m$, the requirement of constant U_* implies that V_m must scale inversely with the square root of the fluid friction factor, a quantity that changes only slightly with pipe diameter. This concept was introduced earlier (Wilson, 1972; see Section 3.6), and alternative ways expressing the velocity scaling law were given in Chapter 3, specifically Eq. 3.13 and 3.16.

In the case of heterogeneous slurry flow, a similarity condition must apply not only to the fluid flow (which will include the suspend-load solids as well as the fluid itself), but also to the 'solids effect' associated with the contact load. The major result of this contact load is to increase the frictional pressure gradient, which is now viewed as the sum of the fluid component i_f, and the solids effect (i_m - i_f). The fluid component i_f scales inversely with the diameter ratio, but the solids effect is independent of pipe diameter, and thus forms a further parameter to be kept constant in the scale-up process. Velocity scaling does not appear to be greatly influenced by the solids effect, and should be based on the friction factor for the fluid component only. This use of the fluid friction factor f_f has been proposed previously (see Wilson, 1972, and the discussion in Section 4.6). The associated shear velocity, $\sqrt{\frac{f_f}{8}}V_m$, will be called the fluid shear velocity, and denoted by U_{*f}.

Figure 6.3 shows, on logarithmic co-ordinates, results obtained at the GIW Hydraulic Laboratory for an aqueous slurry of 0.42 mm masonry sand in a 0.20 m pipe (Clift *et al.*, 1982). In calculating the ratio $(i_m\text{-}i_w)/(S_m\text{-}1)$ the clear water gradient i_w has been used to approximate i_f, and S_f is S_w, i.e. 1.0. The value of i_w was calculated from the Darcy-Weisbach formula using the friction factor $f_f = 0.013$ measured previously for flows of water in this pipe.

The plotted points fall on an essentially straight line on Fig. 6.3. This behaviour corresponds to Eq. 6.3; the slope of the line gives M = 1.7, and the point on the line where $(i_m\text{-}i_w)/(S_m\text{-}1) = 0.22$ gives V_{50}, which is approximately 2.8 m/s.

With V_{50} determined experimentally, Eq. 6.2 can be used to obtain the particle-associated velocity w. For the case plotted on Fig. 6.3, with f_f = 0.013 as noted above, $\sqrt{(8/f_f)}$ equals 24.8 and cosh[60 d/D] is 1.008, i.e. not significantly different from unity. Thus w is 0.040 V_{50} or 0.112 m/s. With w and M obtained for the 0.20 m pipe, scale-up to a larger pipe diameter can be carried out. The larger diameter of 0.44 mm has been selected because data are available with the same sand in this larger pipe (Clift *et al.*, 1982) and thus the scaled-up results can be verified directly.

The measured internal diameters of the two pipes are 203 mm and 440 mm, and the friction factor for clear-water tests in the larger pipe was 0.013, the same as for the smaller pipe. As cosh[60d/D] is also virtually indistinguishable from unity in the larger pipe, and as both pipes have the same value of $\sqrt{(8/f_f)}$, the relation between w and V_{50} is the same for both pipes, as is M. Therefore, in the present instance (but not universally), the equation linking $(i_m\text{-}i_w)/(S_m\text{-}1)$ to V_m which had been established for the smaller pipe should also apply to the larger one. Verification is given by Fig. 6.4, which plots the proposed power-law equation together with the experimental data from both pipe sizes. For the velocity range less than 4.5

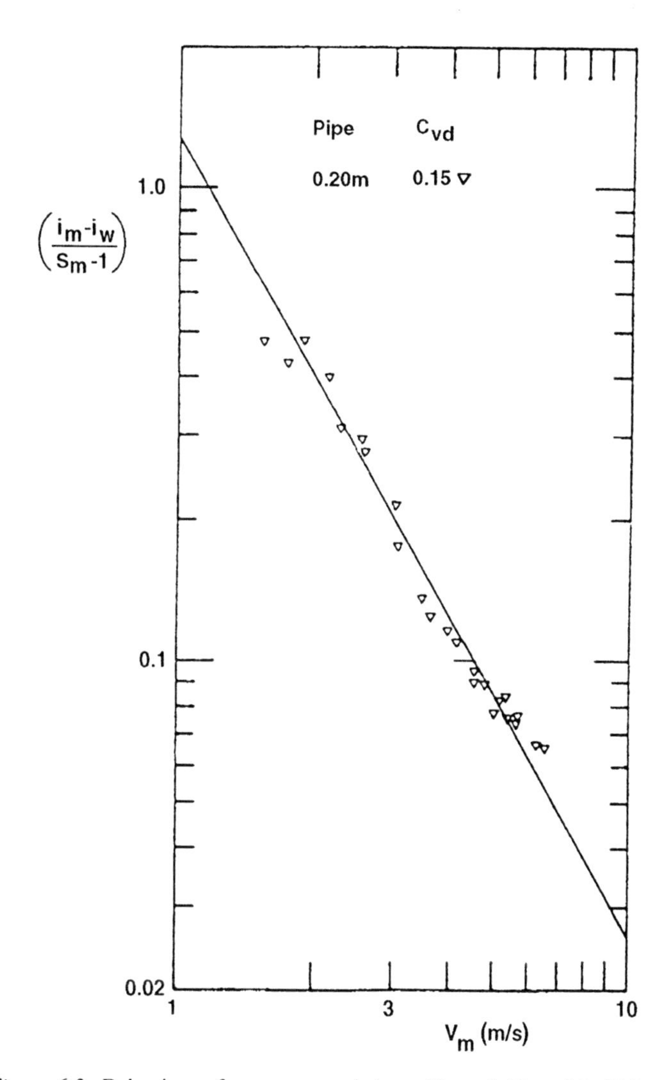

Figure 6.3. Behaviour of masonry-sand slurry (d_{50} = 0.42 mm) in 203 mm pipe

m/s (which includes the velocities of commercial interest), it is clear that both data sets are part of the same population, and can be represented by the same power-law fit line. For higher velocities it can be seen that the data points for the larger pipe droop below the common line, appearing to have an enhanced value of M in this range.

The final step in the scale-up procedure is to prepare curves of i_m versus V_m for various values of C_{vd} in the larger pipe. For the velocity range of commercial interest, the relation for each line of constant C_{vd} is expressed as

$$i_m = \frac{f_w}{2gD} V_m^2 + 0.22 (S_s - 1) V_{50}^M C_{vd} V_m^{-M} \qquad (6.5)$$

The velocity at the minimum points can be obtained by differentiating Eq. 6.5 with respect to V_m, and setting the result equal to zero. For the case dealt with here, with f_w and V_{50} independent of pipe diameter and M = 1.7, it can readily be shown that the value of V_m at the minimum varies with $(DC_{vd})^{0.27}$; i.e. for given pipe size this velocity depends on $C_{vd}^{0.27}$ and for given delivered concentration it scales with pipe size according to $D^{0.27}$. A numerical example involving the minimum point of the system curve is given in Case Study 6.2.

It should be noted here that the constant value of $\sqrt{\frac{8}{f_f}} \cosh[60d/D]$ which was found for the scaling example studied in this section, while not unusual, should not generally be assumed. It is typical to find a moderate decrease in f_w with increasing pipe size, which produces a small increase in V_{50}. Increases in D have the opposite effect on cosh[60d/D], but for small ratios of d/D this term is already indistinguishable from unity, and so cannot counterbalance the influence of the pipe diameter on f_w. In each case of scale- up there is no difficulty in calculating $\sqrt{\frac{8}{f_w}} \cosh[60d/D]$ and using Eq. 6.3 to determine the scaling effect on V_{50}.

Other problems arise when pipeline experiments have not been carried out with the particular slurry of interest. Equations 6.2 and 6.4 can still be used in such cases, provided the parameters w and M can be evaluated. The value of M depends on particle size distribution, and its evaluation will be discussed in Section 6.5; the particle-associated velocity w of Eq. 6.2 will be dealt with first.

As noted above, the particle-associated velocity w can be considered as replacing the terminal settling velocity, which had been used in earlier formulations for estimating the stratification ratio. These earlier approaches had considered the ratio V_m/v_t as a unit, in accord with the perceived mechanism of suspension by turbulent eddies, described by Schmidt (1932) and Rouse (1937). Although this type of fluid suspension dominates the

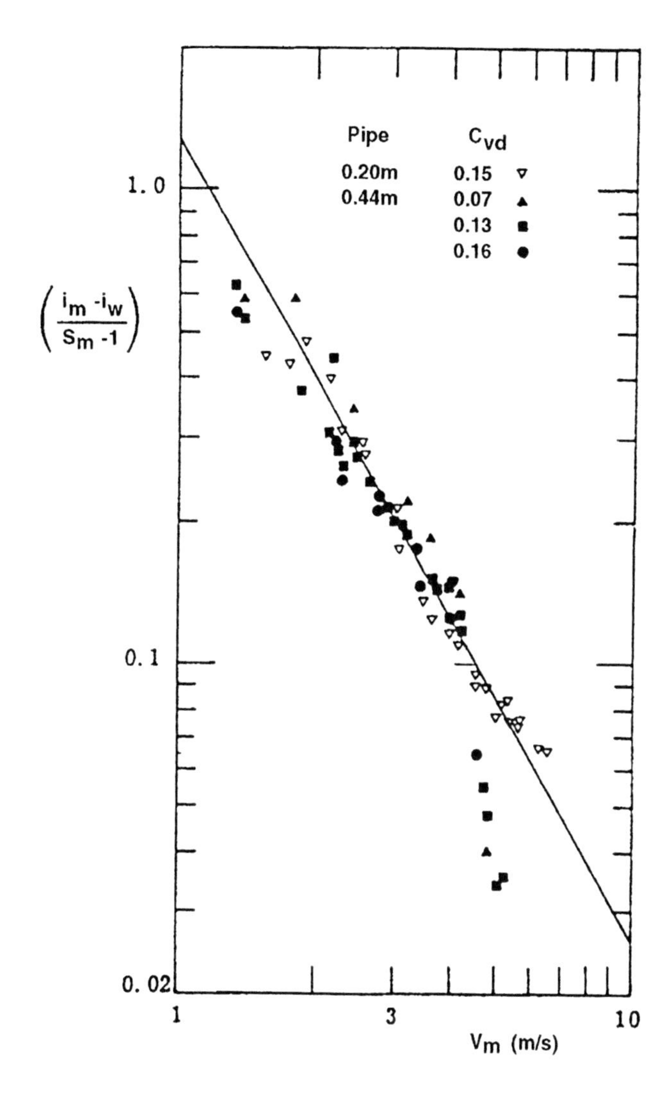

Figure 6.4. Behaviour of masonry-sand slurry (d_{50} = 0.42 mm) in 203 mm and 440 mm pipes, after Clift *et al.* (1982)

fully-turbulent portion of the flow, it was seen in Section 6.2 that a different mechanism of fluid support – hydrodynamic lift – applies in the near-wall layers where velocity gradients are large. In the 2nd Edition of this book, it was proposed that w could be approximated by a two-term expression that contains two coefficients for which numerical values must be determined. Based on experimental results available at the time, these coefficients were evaluated to produce a formula for the particle-associated velocity, which was then used to calculate w as a function of particle diameter. Figure 6.5 shows a curve obtained on this basis for sand particles (S_s=2.65) in water at 20°C.

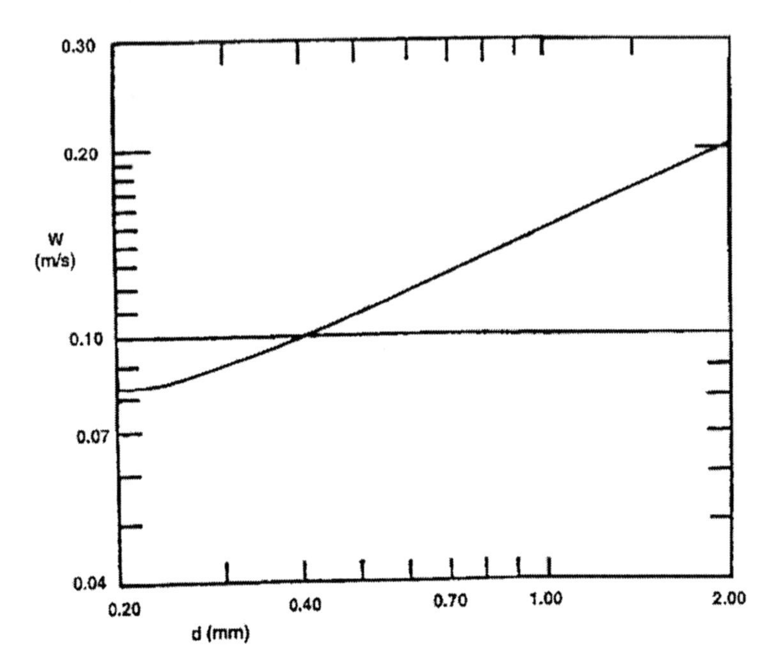

Figure 6.5. Particle-associated velocity w for sand-water slurries

6.4 Effect of Particle Size Distribution

Up to this point, the discussion of slurry flow has been directed to solids with narrow particle size grading, and with no significant fraction of -40 μm. In this section, the effect of relaxing these conditions will be investigated. The first step will be to consider broader particle gradings, but still without -40 μm material. Consideration of the case where finer material is also present will be deferred to Chapter 7.

Plots of $(i_m-i_f)/(S_m-1)$ versus V_m have been shown on Fig. 6.3 and 6.4 for closely graded materials. These figures (after Clift *et al.*, 1982) have M ≈ 1.7, as is often found for narrow size grading. Even materials with broader grading can approximate this value of M, as shown on Fig. 6.6, also from the work of Clift *et al.* (1982), which is a plot of data for slurries of crushed granite in pipes of internal diameter 203 mm and 440 mm. Although, as with Fig. 6.4, the data points for the larger pipe diverge downward at the highest velocities tested, the points in the range of commercial interest are in reasonable accord with the line shown on the figure, which has M = 1.7.

The slurry which provided the data for Fig. 6.6 had solids with a mass-median particle size, d_{50}, of 0.68 mm. Subsequently, tests were performed in the 203 mm pipe using a slurry with the same value of d_{50} but with a still broader range of particle size (Wilson *et al.*, 1990). The results have been plotted on Fig. 6.7 which employs the usual logarithmic axes. Once more a straight line gives a suitable representation of the bulk of the data, with scatter only at extreme values of V_m (low values in this case). However the slope is now much flatter, with $M \approx 0.9$. The influence of particle grading on the distribution parameter M is evident here, and will be discussed below, but the most important point of comparison of the two slurries is the intercept of the fit lines. This intercept has been shown on Fig. 6.7 by adding the fit line from the data plotted on Fig. 6.6. The two lines intercept at $V_m \approx 3.2$ m/s and $(i_m - i_w)/(S_m-1) \approx 0.22$. These intercept values define the central tendency of the distribution. The velocity at intercept, which depends on the properties of the median particle but not on the distribution of particle size, represents V_{50}. The corresponding value of the ordinate indicates the point where half the mass of the particle is supported by fluid suspension. It was noted in Section 6.3 that this condition is expected occur when $(i_m-i_f)/(S_m-1)$ is approximately 0.22. The fact that the same value was found experimentally for the intercept of curves for slurries of the same d_{50} provides a gratifying verification.

As mentioned above, it has been observed that the exponent M in Eq. 6.4 is often near 1.7 for narrow particle grading, and decreases as the grading becomes broader. A plausible reason for this behaviour can be found by considering how M is related to the standard deviation of the integrated log-normal distribution representing the ogee curve. The magnitude of M is the same as the slope of a log-log plot of the integrated probability function. If this slope is evaluated at some fixed value of V_m/V_{50}, it will be found that M is inversely proportional to the standard deviation of $\log(V_m/V_{50})$, where log indicates the logarithm to the base 10. This standard deviation is denoted here by σ_{10}. For the case of uniform particles, a condition approximated by narrow particle grading, V_{50} is constant, so that the variation which accounts for the observed value of σ_{10} must be associated with the inherently stochastic nature of the turbulent flow field. As the value of σ_{10} for this uniform-particle condition depends only on fluid turbulence it is given the symbol σ_f. For fully-developed turbulent flow it may be assumed that σ_f will be sensibly constant.

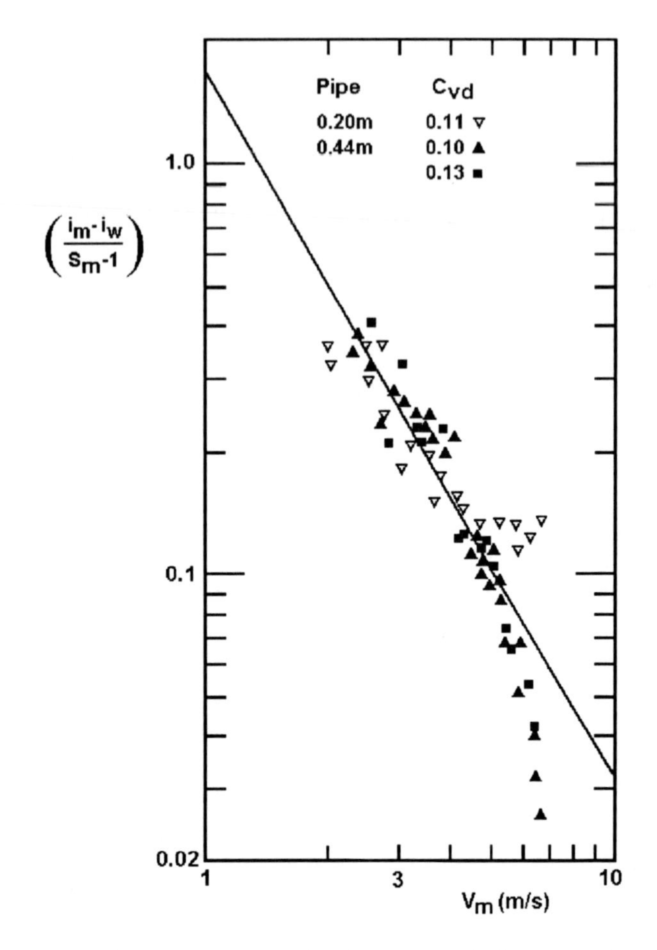

Figure 6.6. Behaviour of crushed-granite slurry (d_{50} = 0.68 mm) in 203 mm and 440 mm pipes, after Clift *et al.* (1982)

When the particles are not uniform, the effect of the solids grading can be expressed by the standard deviation of $\log V_{50}$, denoted σ_s. It can be shown that the standard deviations σ_f and σ_s combine on a summed-squares basis, i.e.

$$\sigma_{10} = \sqrt{(\sigma_f^2 + \sigma_s^2)} \tag{6.6}$$

With M inversely proportional to σ_{10}, and σ_f effectively constant, it follows the expression for M is of the form

$$M = (a + b\,\sigma_s^2)^{-0.50} \tag{6.7}$$

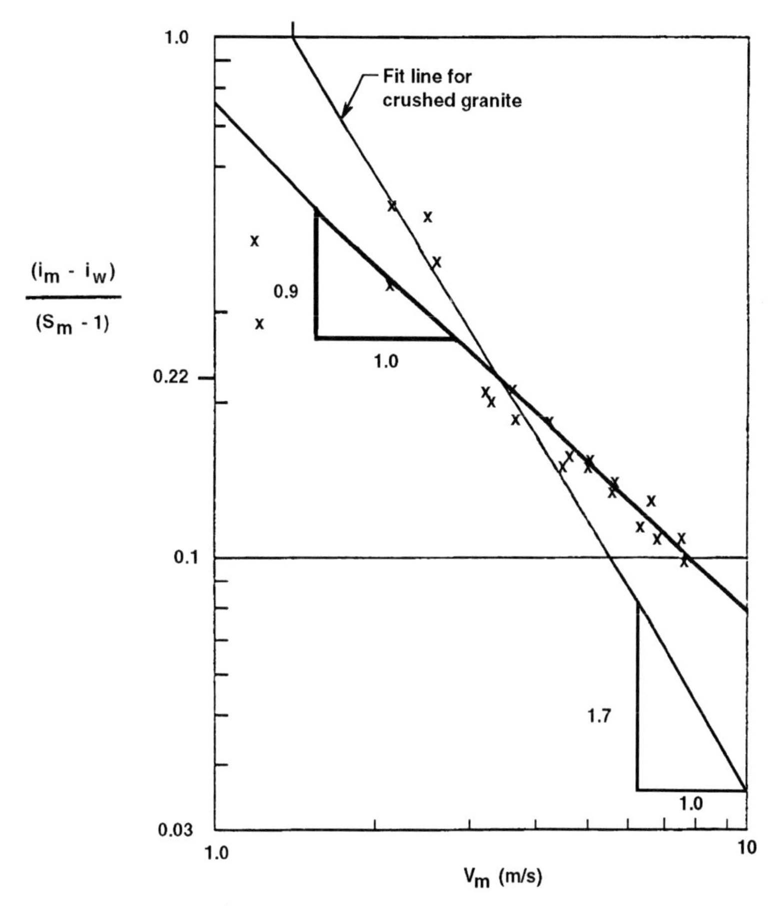

Figure 6.7. Behaviour of slurry of broad-graded material (d_{50} = 0.68 mm) in 203 mm pipe

where a and b are numerical coefficients to be determined empirically.

For this purpose σ_s is required. As $\sqrt{\dfrac{8}{f_f}}$ in Eq. 6.3 is a common factor for particles of all sizes, σ_s represents the standard deviation of log[w cosh(60d/D)]. This quantity can be taken from the fitted slope of a log-normal plot of w cosh(60d/D), but for present purposes it is sufficiently accurate to estimate this slope using only two points. One way of doing this is to use the mass-median particle (with a cumulative probability of 50%) and the particle which lies one standard deviation above d_{50}. The latter represents a cumulative distribution of 84.1% (which in engineering practice is rounded to 85%).

If we use the method with d_{85} and d_{50}, we will already know the values associated with d_{50}. The value of cosh($60d_{85}/D$) is readily calculated, and the associated value of w(i.e. w_{85}) can be obtained from Fig. 6.5. The value of σ_s is then given by

$$\sigma_s = \log\left[\frac{w_{85}\cosh(60\,d_{85}/D)}{w_{50}\cosh(60\,d_{50}/D)}\right] \tag{6.8}$$

where, as noted previously, log specifies the logarithm to the base 10. The numerical coefficients in Eq. 6.7 have been evaluated using values of σ_s calculated from Eq. 6.8, together with the associated values of M determined from pipeline experiments. Based on data obtained for various slurries tested at the GIW Hydraulic Laboratory, the resulting expression is

$$M = (0.25 + 13\,\sigma_s^2)^{-0.50} \tag{6.9}$$

Equation 6.9 includes the *proviso* that a calculated M larger than 1.7 must be reduced to that value. The resulting value of M can be used directly in Eq. 6.4, with V_{50} evaluated for the median (d_{50}) particle. Case Study 6.1 includes calculations of M for slurries of two different size gradings (see Section 6.8).

Recent experimental results obtained by Sundqvist *et al.* (1996) in pipes 0.2 - 0.3 m in diameter give support to Eq. 6.5 and Eq. 6.9 for C_{vd} less than about 20%. Slurries of particles with grading intermediate between narrow and broad showed general agreement up to C_{vd} of 30%. However Sundqvist *et al.* (1996) found different behaviour with narrow and broad gradings at concentrations above about 20% (see Wilson et al. 2002); this feature will be discussed in Chapter 7. A recent paper by Whitlock *et al.* (2004) dealt with friction-loss modelling as part of a study of energy requirements for pumping sand slurries. Using data from the GIW Hydraulic Laboratory, and also those obtained from a large dredge pipe by (1.0 m in diameter) by van den Berg et al. (1999) and Vercruijsse *et al.* (2002), they obtained results for moderate solids concentrations that were in basic accord with the formulations presented in this chapter. For higher concentrations they found relatively lower losses in both the 0.5 m and 1.0 m pipes.

6.5 Simplified Approach for Estimating Solids Effect

In many cases of practical importance, information on the size and grading of the material to be pumped is limited, but estimates of the solids effect must be made. For example, consider the case of an ore that is to be crushed and then transported by pipeline. At the initial stage of the design there may be no adequate sample of the crushed ore, but estimates of the effect of the solids on the head loss must be made: for feasibility studies, preliminary designs and cost estimates, and even for justifying the expenses of laboratory or pilot-plant testing.

To estimate the solids effect, two parameters are required in Eq. 6.4: the power M and the velocity V_{50} . The value of M has a lower limit of 0.25 (for fully-stratified flow, see Eq. 5.13) and, as shown earlier in this chapter, tends to approach 1.7 for slurries with narrow particle grading (at least for pipes of moderate size, and particles between 300 μm and 800 μm). If only a rough idea of the grading is available, it could be difficult to apply the detailed evaluation method described in Section 6.4. Instead, it may be adequate to use the following approximation, which requires only an estimate of the particle diameter ratio σ_g, which can be evaluated as d_{85}/d_{50}. Using this evaluation, the approximation for M is written

$$M \approx [\ell n(d_{85} / d_{50})]^{-1} \tag{6.10}$$

Here ln is the natural logarithm. When using this approximation, M should not be allowed to exceed 1.7 or fall below 0.25. Thus if d_{85}/d_{50} is less than 1.8, M will be set to 1.7.

The next step is to obtain a commensurate approximation for V_{50}. This is done by ignoring variations of f_f and d/D in Eq. 6.2, and using a power-law equation in d_{50}, $(S_s\text{-}1)$ and relative viscosity ν_r to approximate the particle-associated velocity w. The resulting evaluation formula reads

$$V_{50} \approx 3.93\, d_{50}^{0.35} \left[(S_s - 1)/1.65\right]^{0.45} \nu_r^{-0.25} \tag{6.11}$$

Here d_{50} is in mm. The coefficient 3.93 applies for velocities in m/s. With sand-weight solids $(S_s\text{-}1)$ equals 1.65 and the bracketed portion of Eq. 6.11 equals 1.00. The relative viscosity ν_r represents the ratio of the actual viscosity of the carrier fluid to that of water at 20 C. The value of V_{50} obtained from Eq. 6.11 is substituted into Eq. 6.4 to obtain the solids effect

(i_m - i_f), where i_f is the gradient for an equal flow of carrier fluid. Note, for sand-weight solids Eq. 6.11 is only applicable for 0.15 mm < d_{50} < 1.5 mm.

6.6 Effect of Large Pipe Diameter

In Sections 6.3 and 6.4 it was noted that large pipes could show a drop-off in the data points for stratification ratio R at high velocities (see Fig. 6.4 and Fig. 6.6), and a somewhat similar drop-off has been noted for other pipes of large diameter (see Whitlock *et al.* 2004). The mechanisms that produce this drop-off are not fully understood at present, but some light has been shed by research now underway at Queen's University (Wilson, 2005). For flows above a stationary bed of granular solids, this bed is topped by a sheared layer of moving solids, in which the particle support is provided mainly by intergranular contacts, shading into turbulent suspension in the upper parts of the layer. Within this sheared layer, the Richardson number, based on the gradient of the mixture density, exerts a strong stabilising effect, inhibiting rapid increases in suspended load in the upper portion of the flow. The thickness of the sheared layer is proportional to the shear stress in that layer, which in turn is, in a general sense, depends on the square of the velocity of the flow as a whole.

When the bed itself is in motion (the case for heterogeneous flow) the relationships are less precise, but still conceptually useful. As the flow velocity is increased, the stratification ratio decreases, but stability is maintained until the sheared layer (in a conceptual sense) comes to comprise the whole of the moving bed. With the shear layer thickness essentially dependent on V_m^2, then for larger pipe sizes (and the same value of V_m) this layer forms a reduced fraction of the pipe cross-section. This finding implies that there are differences in the geometry for the moving 'beds' in the large and small pipes, with the bed in the small pipe subtending a large angle at the pipe axis, so that the top of this layer meets the pipe wall where the wall already has a rather steep angle with the horizontal. On the other hand, for the large pipe the 'bed' subtends a smaller angle at the pipe axis, lying in the bottom of the pipe where the wall is not far from horizontal. It follows that, in the larger pipe, the off-wall hydrodynamic lift in the area near the bed is closer to vertical, and hence more effective in particle support. If the moving bed is already only marginally stable (because of the Richardson-number effect), then the additional off-wall force may be sufficient to destabilise it, producing the rapid diminution of R with increasing V_m that was noted for the large-pipe data mentioned above. This line of reasoning suggests that, for given particles, the value of V_m at which destabilisation begins might be taken as inversely proportional to the square root of pipe diameter.

It should be remarked that this proposed mechanism also depends on particle size. For large particles the off-wall lift is less effective, and the velocity for bed destabilisation should be larger. It follows that significant reductions in pressure drop can only be expected for large pipes and modest particle diameters (perhaps between 150 and 450 μm).

6.7 Effect of High Concentration of Solids

Results for pipeline flows of highly-concentrated particulate slurries have recently been obtained at several major laboratories. An article by Wilson (2004) considers specific energy consumption (SEC) based on the new results for particulate slurries, which are those having a solid phase comprised of individual particles larger than 40 μm, with no rheologically-active finer material. It is usually economically desirable to design and operate a slurry pipeline with the lowest practicable SEC. At moderate volumetric solids concentrations (say $C_v < 0.25$), SEC decreases with increasing concentration, but it will be shown that this is no longer the case at high concentrations. When the volumetric concentration of solids delivered by a slurry pipeline exceeds 0.25, there is a tendency for the hydraulic gradient and the stratification ratio R to increase at a greater-than-linear rate with further increases in concentration. This behaviour produces a minimum in the specific energy consumption, which occurs where C_v is in the vicinity of 0.35.

Figure 6.8 shows plots of results obtained by Korving (2002) for aqueous slurries of 100 μm sand in a 158 mm pipe, with volumetric solids concentration C_v up to 0.48. It plots SEC *versus* volumetric solids concentration, based on throughput velocities of 2.0 m/s and 4.0 m/s (velocities intermediate between these values give intermediate energy consumptions). It should be noted that, for both plots, as concentration increases there is a slow initial decline in SEC to a minimum near $C_v = 0.36$, followed by a moderate rise to C_v about 0.42, and finally a very pronounced steep rise in SEC. It appears that the SEC variation can be connected to the shape of curves of local concentration *versus* height. Such curves are not available for the specific sand of Korving's experiments, but curves for a similar case were obtained by Schaan et al. (2000) using 90 μm sand in a 150 mm pipe for a throughput velocity of 3.0 m/s. They found that the profile for the lowest concentration ($C_v = 0.15$) shows an essentially monotonic decrease of concentration with height, but for the two higher concentrations ($C_v = 0.32$ and 0.39) the concentration was virtually uniform throughout the pipe. The flow data for the runs of Schaan et al. (2000) are similar to Korving's data, in that SEC is initially high, passes through a minimum when C_v is somewhat larger than 0.3, and is already rising rapidly as C_v approaches 0.4.

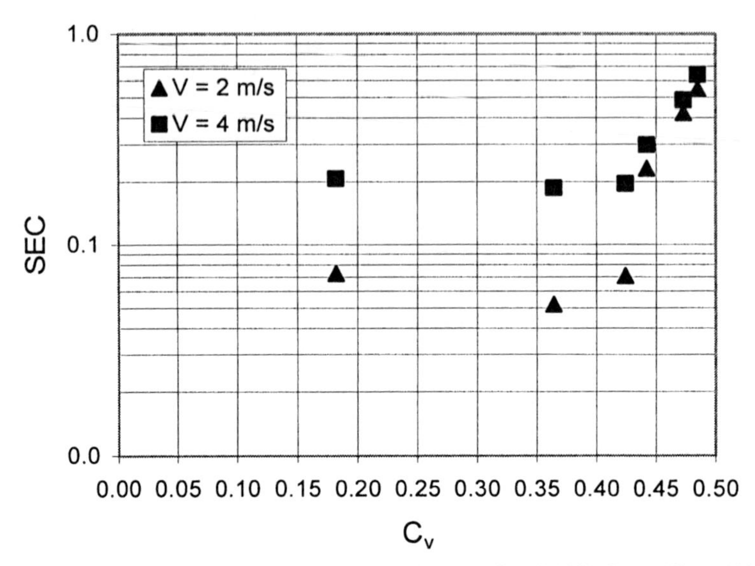

Figure 6.8. SEC plots for 100 μm sand tested by Korving (2002), fromWilson (2004)

For heterogeneous slurries curves of concentration *versus* height differ significantly from those for fine-particle slurries. Figure 6.9 shows concentration curves from tests made at the Saskatchewan Research Council (Gillies & Shook, 1996) for a slurry of 180 μm sand in a pipe of 105 mm internal diameter, at a throughput velocity of 4.0 m/s. For the lower throughput concentration (C_v = 0.36) the local concentration has a maximum near y/D = 0.15, and above that level it diminishes with height at a rate greater than that for the low-concentration fine sand shown on Fig. 2. On the other hand, near the bottom of the pipe, the concentration profile for the 180 μm sand bends backward, a feature that has been observed for many heterogeneous slurries and is associated with an off-wall hydrodynamic lift that drives particles away from a boundary (Wilson *et al.*2000, Wilson & Sellgren 2003, Whitlock *et al.*2004). For the intermediate concentration observed by Gillies & Shook (C_v = 0.42), the profile is essentially constant in the lower half of the pipe, and then diminishes with height in a fashion similar to that for the lower concentration. For the largest concentration (C_v = 0.47), there is little change of concentration with height.

The three runs for the 180 μm sand give the lowest value of SEC at C_v = 0.36, and a progressive rise for larger concentrations, and similar results were found for other data sets (Wilson 2004). In summary, once C_v

substantially exceeds 0.4, the solids are almost uniformly distributed over the pipe section. This observation is consistent with the close particle spacing at high concentration which will reduce the effectiveness of turbulent suspension, so that the particles' submerged weight must be carried by granular contact. Further increases in C_v lead to larger hydraulic gradients and a pronounced rise in specific energy consumption.

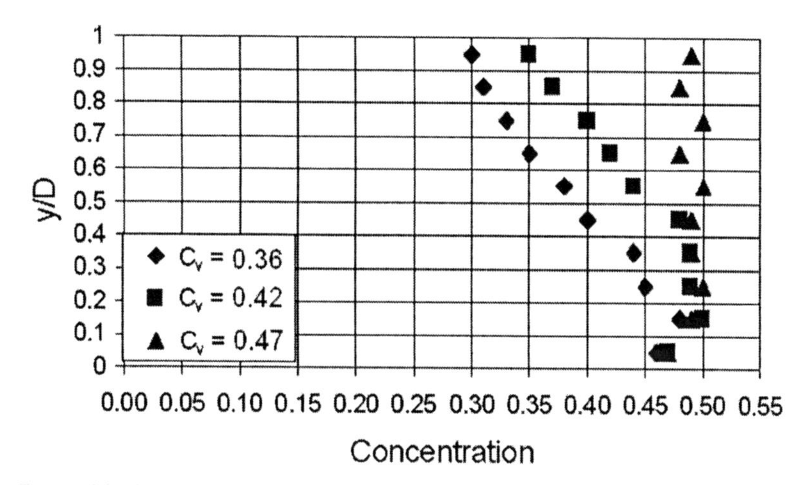

Figure 6.9. Concentration profiles for 180 μm sand tested by Gillies & Shook (1996)

The next step in describing slurry flows will be to discuss complex slurries – those with features such as very broad particle grading or non-Newtonian carrier fluids. Complex slurries are the subject of Chapter 7.

6.8 Case Studies

Case Study 6.1 - Effect of Particle Size and Grading on Sand Transport

In Case Study 5.1 a preliminary pipe selection was made for a system to transport 4000 tonnes/hour of coarse sand. One of the pipe sizes selected was D = 0.65 m, for which the combination V_m = 6.30 m/s and C_{vd} = 0.20 gives the required solids flow rate.

In that case study the mean size and grading of the particles was not specified; and thus the hydraulic gradient and the specific energy consumption could not be calculated. The present case study will investigate these points.

The first material to be considered is a clean sand (negligible fraction smaller than 40 μm) with S_s = 2.65, d_{50} = 0.70 mm and d_{85} = 1.00 mm. We will consider an aqueous slurry ($S_f = 1.00$) in the pipe with D = 0.65 m. The conditions found in Case Study 5.1 ($C_{vd} = 0.20$, $V_m = 6.30$ m/s) will be investigated initially, and the first requirement will be:

(a) Find the hydraulic gradient i_m and the specific energy consumption for this case. For these particles the associated velocity w can be read directly from Fig. 6.5, giving 0.127 m/s for the 0.70 mm particle and 0.148 m/s for the 1.00 mm one. When multiplied by cosh(60d/D), these become 0.1273 m/s and 0.1486 m/s, giving a ratio of 1.17. From Eq. 6.8, the associated value of σ_s is 0.069, and when this is substituted into Eq. 6.9, M is given as 1.79. As this is larger than 1.7, M = 1.7 will be used. (This result is not surprising, since d_{85}/d_{50} is less than 1.5, indicating a narrow particle grading).

With M evaluated, we can turn to V_{50}, equal to $w\sqrt{\frac{8}{f_w}}\cosh(60d/D)$, based on d_{50}. For $f_w = 0.012$ (found from the Stanton-Moody diagram for commercial steel pipe of this size and velocity range), the resulting value of V_{50} is 3.28 m/s. If the simplified approach of Section 6.5 had been used, M would remain at 1.7 and V_{50} would be 3.47 m/s. On the basis of mixture velocity $V_m = 6.30$ m/s, and $(S_m - S_f) = (1.65)\,0.20 = 0.330$, Eq. 6.4 gives the solids effect $(i_m - i_f)$ as $(0.330)(0.22)[1.921]^{-1.7}$ or 0.0239 (m water/m pipe).

With no fines, the fluid is water and $i_f = i_w$, given by $0.012\,(6.3)^2/(19.62)(0.65)$, or 0.0373. The value of i_m is the sum of i_w and the solids effect or 0.0612 (m water/m pipe).

The specific energy consumption is proportional to i_m/S_sC_{vd}, which is 0.115, giving SEC of 0.315 kWh/tonne-km.

(b) Next, it is of interest to vary some of the quantities and as input, to find the effect on i_m and SEC. The initial case to be investigated is a variation of V_m only, with C_{vd} kept constant. The second case involves adjusting C_{vd} as well as V_m, so that the solids throughput Q_s is maintained at a constant value.

The effect of varying V_m alone could be obtained by differentiation, but it is probably clearer simply to change V_m by about 10% in each direction, repeating the calculations for, say, V_m = 6.9 m/s and 5.7 m/s. As i_w varies with V_m^2 (ignoring any changes in f_w) and $(i_m - i_w)$ varies with $V_m^{1.7}$, the calculations are straightforward. For $V_m = 6.9$ m/s, $i_m = 0.0653$ and SEC = 0.336

kWh/tonne-km. For V_m = 5.7 m/s, i_m = 0.0589 and SEC = 0.304 kWh/tonne-km.

If the same values of V_m are employed, but C_{vd} is adjusted to maintain Q_s, the values shown in Table 6.1 are obtained.

Table 6.1.

V_m (m/s)	C_{vd}	i_m	SEC (kWh/tonne-km)
5.70	0.221	0.0619	0.289
6.30	0.200	0.0612	0.315
6.90	0.183	0.0636	0.358

As with the figures for the constant-C_{vd} case, this table indicates a significant increase in SEC when V_m is increased to 6.90 m/s. As before, the specific energy consumption is diminished when V_m is decreased to 5.70 m/s, but once again, the change is smaller for a decrease in velocity. Thus it would appear that the lower velocity of 5.7 m/s could be attractive. On the other hand, the lower velocity is less than 20 percent above the deposition limit (V_{sm} = 4.84 m/s by Eq. 5.11, see Case Study 5.1), and considerations of system stability may be involved. In order to resolve these questions, it will be necessary to know the pump characteristic as well as the system characteristic. Pump characteristics will be discussed in Chapter 9, and the intersection of pump and system characteristics in Chapter 13.

(c) It is also of interest to find the effect on i_m of changes in particle grading. As an initial example, suppose that the coarse sand specified above is replaced by a fine, clean, narrow-graded sand with d_{50} = 0.20 mm. From Fig. 6.5, w for this sand is 0.080 m/s. In the pipe of 0.65 m diameter the resulting value of V_{50} from Eq. 6.2 is 2.08 m/s. For V_m = 6.3 m/s, and C_{vd} = 0.20, the solids effect as calculated from Eq. 6.4 is 0.0110, giving i_m = 0.0483 m water/m pipe. (If the simplified approach had been used the value of i_m would be 0.0498 m water/m pipe). Clearly, i_m is considerably less than the value of 0.0612 found in part (a) for the coarse sand. It may be compared with the result of the 'equivalent fluid' model, by which i_m is estimated as the product of S_m and i_w. As S_m = 1.330, and i_w = 0.0373, the equivalent fluid model gives i_m = 0.0496, rather close to the result of the new model for this specific instance of fine-sand slurry flow.

Consider next a coarse sand with d_{50} = 0.70 mm as before, but with a broader particle grading so that d_{85} = 1.50 mm. For the latter size, w = 0.18 m/s, and for the 0.65 m pipe cosh(60d/D) is 1.01. Substitution into Eq. 6.8 shows that the value of σ_s for this material is 0.145. When this is substituted into Eq. 6.9, M is found to be 1.38. (The simplified approach gives M = 1/ln(1.50/0.70) = 1.31.)

The value of V_{50} remains at 3.28 m/s and for V_m = 6.30 m/s and C_{vd} = 0.20, the solids effect becomes $(0.330)(0.22)[1.921]^{-1.38}$ or 0.0295. Adding i_w of 0.0373 gives i_m = 0.0668 m water/m pipe, which is about 9% higher than the value obtained previously for the narrow-graded particle distribution with the same median size. (A similar increase, about 11%, is obtained if the simplified approach is used throughout.)

Case Study 6.2 - Transport of Coal Slurry

The material for this study is coal, with S_s = 1.4, μ_s = 0.44, C_{vb} = 0.60. The coal is clean with few fines, and d_{50} = 2.0 mm and d_{85} = 2.8 mm (as the ratio of d_{85}/d_{50} is less than 1.5, the particle grading is narrow, and M will be taken as 1.7).

The pipe to be used has D = 0.440 m and f_w = 0.013. Solids concentration C_{vd} can be taken initially as 0.25.

The minimum in the curve of constant C_{vd} is to be investigated, comparing the velocity at this point with V_{sm} and V_{50}, and finding the minimum specific energy consumption and how this is influenced by changes in C_{vd}.

(a) the limit of deposition will be determined first. For d = 2.0 mm and D = 0.440 m, Fig. 5.3 shows that V_{sm} would be 4.1 m/s for material with S_s = 2.65, and for S_s = 1.4, the right-hand side of that figure gives V_{sm} = 1.90 m/s. Equation 5.11 gives a value of 1.94 m/s, but the smaller number should be taken, i.e. V_{sm} = 1.90 m/s. (Note that for these 2.0 mm coal particles the nomographic chart shows a smaller value than Eq. 5.10; the reverse of the condition found in Case Study 5.1 for 0.7 mm sand).

(b) To obtain V_{50} it is sufficient to use the simplified approach of Eq. 6.11 which gives 2.64 m/s. For this narrow- graded material M = 1.7, and with $(S_s - 1)C_{vd}$ = 0.4 (0.25) = 0.10, the expression for solids effect (Eq. 6.4) becomes:

$$i_m - i_w = 0.022\left(\frac{V_m}{2.64}\right)^{-1.7} = 0.115\,V_m^{-1.7}$$

Also, i_w is $\frac{0.013 V_m^2}{(19.62)(0.44)} = 0.00151 V_m^2$. The sum of these two terms is i_m, and differentiation gives the following relation for the velocity at minimum i_m:

$$1.7(0.115)\,V_m^{-2.7} = 2(0.00151)\,V_m$$

from which V_m equals 3.1 m/s. At this point i_m is calculated to be 0.0313 m water/m pipe and i_m/S_sC_{vd} is 0.089, giving SEC = 0.244 kWh/tonne-km. (This is the quantity which must be provided by the pumps to the slurry, the energy which must be supplied to the pumps will, of course, be larger).

(c) In this case, the value of V_m at the minimum point is comfortably above the deposition limit of 1.9 m/s, so no problem with deposition should arise. However, stability problems would occur near the minimum, as shown in Chapter 13, and thus the operating velocity should be larger than 3.1 m/s. It is also worth noting that in the present study, the flow is definitely heterogeneous near the minimum point. At the minimum point V_m/V_{50} equals 1.17 and the stratification ratio, which is given by 0.5 $(V_m/V_{50})^{-1.7}$ equals 0.38, implying that about 38% of the submerged weight of solids is carried by intergranular contact, with fluid support accounting for the rest.

(d) Finally, it is of interest to consider the effect of a change in concentration on the minimum point. As shown in the discussion following Eq. 6.5, the velocity at minimum i_m varies *cœteris paribus* with $C_{vd}^{0.27}$. Thus, if C_{vd} increases from 0.25 to 0.30 the value of V_m at the minimum point will increase from 3.1 m/s to 3.25 m/s; not a large change. It can readily be calculated that the equivalent value of i_m is 0.0345 m water/m pipe. This gives SEC = 0.224 kWh/tonne-km, or about 92% of the value obtained previously for C_{vd} = 0.25.

REFERENCES

Carstens, M.R. & Addie, G.R. (1981). A sand-water slurry experiment. *Jour. Hydr. Div.* ASCE, Vol. 107, No. HY4, p. 501-507.

Clift, R., Wilson, K.C., Addie, G.R., & Carstens, M.R. (1982). A mechanistically-based method for scaling pipeline tests for settling slurries. *Proc. Hydrotransport 8*, BHRA Fluid Engineering, Cranfield, UK, pp. 91-101.

Hsu, S.T., Van der Beken, A., Landweber, L., & Kennedy, J.F. (1980). Sediment suspension in turbulent pipeline flow. *Jour. Hydr. Div.* ASCE Vol. 106, No. HY11, pp. 1783-1792.

Matoušek, V. (2002). Non-stratified flow of sand-water slurries, *Proc. Hydrotransport 15,* BHR Group, Cranfield, UK, pp 563-573.

Newitt, D.M., Richardson, J.F., Abbot, M., & Turtle, R.B. (1955). Hydraulic conveying of solids in horizontal pipes. *Trans. Inst. of Chem Engrs.,* Vol. 33, London, UK.

Rouse, H. (1937). Modern conceptions of the mechanics of fluid turbulence. *Trans. Amer. Soc. Civil Engrs.* Vol. 102, pp. 436-505.

Schmidt, W. (1932). Der Massenaustausch in freier Luft und werwandte Erscheinungen. *Die Wasserwirtschaft,* No. 5-6 (Mitteilung 10, Inst. für Wasserbau, Tech. Hochschule, Berlin).

Shook, C.A., Gillies, R., Haas, D.B., Husband, W.H.W. & Small, M. (1982). Flow of coarse and fine sand slurries in pipelines. *Jour. Pipelines,* Elsevier, Vol. 3, pp. 13-21.

Sundqvist, Å., Sellgren, A. & Addie, G.R. (1996). Pipeline friction losses of coarse sand slurries; comparison with a design model. *Powder Technology*, Elsevier, Vol. 89, pp.9-18.

van den Berg, C.H., Vercruijsse, P.M. & van den Broeck, M. (1999). The hydraulic transport of highly concentrated sand-water mixtures using large pumps and pipeline diameters, *Proc. Hydrotransport 14*, BHR Group, Cranfield, UK, pp 445-453.

Vercruijsse, P.M., de Gruijter, A, & Corveleyn, F. (2002). The solids effect on pump and pipeline characteristics – Keeping up with present trends n the dredging industry, *Proc. Hydrotransport 15*, BHR Group, Cranfield, UK, pp 711-723.

Whitlock, L., Sellgren, A. & Addie, G.R. (2004). Energy requirement for pumping sand slurries: comparison of large-scale loop and field results with a design model, *Proc WEDA XXIV Dredging Conference and Texas A&M 36th Annual Dredging Seminar,* Wyndham Palace Resort & Spa Lake Buena Vista, Florida, July 2004.

Whitlock, L., Wilson, K.C. & Sellgren, A. (2004). Effect of near-wall lift on frictional characteristics of sand slurries, *Proc. Hydrotransport 16,* BHR Group, Cranfield, UK, pp. 443-454.

Wilson, K.C. (1973). A formula for the velocity required to initiate particle suspension in pipeline flow. *Proc. Hydrotransport 2,* BHRA Fluid Engineering, Cranfield, UK, pp. E2-23 - E2-36.

Wilson, K.C. (2004). Energy consumption for highly-concentrated particulate slurries, *Proc. 12th Intern'l Conf. on Transport and Sedimentation of Solid Particles.* Prague, Czech Republic.

Wilson, K.C. (2005). Rapid increase in suspended lad at high bed shear, *J. Hydr. Engrg,* ASCE, Vol. 131, No. 1, pp.46-51.

Wilson, K.C. & Pugh, F.J. (1988). Dispersive-force modelling of turbulent suspension in heterogeneous slurry flow. *Canad. Jour. Chem Engrg.*, Vol. 66, pp. 721-727.

Wilson, K.C. & Sellgren, A. (2002). Effect of particle grading on pressure drops in slurry flows. *Proc. 11th Conf. on Transport and Sedimentationof Solid Particles.* Ghent, Belgium.

Wilson, K.C. & Watt, W.E. (1974). Influence of particle diameter on the turbulent support of solids in pipeline flow. *Proc. Hydrotransport 3,* BHRA Fluid Engineering, Cranfield, UK, pp. E1-1-E1-13.

Wilson, K.C., Clift, R. & Sellgren, A. (2004). Operating points for pipelines carrying concentrated heterogeneous slurries. *Powder Technology*, Vol. 121, pp 19-24.

Wilson, K.C., Sellgren, A. & Addie, G.R. (2000). Near-wall fluid lift of particles in slurry pipelines, *Proc. 10th Conf. on Transport and Sedimentation of Solid Particles,* Wrocław, Poland.

Wilson, K.C., Clift, R., Addie, G.R., & Maffett, J. (1990). Effect of broad particle grading on slurry stratification ratio and scale-up. *Powder Tech.*, Elsevier, Vol. 62, No. 2, pp. 165-172.

Chapter 7

COMPLEX SLURRIES

7.1 Introduction

The term 'complex slurries" is used here to indicate slurries with a range of particle size that overlaps more than one of one of the categories covered in the previous four chapters – non-Newtonian, pseudo-homogeneous, heterogeneous and fully-stratified. In Section 7.2, we will begin with essentially particulate slurries where the fraction of fine particles (less than 40 μm in size) is reasonably small, so that it may increase the viscosity of the carrier fluid but does not cause it to become strongly non-Newtonian. For such cases turbulent flow can be expected.

Subsequently, Sections 7.3 to 7.5 cover the other limiting case where the fine-particle fraction is of major importance, producing a strongly non-Newtonian carrier, and a flow that is typically laminar. If a fraction of larger particles is included in this flow, there are questions as to whether, and under what conditions, these particles will settle, and the effect of any settling on pressure gradient. Section 7.3 gives an overview of the 'Stabflo' debate, Section 7.4 reviews particle settling in non-Newtonian fluids, and Section 7.5 (together with Case Study 7.1) deals with settling in the pipeline environment.

Finally, Section 7.6 considers cases where the fine-particle fraction is sufficient to produce a strongly non-Newtonian carrier fluid, but the flow remains in the turbulent régime.

7.2 Broad-graded Slurries with Newtonian Carrier Fluids

As noted above, for broad-graded slurries with a modest fraction of fine particles, i.e. finer than 40μm, the combination of the fines and the liquid form a carrier fluid which typically approximates Newtonian behaviour, and the flow will generally be turbulent. It is desirable to carry out pilot-plant testing of all broadly-graded slurries, but in many cases this will not be practicable, particularly in the early stages of design. Thus, there is a place for computational techniques, even if approximate. The technique that follows is an extended version of that presented by Wilson & Sellgren in the Third Edition of the McGraw-Hill Pump Handbook (2001).

The mass fraction of the solids smaller than 40 μm is designated X_f, and the remaining fraction (1- X_f) can be divided into three segments – pseudo-homogeneous, X_p, heterogeneous, X_h, and fully-stratified, X_s. The liquid plus the fraction X_f is acts as the carrier fluid, and the pseudo-homogeneous fraction X_p (called the 'middlings') is considered to combine with the carrier fluid produce an equivalent fluid of the sort described by Eq. 4.3. The next-larger fraction, X_h (the 'grits'), follows the heterogeneous behaviour described in Chapter 6, while the largest fraction (the 'clunkers') represents the fully-stratified load described in Section 5.5.The fraction X_p comprises particles that are small enough to be engulfed in the viscous sub-layer (see Section 6.2). The upper size limit for X_p can be taken as 150 μm for cases where the properties of the carrier fluid do not differ significantly from those of water. If the carrier fluid has a Newtonian viscosity greater than that of water, the value of 150 μm is to be multiplied by the viscosity ratio ν_r (the ratio of the kinematic viscosity of the carrier fluid to that of water at 20°C). If the carrier fluid has non-Newtonian properties, the ν_r should be based on its secant viscosity, evaluated near the expected wall shear stress, and a further factor, α, is required (see Section 3.3), which is to be evaluated at the same wall shear stress. The heterogeneous fraction spans particle diameters between 150 μm (or $150\alpha\nu_r$ μm) and a multiple of the internal pipe diameter D. As noted in Section 5.5, flow tends to shift from heterogeneous to fully-stratified at values of d/D between 0.015 and 0.018. Here the value of 0.015D will be used to define the upper particle diameter for heterogeneous flow, with larger particles comprising the fully-stratified solids fraction X_s.

The hydraulic gradient for the mixture as a whole, i_m (m water/m pipe), can be considered as having three components, associated with the equivalent fluid, the heterogeneous load and the fully-stratified load. As indicated by Eq. 4.3, the hydraulic gradient for pseudo-homogeneous flow of an aqueous slurry is given by $i_w[1 + A'(S_m - 1)]$, where A' is a coefficient that equals unity for the equivalent-fluid case. However, $(S_m - 1)$, i.e. $C_v(S_s - 1)$, refers to the excess relative density of the slurry as a whole, and as only the fraction X_p is part of the homogeneous flow, $(S_m - 1)$ must now be replaced by $C_vX_p(S_s - S_f)$, where S_f is the relative density of the carrier fluid. Likewise, the hydraulic gradient for water, i_w, must be replaced by that for the carrier fluid, i_f. Here i_f will be somewhat larger than i_w because of increases in density and viscosity produced by the fine particles (Gillies & Shook, 1991), but a reasonable estimate of any such increase is often sufficient. Thus the equivalent fluid component of the hydraulic gradient, i_e (m water/m pipe), is given by:

$$i_e = i_f[1 + C_vX_p(S_s - S_f)] \tag{7.1}$$

In dealing with the component due to heterogeneous flow, we begin with Eq. 6.4, which gives this type of solids effect as $0.22\,(S_m - 1)\left[\frac{V_{50}}{V_m}\right]^M$. This term must now be multiplied by the fraction of homogeneous particles X_h, with $(S_m - 1)$ replaced by $C_v(S_s - S_{fp})$ where S_{fp} represents the relative density of the combination of carrier fluid and the pseudo-homogeneous fraction, typically the relative density of the mixture of fluid and all particles less than 150 μm in size. This expression will be denoted Δi_h , i.e.

$$\Delta i_h = 0.22\, C_vX_h(S_s - S_{fp})[V_{50}/V_m]^M \tag{7.2}$$

As the technique outlined here is necessarily an approximate one, the power M will be taken as 1.0 [following the article by Wilson & Sellgren (2001) in the McGraw-Hill Pump Handbook]. In order to estimate the appropriate value of V_{50} for the heterogeneous fraction it is necessary to have a representative particle for this fraction, say of diameter d_h. It is suggested that this is the particle that corresponds to the mean of the percentages passing the upper and lower limits of the heterogeneous fraction. This point is illustrated by Example 7.1 in section 7.7. Wilson and Sellgren (2001) had presented a curve for the effect of relative viscosity on V_{50}, but in most cases it should be sufficient to evaluate V_{50} by using Eq. 6.11, i.e.

$$V_{50} \approx 3.93\, d_h^{0.35}\left[(S_s - S_{fp})/1.65\right]^{0.45} \nu_r^{-0.25} \tag{7.3}$$

Reference may also be made here to the effect of particle shape, studied by Schaan et al. (2000).

Finally, there is the fully-stratified load, which will produce an additional solids effect Δi_s. It will be recalled that, in the absence of other solids, this is given by Eq. 5.13 as $\Delta i_s = (S_m - 1)B'[V_m/0.55V_{sm}]^{-0.25}$, where V_{sm} is the velocity at the limit of stationary deposition (see Fig. 5.3), and B' is a coefficient that may be as high as unity but is lowered by the effects of finer particles (see below). It will be taken here as 0.5. For the case dealt with here, $(S_m - 1)$ must be replaced by $C_vX_s(S_s - S_{fph})$, where S_{fph} is the relative density of the mixture of fluid, pseudo-homogeneous and heterogeneous particles. Thus:

$$\Delta i_s = C_v X_s (S_s - S_{fph}) B' [V_m / 0.55 V_{sm}]^{-0.25} \tag{7.4}$$

The sum of the terms given by Eq. 7.1 to 7.4 gives the gradient for the mixture as a whole, i_m (m water/m pipe), thus:

$$i_m = i_e + \Delta i_h + \Delta i_s \tag{7.5}$$

At low total solids concentrations, the total solids effect will be approximately linear in C_v, but as the concentration is increased S_{fp} and S_{fph} also increase, and the effect of increasing C_v is less than linear for broadly-graded solids. In addition, it appears that both V_{50} and B' decrease with S_{fp} and S_{fph}, respectively. Maciejewski et al. (1993) compared pipe transport of very coarse particles (about 100 mm in size) in clay suspensions and in oil-sand tailings slurries (particle size below 0.8 mm). They found that the sand slurry was more effective than the clay in reducing the solids effect of the coarse particles, equivalent to reducing B'. The important role of particles with sizes 0.1 to 0.8 mm in diminishing friction was also demonstrated by Sundqvist et al. (1996a, 1996b) for products with d_{50} near 0.6 mm. Wilson & Sellgren (2001) made suggestions for approximate methods of estimating the reductions in V_{50} and B' mentioned above, but for design purposes it is recommended that laboratory tests be undertaken to evaluate friction losses for broadly-graded slurries, especially if the solids concentration is high. An example of experimental results is shown on Fig. 7.1 (from Wilson et al. 2002). It gives plots of solids effect $(i_m - i_w)$ *versus* $(S_m - 1)$ based on Sundqvist's sand data with $V_m \approx 4.3$ m/s. All three sands had d_{50} close to 0.6 mm; Sand 1 has a very narrow grading ($d_{85}/d_{50} = 1.2$), for Sand 2 this ratio is 2.3 and for Sand 3 it is 7.1 (a very broad grading).

This figure shows that the slurry of intermediate grading has a solids effect that is directly proportional to $(S_m - 1)$, i.e. to C_v, up to the maximum C_v of about 0.37. As expected in view of the analysis put forward in this section, Sand 3, with its very broad grading, shows a pronounced flattening of the solids-effect relationship at high C_v, a feature that has considerable commercial potential. Sand 1, with its very narrow grading, exhibits the opposite (and commercially undesirable) feature – a disproportionate increase in the solids effect at high concentrations. Similar results were found by Gillies & Shook (2000).

Wilson & Sellgren (2002) suggested reasons for the effects discussed above, noting that solid particles in a turbulent fluid flow influence the energy spectrum of turbulence, typically plotted as energy density versus turbulent frequency (which is inversely proportional to eddy size). For flows behind stationary cylinders in a wind tunnel, it is known that spectral peaks occur at vortex-shedding frequencies. In that configuration, turbulent energy is produced, but it is expected that moveable particles will have the opposite effect, damping eddies of approximately their own size. The result should be a 'blip' of reduced energy density at the corresponding frequency. If the particle size is uniform, and the concentration is sufficiently high, this 'blip' will tend toward a gap or 'hole' in the spectrum.

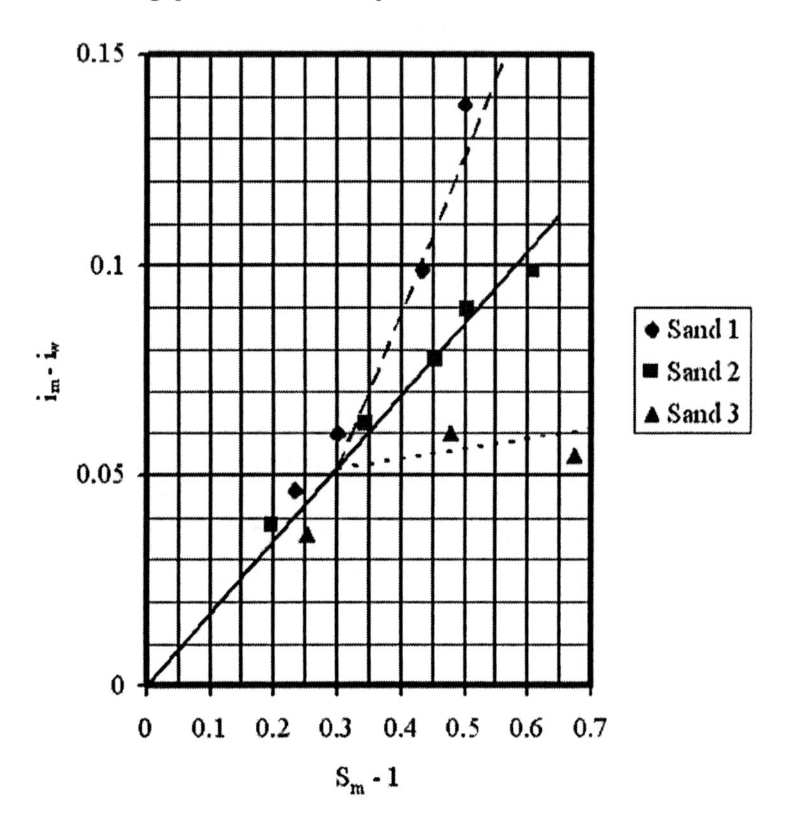

Figure 7.1. Plots of solids effect for Sundqvist's sands for average mixture velocity 4.3 m/s (from Wilson et al. 2002)

By the 'energy cascade' concept, the energy of the small high-frequency eddies is maintained by an energy flow from the larger eddies. A gap in the spectrum will interfere with the cascade of energy to eddies smaller than the gap, reducing the intensity of the smallest eddies, and thus tending to increase the thickness of the viscous sub-layer. Such an increase can give some benefit in reducing fluid friction, but also tends to reduce the off-wall lift force, increasing contact load and Coulombic friction. Thus it is not unexpected that large concentrations of uniform particles (which produce spectral holes) are found to have the disproportionately high pressure drops noted above.

On the other hand, for broadly-graded solids there will be a series of 'blips' in the turbulent spectrum that combine to form a broad 'trough' rather than a narrow 'hole'. Such a trough will be less likely to interfere with the cascade of energy to the smaller eddies, so significant reductions of off-wall lift are not expected in this case. Also, if the solids include both particles directly susceptible to the lift force and somewhat larger particles, a benefit may be obtained, at least at large overall solid concentrations. Instead of a particle subject to lift being completely removed from the near-wall area, it will tend to rise only until it hits a larger particle, thus contributing to the support of the larger particle and simultaneously reflecting the smaller particle back into the high-lift zone near the wall, where the process can be repeated. It is believed that this is the major mechanism contributing to the reduction of B' for broad-graded slurries. As a result of this mechanism, possibly in combination with the effect of large pipe diameter noted in Section 6.6, the technique just discussed for calculating pressure drops for broad-graded slurries can produce significant over-estimates for some particle gradings. Additional research is underway, but the present technique may be considered to be conservative. When maximum accuracy is required, the best approach is to perform pipe-loop tests with the actual slurry of interest.

7.3 The 'Stabflo' Debate

Many slurries that are transported in pipelines are comprised principally of a homogeneous aqueous mixture of fine particles. As shown in Chapter 3, such mixtures are often strongly non-Newtonian, and typically are viscoplastic. If a material of this sort is subjected to a shear stress that is gradually increased, no strain rate is produced until the applied shear stress τ exceeds the yield stress τ_y. In addition, there is often a fraction (perhaps only a small one) of larger particles, and if such a discrete particle is placed in a quiescent viscoplastic medium it will not settle unless it is heavy enough to produce a shear stress in the medium that exceeds τ_y. It is important to know if these larger particles will be supported by the non-Newtonian medium, or will settle out; and if they settle, what their terminal fall velocity will be.

In 1970, the first conference of the well-known Hydrotransport series was held, and at that conference Elliot & Gliddon (1970) presented a paper describing their work on pumping highly-concentated mixtures of coarse and fine coal under laminar flow conditions. This type of flow was later called "stabilised flow" or 'Stab-Flo' for short. Central to the concept of stabilised flow was the belief that the laminar-flow pressure gradients could be scaled up to larger pipe sizes in accord with the Rabinowitsch-Mooney relationship mentioned in Chapter 3. On this basis stabilised flow appeared to offer ever-reducing pressure gradients with increasing pipe size, implying major economies for transport in larger pipes. A contrary viewpoint was expressed by Thomas (1978, 1979a), who considered that deposition under laminar-flow conditions might affect the scaling of pressure gradients, and that settling might occur progressively along the pipeline length.

In the 1980's large-scale pipe-loop tests were conducted on stabilized coal mixtures by Duckworth and his group at the CSIRO laboratories in Australia (e.g., Duckworth et al. 1983, 1986). These publications specifically noted that in some circumstances particles did not settle even if left for long periods of time. In the most basic sense, the question of whether such particles settle or not can be obtained by measurement, but it is also convenient to have an equation that can predict the presence or absence of particle settling in terms of the rheological characteristics of the carrier. What is required is a force balance in the vertical direction, involving the driving force provided by the submerged weight of the particle and the resisting force offered by the fluid (see Fig. 2.13).

For a spherical particle of diameter d and solids density ρ_s, immersed in a fluid of density ρ_f, the submerged weight force F_w is given by:

$$F_w = \left(\pi d^3/6\right) g\left(\rho_s - \rho_f\right) \tag{7.6}$$

where g is gravitational acceleration. The resisting force must be related to the rheological behaviour of the carrier fluid, and if the fluid has a yield point, τ_y, then no movement is produced until $\tau > \tau_y$. The submerged weight of the particle sets up a stress field in the surrounding medium. The largest shear stress occurs at the surface of a spindle shape that encapsulates the particle (and necessarily has an area somewhat larger than the particle surface), and the shear stress diminishes with increasing distance from the particle. The particle settles only if the shear stress at the surface of the spindle equals or exceeds the yield stress τ_y. Since the area of this surface is proportional to d^2, the following static shear stability criterion must be met if the particle is not to settle:

$$\tau_y \geq kgd(\rho_s - \rho_f) \tag{7.7}$$

The coefficient k must be determined empirically. Several authors have proposed values that are effectively equal to 0.10.

On the other hand, visualization studies by Pullum et al. (2001), where the carrier fluid is replaced by a clear gel with appropriate rheological properties, show clearly that the solids settle rapidly under shear, even when the yield stress of the carrier fluid exceeds the static criteria given by Eq. 7.7 by more than an order of magnitude. In moving downstream, the settled particles are seen to form a sliding bed. This bed (which is traveling at a slower velocity than that of the flow above it) distorts the velocity profile, as shown on Fig. 7.2.

Once a bed starts to form, it tends to grow as time progresses until, eventually, virtually all the particulate solids are conveyed in the sliding bed. In this case, modern instrumentation has produced new insights into laminar-flow transport of coarse particles. As a result of these studies, and the work of various other writers that has been summarised in earlier sections of the present paper, today very few engineers would recommend stabilised flow for long-distance transport. It still has advantages for relatively short-distance transport, for example to tailings impoundments. Here the ability to stop a line and restart it at will, and to deliver very-high-concentration 'pastes' still makes high-concentration flow an attractive means of conveying short distances with limited water consumption.

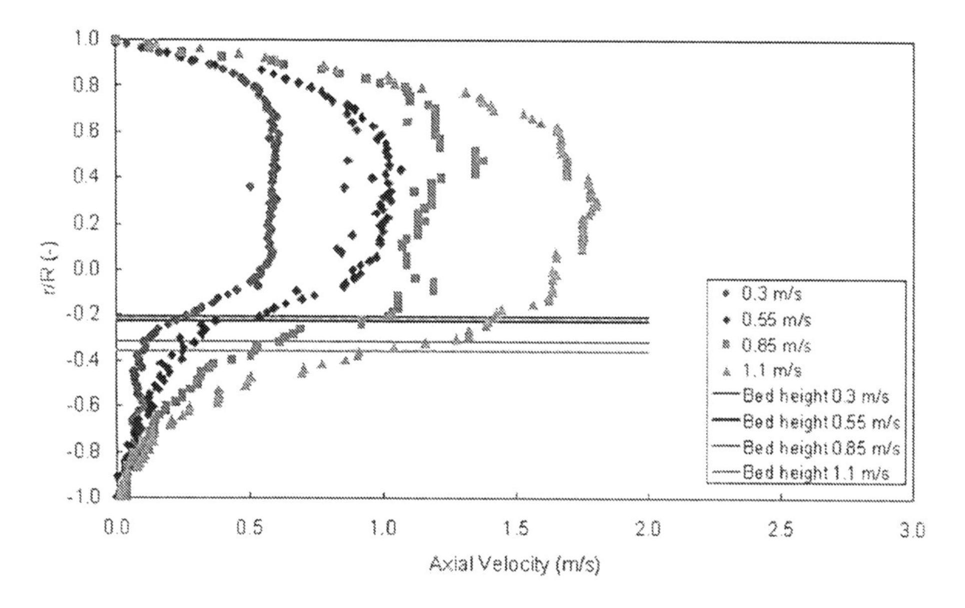

Figure 7.2. Fluid velocity profiles captured by MRI 200D downstream of the visual observation window for a statically stable "Stab-Flo" suspension, from Thomas et al. (2004)

It has been shown above that flows of coarse particles in a non-Newtonian carrier fluid do exhibit stratification, a fact that points to the need for these flows to be re-analysed using non-Newtonian stratified-flow concepts; although this is likely to present difficulties that did not occur in the earlier analyses of turbulent flows with Newtonian carrier fluids. However, the hydrodynamic lift force that was found to be significant for turbulent flow (see Section 6.1) is expected to be much smaller for laminar flow, and is unlikely to be significant laminar flows of materials normally associated with the mineral industry.

7.4 Particle Settling in non-Newtonian Fluids

Recent publications based on experiments at Curtin University, Western Australia, (Wilson et al. 2003, Wilson & Horsley 2004) proposed a technique for predicting fall-velocities that was applied to Newtonian fluids in Chapter 2, but is also appropriate for particles in non-Newtonian materials, and is especially useful for the non-Newtonian case.

The new method does not use the traditional variables of Reynolds number and drag coefficient, but instead has developed a pair of dimensionless variables on the basis of concepts developed in the pipe-flow analysis of Prandtl (1933) and Colebrook (1939). As shown in Chapter 2, this analysis expresses the velocity ratio (mean velocity to shear velocity) as a function of the shear Reynolds number (based on shear velocity rather than mean velocity). The shear stress set up on the surface of a spherical particle is non-uniform, but the mean surficial shear stress, denoted by $\bar{\tau}$, is used as the basis for the shear velocity. This stress is given by the submerged weight force divided by the surface area of the sphere, and equals $\rho_f g(S-1)d/6$. As shown in Chapter 2, the resulting expression for the shear velocity V* is

$$V^* = \sqrt{(S_s - 1)gd/6} \tag{7.8}$$

For falling particles the velocity ratio is based on the terminal fall velocity of the particle v_t, i.e the velocity ratio is v_t/V^*. The shear Reynolds number Re^* has the form:

$$Re^* = \rho_f V^* d/\mu \tag{7.9}$$

Here d is the particle diameter (analogous to the pipe diameter for the pipe-flow case) and, for a Newtonian fluid, μ is the viscosity.

The form of the settling curve for Newtonian fluids for the v_t/V^* and Re^* axes is shown on Fig. 2.15. In terms of these variables, the drag coefficient equals $8/[v_t/V^*]^2$ and the conventional Reynolds number Re equals the product of Re* and v_t/V^*. For Re larger than about 1100 the drag coefficient can be taken as effectively constant at 0.445, equivalent to $v_t/V^*=4.24$ for $Re^*>260$. At the other end of the Reynolds number range, settling obeys Stokes law which can be expressed as $v_t/V^*=Re^*/3$. In the intermediate region, piecewise functions were developed (Wilson et al., 2003), and presented earlier as Eq. 2.64 and 2.65. For Re^* less than or equal to 10, the fit equation is:

$$V_t / V^* = Re^* /[3(1+0.08\, Re^{*1.2})] + 2.80/[1+3.0(10^4)(Re^{*-3.2})] \tag{7.10}$$

In the range $10 < Re^* < 260$ a different fit equation is used, based on x = $\log(Re^*/10)$ and y = $\log(v_t/V^*)$. It is:

$$y = 0.2069 + 0.500\, x - 0.158\, x^{1.72} \tag{7.11}$$

The curve represented by these equations is shown, on logarithmic coordinates, on Fig. 2.15. As mentioned above, for $Re^* > 260$, $v_t/V^* = 4.24$.

As mentioned in connection with Fig. 3.3, for any specific reference point on a rheogram, (as determined by either a given shear rate or a given stress). The viscosity μ (sometimes called the 'secant' or 'equivalent' viscosity) is defined as

$$\mu = \tau /(dU / dy) \tag{7.12}$$

It can be determined directly from the data used to plot the rheogram, without the intermediary of a rheological model. In applying this concept to the fall of particles in non-Newtonian materials it is necessary to have a method of determining the appropriate reference point on the rheogram. This can be based on either a specified strain rate τ or a specified shear stress γ'. Since the strain rate will not be known initially, it is more convenient to specify the shear stress. For the simpler problem of pipeline flow of non-Newtonian materials, the stress of the pipe wall was found to give an appropriate reference (see Wilson & Thomas, 1985; Thomas & Wilson, 1987). In the present case it is logical to base the reference point for the rheogram on mean surficial shear stress of the particle $\overline{\tau}$, as given prior to Eq. 7.8. It was found (Wilson et al., 2003) that the use of $\overline{\tau}$ itself produced a large scatter, and the calculated points generally lay considerably to the right of the Newtonian line of Fig. 2.15. Thus it was decided to select a reference shear stress of $\zeta\overline{\tau}$, where ζ is a multiple less than unity. [It is worth noting here that this proposed evaluation of reference shear stress would cause no problem if the fluid should happen to be Newtonian. For a Newtonian fluid, α is always equal to unity and μ is the same for any reference shear stress, i. e. for any value of ζ.]

For various values of ζ, the stress $\zeta\overline{\tau}$ was calculated for each of the available data points, and the corresponding value of μ (denoted μ_{ep}) was found from the appropriate rheogram. [Note that the stress $\zeta\bar{\tau}$ was used in determining μ_{ep} (and α), but V* is always based on $\sqrt{\bar{\tau}/\rho_f}$] As ζ was decreased from 1.0 to 0.3 it was found that the plotted points generally approached the Newtonian line, and, equivalently, the standard error of estimate diminished. For ζ less than 0.3, it was found that the standard error increased. Details are given in Wilson et al. (2003) and Wilson & Horsley (2004). Thus $\zeta = 0.3$ was selected. For this value the points lay closer to the Newtonian line than was the case for other values of ζ, and the divergence of the points from the Newtonian line appears to be random, i.e., no right-left trend is evident. The scatter, which is small to moderate for $\alpha < 1.7$ is significantly larger (though without a preferred direction) for the larger values of α. It should be noted that for $\alpha > 1.7$ the reference point on the rheogram at $\tau = 0.3\bar{\tau}$ is associated with rather small strain rates, typically in the order of 10 s^{-1}, and here viscometers do not give rheograms of high accuracy.

7.5 Particle Settling in Laminar non-Newtonian Pipeline Flow

The next step is to move from the static stability criterion described earlier to the case where the carrier fluid is itself sheared by an external agency, as is typical of pipeline flow. For the static case of an unsheared medium, a reference particle shear stress can be defined, and, as shown by Wilson et al. (2003), this can be used to determine an equivalent strain rate for the particle, say γ'_{ep}. This strain rate serves to define the equivalent viscosity μ_{ep} that would apply to a Newtonian fluid producing the same particle fall velocity. The combination of a particle in a stationary medium (with equivalent strain rate γ'_{ep}) and the medium itself being subjected to an external strain rate γ'_x gives rise to a resultant (in the vector sense) strain rate, γ'_r. It has been proposed (Wilson & Horsley, 2004, Thomas et al., 2004) that the magnitude of this strain rate can be given by:

$$\gamma'_r = [(\gamma'_{ep})^2 + (\gamma'_x)^2]^{0.5} \tag{7.13}$$

Several basic checks can be used to verify the general applicability of this relationship. Thus, for γ'_x of zero, $\gamma'_r = \gamma'_{ep}$, as expected. Also, for a fluid with a yield shear stress τ_y, if $\tau_r < \tau_y$ the particle will not settle. This condition implies that $\gamma'_r = 0$, and thus both γ'_{eq} and γ'_x are zero. Also, if the reference shear stress of the particle is well below τ_x, then $\gamma'_r \approx \gamma'_x$ and the associated value of μ_{ep} for the particle is essentially the same as for the sheared fluid.

Note that in a sheared fluid ($\gamma'_x > 0$), Eq. 7.13 predicts that γ'_r is larger than γ'_{ep}. For typical rheological behaviour, such as that given by the Bingham or Casson models, a larger γ' is associated with a smaller μ. Thus the resultant viscosity μ_r is smaller than μ_{ep}. With a smaller viscosity, the particle falls more rapidly. In other words, the proposed relation predicts that a particle in a sheared medium settles, and its fall velocity is greater than that which the same particle would have in the unsheared medium. This result corresponds in general terms with the findings of Thomas' (1979b) experiments with particles settling in a clay slurry sheared in the annulus between a fixed cylinder and a rotating one. Particles which were statically stable in a clay slurry were found to settle when the slurry was sheared in the gap between the rotating cylinders. In recent years other workers have conducted similar tests with substantially similar results (Wilson 2000, Cooke 2002).

It follows from the remarks above that particles is a sheared laminar flow will generally settle out in the upstream section of a pipeline. An example of this behaviour is given below as Case Study 7.1. The expected result is a stratified or sliding-bed configuration as described in Section 7.3 and illustrated on Fig. 7.2. Reference may also be made to experiments with a sand-clay mixture by Wilson & Addie (2002).

7.6 Turbulent Flow of Slurries with non-Newtonian Carrier Fluids

The previous section dealt with particles settling in the laminar flow of a non-Newtonian fluid. Turbulent flows, unlike laminar ones, have mechanisms (described in Chapter 6) that can support some fraction of the particles. When the remaining, contact-load, fraction is present, the pressure gradient has two components. The first, associated with the carrier fluid, scales inversely with pipe diameter (Eq. 3.11); the second, associated with the contact-load solids, is independent of pipe diameter. For this purpose the carrier fluid (indicated by the superscript f) comprises all of the slurry except the coarse particles which may be subject to settling. The carrier fluid has density ρ_f and its rheological properties are often non-Newtonian. The component of the pressure gradient associated with the carrier fluid can be expressed in terms of a fluid-associated shear velocity, U_{*f}, i.e.

$$\left(\frac{\Delta p}{\Delta x}\right)_f = 4\,\rho_f\,\frac{U_{*f}^{\;2}}{D} \qquad (7.14)$$

For non-Newtonian flows without contact load, the pressure gradient is entirely of this type; analysis of this case was presented in Chapter 3.

An earlier publication (Wilson 1988) showed how flows without settling particles can be employed to establish a relation for U_*, which in turn can provide a value of U_{*f} for use when contact load is present. This method is applicable if the range of turbulent-flow experiments includes points for which solids settling is absent. In this case, exemplified by the data of Fig. 7.3 which will be discussed below, these points can be used for the evaluation U_{*f}. Rheological modelling is not required; it is sufficient to calculate an equivalent viscosity, μ_{eq}, for each data point, and then to express μ_{eq} in terms of shear velocity. This point-wise calculation of μ_{eq} is based on the relation that the mean velocity of turbulent Newtonian pipe flow equals 2.5 U_* ln (ρU_* D/μ_{eq}). The results of such calculations for coal slurries at various concentrations are shown on Fig. 7.3 As a logarithmic plot of μ_{eq} against U_*. The steep left-hand portion of the graphs shows the effect of particle settling, which is absent from the gently-sloping right-hand portion.

These gently-sloping graphs are virtually straight lines on Fig. 7.3, indicating a power-law relation of the type

$$\mu_{eq} = b\, U_*^{-\beta} \tag{7.15}$$

where the numerical values of the coefficients b and β can be found from the graph for any given slurry. A power-law relation such as Eq. 7.14 generally fits the data adequately over the portion of the experimental range free of contact load, and hence can also be used to link μ_{eq} to U_{*f} when contact load is present.

For an experimental point with mean velocity $(V_m)_1$ in a pipe of diameter D_1, it follows from substitution of Eq. 7.15 into the turbulent pipe-flow equation that the value of U_{*f} can be found by numerical solution of the equation

$$(V_m)_l / U_{*_f} = 2.5\, \ell n(\rho_f\, U_{*_f}^{(1+\beta)}\, D_l / b) \tag{7.16}$$

In Section 7.7, the use of this equation will be illustrated by example problems. The fraction of contact load decreases with increasing flow velocity, and in these examples becomes negligibly small above a certain velocity. This type of behaviour, together with that discussed in Section 7.5, shows that the presence of particulate solids can transition from laminar to turbulent flow of non-Newtonian fluids, making it gradual rather than abrupt.

This section completes the description of slurry flow in horizontal pipes. The following chapter, dealing with slurry flows in vertical and inclined pipes, will end the first half of the book.

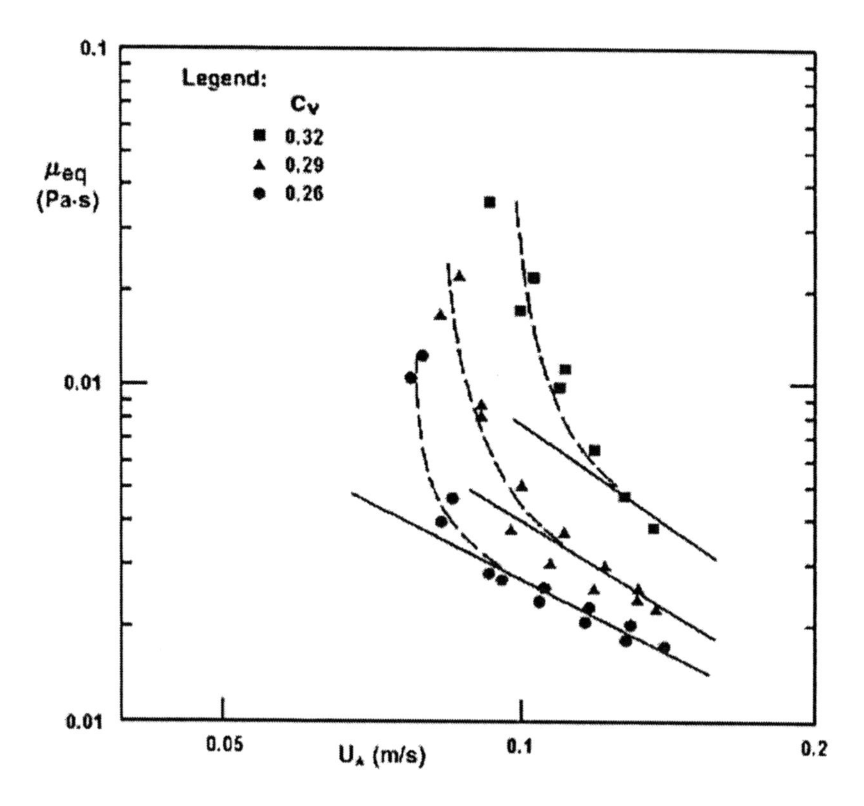

Figure 7.3. Equivalent-viscosity plot for coal slurries (after Wilson, 1988)

7.7 Example Problems and Case Study

Example 7.1

A broadly-graded granular material with S_s=2.65 has a grading curve that passes through the points listed below. An aqueous slurry of this material having C_v=0.25 is to be pumped through a horizontal pipe with internal diameter 300 mm. The throughput velocity V_m will be 3.2 m/s. Estimate the hydraulic gradient i_m using the method of Section 7.2.

Percent Finer	Diameter (μm)
20	40
44	150
50	200
70	400
85	1400
95	4500

We can begin with the hydraulic gradient for water, i_w, which is given by $fV^2/(2gD)$. With a typical f of 0.013, i_w=0.0226. From the grading table it is seen that the -40 μm fraction, X_f, is 0.20, giving S_f=1.0+0.25(0.20)(1.65), which is 1.082. For these granular fines, at relatively modest concentration, it can be assumed that ν_r and α are both equal to 1.0, placing the upper boundary of X_p at 150 μm, and giving X_p as 0.44-0.20=0.24. Also, i_f can be estimated as equal to $S_f.i_w$ or 1.082(0.0226).

By Eq. 7.1, i_e is given by 1.082(0.0226)[1.0+0.25(0.24)(2.65-1.082)] = 0.0268 (m water/m pipe). The lower limit of the heterogeneous fraction is 150 μm, as seen above, and its upper limit is 0.015D, i.e. 4500 μm. From the grading data provided, these boundaries of X_h represent 44% finer and 95% finer, with an average of 69.5%, which gives the representative diameter for this fraction, d_h, as about 400 μm. Substitution into Eq. 7.3 (with ν_r=1.0 and S_{fp}=1.182) gives V_{50}=2.71 m/s. When this value (together with relative densities and other properties) is substituted into Eq. 7.2, it is found that Δi_h=0.0349 (m water/m pipe).

Finally, there is the 5% of the solids that travel as fully-stratified load, producing an additional component of hydraulic gradient, Δi_s, in accord with Eq. 7.4. Here X_s=0.05, S_{fph}=1.392 and B' is taken as 0.5. The deposition limit V_{sm} for this fraction can be estimated from Fig. 5.3 for d=4.5 mm as 2.7 m/s, producing Δi_s=0.0129. Summing i_e, Δi_h and Δi_s [Eq. 7.5] gives $i_m \approx 0.068$ (m water/m pipe). [It may be of interest that the calculations for the same example using the simplified single-fraction approach of Section 6.5 give i_m some 30% higher than that obtained with the present method.]

Example 7.2

The coal slurry used for plotting Fig. 7.3 is to be pumped through a horizontal pipe with internal diameter 0.44 m. The volumetric concentration C_v will be 0.26, giving a slurry density of 1130 kg/m^3. For a pressure gradient ($\Delta p/\Delta x$) of 140 Pa/m, find j, τ_o, U_*, μ_{eq}, V_m and Q. The lowest line on the figure indicates that, for the design concentration, μ_{eq} equals 6.6 (10^{-5}) $U_*^{-1.6}$.

As $\Delta p/\Delta x$ equals $\rho_m gj$, j is given by 140/(1130)(9.81) = 0.0126 m of slurry per m of pipe. From Eq. 2.17, τ_o=($\Delta p/\Delta x$)D/4 or 140(0.44)/4=15.4 Pa. U_* equals $\sqrt{\tau_o/\rho_m} = \sqrt{15.4/1130}$ or 0.117 m/s. Using U_*, the value of μ_{eq} can be obtained graphically from Fig. 7.3 or by means of the fit equation $6.6(10^{-5})\,U_*^{-1.6}$. This gives μ_{eq} = 0.00205 Pa.s. As the points in this region are essentially on the fit line, there is no need to distinguish U_{*f} from U_*. Substitution into Eq. 7.15 gives V_m= 2.98 m/s (corresponding to f = 0.0122). Flow rate is the product of V_m and $\pi D^2/4$, giving Q = 0.453 m^3/s.

Example 7.3

For the pipe of the previous example, it has been decided to use a design flow rate of 0.60 m^3/s. Find the corresponding values of U_*, μ_{eq}, $\Delta p/\Delta x$ and j.

From the previous example, it can be expected that the fit line from Fig. 7.3 will also apply here, and there is no need to distinguish U_* and U_{*f}. It will be necessary to substitute V_m, pipe diameter, and b into Eq. 7.16. The value of V_m is readily found by dividing flow rate by pipe area, yielding 3.95 m/s. Substitution then gives

$$3.95 = 2.5\,U_*\,\ell n\left(\frac{(1130)(0.44)\,U_*^{2.6}}{6.6\,(10^{-5})}\right)$$

On iteration it is found that the required value of U_* is 0.146 m/s. The corresponding value of μ_{eq} is 0.00143 Pa.s, and that of f is 0.011. The shear stress equals $\rho_f U_*^2$, *i.e.* 24.1 Pa, from which $\Delta p/\Delta x$ is 219 Pa/m and j is 0.0198 metres of slurry per metre of pipe.

In the examples just presented, there was no contact load. However, if the value of U_{*f}, (determined from Eq. 7.16) is smaller than U_* (calculated from the pressure gradient in the usual fashion), then the pressure gradient includes a contact-load component. This component does not effect velocity scaling, and hence U_* in Eq. 3.12 must be replaced by U_{*f}, giving

$$(V_m)_2 = (V_m)_1 + 2.5\,U_{*_f}\,\ell n\,(D_2/D_1)$$

The scaling relation for pressure gradient must also be changed, as now only the fluid-phase component obeys the inverse-diameter law. The replacement for Eq. 3.10, calculated on this basis (Wilson 1988) is given by

$$(\Delta p/\Delta x)_2 = (\Delta p/\Delta x)_1 - 4\,\rho_f\,U_{*_f}^{\,2}\,(D_2 - D_1)/(D_1 D_2)$$

The scale-up process is greatly facilitated if the data taken in the test loop include measurements of pressure differences in ascending and descending vertical flow. If the average frictional gradient in ascending and descending vertical flow, denoted as $\bar{i}_v$ (see Sections 4.6 and 8.2) is essentially equal to that measured for horizontal flow, i_m, then it follows that particle settling is not important and the scale-up techniques of Section 4.3 are directly applicable.

On the other hand, if the friction gradient in the horizontal pipe, i_m, is significantly in excess of $\bar{i}_v$, it is clear that settling is important and that contact load is present in the horizontal pipe. In this case, the pressure gradient due to the carrier fluid and the excess gradient due the contact-load solids effect must be separated for each test point, and then scaled up individually and recombined to give the prototype value of i_m. As long as the carrier fluid (water plus fines) has near-Newtonian properties, this process is quite simple. The fluid gradient i_f is set equal to the measured value of $\bar{i}_v$, and scaled using the diameter ratio in the same fashion as for fluid flow (Eq. 4.10). The value of $\bar{i}_v$ is also used to obtain the value of the fluid shear velocity, U_{*f}. The gradient difference i_m - $\bar{i}_v$ is used to approximate the contact-load solids effect, which does not depend on pipe size.

$$(i_m)_2 = (i_m)_1 - (\bar{i}_v)_1 \left(1 - \frac{D_1}{D_2}\right)$$

It is expected that in most cases the method described above will be adequate. If v'_t/V_m is not small, a correction may be required. It is expressed in terms of the factor λ, which will be introduced in Section 8.2 in the chapter on vertical and inclined flows. The mean velocity must also be scaled up, using Eq. 6.25, which is based on Eq. 4.12 with U_* set equal to U_{*f} (evaluated from $\bar{i}_v$).

Case Study 7.1

Consider as an example a pipe of internal diameter D_{pi} = 400 mm, with a non-Newtonian carrier fluid that is a clay-water mixture with density 1000 kg/m^3. It is expected that a fluid of this type will have a rheogram similar in form to Fig. 3.3. For convenience the rheogram will be approximated by a Casson fluid with yield stress τ_c = 9.0 Pa and Casson viscosity η_c = 0.0262 Pa.s. For the pipeline a throughput velocity of 1.8 m/s has been selected. A small quantity of crushed rock (ρ_s = 2650 kg/m^3) will be included in the slurry, with maximum size 0.125 inch (about 3 mm). It is known that in a Newtonian fluid the fall velocity of angular particles such as crushed rock is significantly smaller than that of spheres of equal mass (Clift et al., 1978). On the assumption that a similar effect applies to particles in non-Newtonian fluids, the rock particles are considered to be equivalent to spheres with diameter 2.0 mm. The problem is to determine if these particles settle, and, if so, how rapidly.

Before dealing with the settling calculations, it is necessary to find the shear at the pipe wall that is required to produce the selected throughput velocity of 1.8 m/s. The Casson fluid obeys the general equation:

$$\sqrt{\tau} = \sqrt{\tau_c} + \sqrt{(\eta_c \gamma')} \tag{7.17}$$

In order to obtain the relation between the throughput velocity V and the shear stress at the wall τ_w , the Casson equation must be integrated twice across the pipe section, giving:

$$V = (D_{pi}\ \tau_w/8\ \eta_c)[\ 1\ -(16/7)\xi^{0.5}\ +(4/3)\xi\ -\ (1/21)\xi^4\ ]\quad(7.18)$$

where $\xi = \tau_c/\tau_w$. An iterative solution based on V = 1.8 m/s, D_{pi} = 0.40 m, τ_c = 9.0 Pa, and η_c = 0.0262 Pa.s gives τ_w = 24.6 Pa. This wall stress corresponds to $\gamma' = 147\ s^{-1}$ and μ (i.e. τ_w/γ') of 0.168 Pa.s.

The problem of fall velocity of the 2 mm sphere can now be considered. The mean surficial shear stress on the particle, $\bar{\tau}$, equals (2650-1100)(9.81/6)(0.002) or 5.1 Pa, giving V* = 0.068 m/s (by Eq. 7.8). The reference shear stress $\zeta\bar{\tau}$ equals 0.3(5.1) or 1.5 Pa. For a quiescent fluid this value is much less than the yield shear stress of 9.0 Pa, and the particles would not settle, implying that the equivalent particle strain rate γ'_{ep} in Eq. 7.13 is zero. However, in the pipeline the carrier fluid is subjected to an externally-imposed strain rate γ'_x. The maximum strain rate, found near the pipe wall, is 147 s^{-1}, as noted above, and Eq. 7.13 shows that the resultant strain rate γ'_r is also 147 s^{-1}which, by Eq. 7.12, is equivalent to μ_r = 0.168 Pa.s.

Substituting this value of μ into Eq. 7.9 shows the shear Reynolds number of the particle to be (1100)(0.068)(0.002)/(0.168) or 0.89. By Eq. 7.10 the resulting value of V_t/V^* equals 0.27, for a fall velocity of (0.27)(0.068) or 18 mm/s. This is a maximum value, not generally applicable throughout the pipe. A detailed model of settling at various locations will not be attempted here. Instead, a typical settling rate will be estimated based on half the maximum shear stress , i.e. 12.3 Pa. This is equivalent to a strain rate of 9.8 s^{-1}, which, following the previous train of reasoning, gives μ_r = 1.25 Pa.s, Re* = 0.12, V_t/V^* = 0.04 and V_t approximately 3 mm/s. On this basis a particle would settle a distance equal to the pipe diameter in about 130 s, and in this time would be carried downstream by the velocity of 1.8 m/s, giving a settling length between 200 and 300 m.

REFERENCES

Colebrook, C.F. (1939). Turbulent flow in pipes, with particular reference to the transition region between the smooth and rough pipe laws, *J. Inst. of Civ. Engrs.*, **11**, Feb. 1939, pp. 133-156.

Clift, R., Grace, J. R. and Weber, M. E. (1978). *Bubbles, Drops, and Particles*, Academic Press, New York.

Cooke, R. (2002). Laminar flow settling: the potential for unexpected problems, *Hydrotransport 15*. BHR Group, Banff, AB, Canada, 121-133.

Duckworth, R. A., Pullum, L., Lockyear, C. F. and Lenard, J. A. (1983b). Hydraulic transport of coal. *Bulk Solids Handling* **3**.

Duckworth, R. A., Pullum, L. and Lockyear, C. F. (1986b). Pipeline transport of coarse materials in a non-Newtonian carrier fluid. *Hydrotransport 10*. BHRA, Innsbruck, Austria.

Elliot, D. E., and Gliddon, B. J. (1970). Hydraulic transport of coal at high concentrations. *Hydrotransport 1*, BHRA, Cranfield, UK. paper G2.

Gillies, R.G. and Shook, C.A. (1991). A deposition velocity correlation for water slurries, *Canad. J. Chem. Engrg*, Vol. 69, pp 1225-1227.

Gillies, R.G. and Shook, C.A. (2000). Modelling high-concentration settling slurry flows, *Can. J. Chem. Eng.*, Vil. 33(4), pp. 709-716.

Horsley, M.R., Horsley, R.R. and Wilson, K.C. (2003). Non-Newtonian effects on fall velocities of pairs of vertically-aligned spheres, *Internat'l. Conf. on Non-Newtonian Rheometry*, Inst. of Non-Newtonian Fluid Mechanics, University of Wales, Cardiff, UK.

Maciejewski, W., Oxenford, J. and Shook, C.A. (1993). Transport of coarse rock with sand and clay slurries, *Proc. Hydrotransport 12,* Brugge, Belgium, BHR Group, Cranfield, UK, pp. 705-724.

Prandtl, L. (1933). Neuere Ergebnisse Turbulenzforschung, *Zeitschrift des Vereines Deutscher Ingenieure*, Berlin, Germany **77** (5), pp. 105-114.

L.Pullum, L., Rudman, M., Graham, L.J.W., Downie, R.J., Bhattacharya, S.N., Chryss, A. and Slatter, P.T. (2001). AMIRA P599,High concentration suspension pumping, 2nd progress report. Melbourne Australia

Schaan, J., Sumner, R.J., Gillies, R.G. and Shook, C.A. (2000). Effect of particle shape on pipeline friction for Newtonian slurries of fine particles, *Canad. J. Chem. Engrg*, Vol 78(4), pp 717-725.

Sundqvist, Å., Sellgren, A. and Addie, G.R. (1996a). Pipeline friction losses of coarse sand slurries, *Powder Technology*, Vol. 89, pp. 9-18.

Sundqvist, Å., Sellgren, A. and Addie, G.R. (1996b). Slurry pipeline friction losses for coarse and high-density industrial products, *Powder Technology*, Vol. 89, pp 19-28.

Thomas, A.D. (1978). Coarse particles in a heavy medium – turbulent pressure drop reduction and deposition under laminar flow, *Hydrotransport 5*, BHRA, Hannover, Germany, paper D5.

Thomas, A. D. (1979a). Pipelining of coarse coal as a stabilized slurry : another viewpoint. *4th International Technical Conference on Slurry Transportation*. Slurry Transportation Association, Las Vegas, USA.

Thomas, A. D. (1979b). Settling of particles in a horizontally sheared Bingham plastic, *1st National Conference on Rheology*, Melbourne, Australia.

Thomas, A.D. and Wilson, K.C. (1987). New analysis of non-Newtonian turbulent flow – Yield-power-law fluids, *Canad. J. Chem.Engrg.*, **65**, pp. 335-8.

Thomas, A.D., Pullum, L. and Wilson, K.C. (2004). Stabilised laminar slurry flow: review, trends and prognosis, *Proc. Hydrotransport 16*, BHR Group, Cranfield, UK, pp. 701-716.

Turton & Levenspiel (1986). Turton, R. & Levenspiel, O. (1986). A Short Note on Drag Correlation for Spheres, *Powder Technol.*, **47**, p. 83.

Wilson, K.C. (1988). Effect of non-Newtonian slurry properties on drag reduction and coarse particle suspension. *Proc. 10th Intern'l Conf. on Rheology,* Sydney, Australia, Vol. 1, pp. 110-115.

Wilson, K. C. (2000). Particle motion in sheared non-Newtonian media.*3rd Israeli Conf. for Conveying and Handling of Particulate Solids*, Dead Sea, Israel, 12.9-12.13.

Wilson, K.C. and Addie, G.R. (2002). Pipe-flow experiments with a sand-clay mixture, *Proc. Hydrotramsport 15*, BHR Group, Cranfield, UK, pp. 577-588.

Wilson, K.C. and Horsley, R.R., (2004). Calculating fall velocities in non-Newtonian (and Newtonian) fluids: a new view. *Proc. Hydrotransport 16*, BHR Group, Cranfield, UK, pp. 37-46.

Wilson, K.C. and Sellgren, A. (2001). Hydraulic transport of solids. *Pump Handbook, 3rd Edition*, pp. 9.321-9.349.

Wilson, K.C. and Sellgren, A. (2002). Effect of particle grading on pressure drops in slurry flows, *Proc. 11th Intern'l Conf. on Transport and Sedimentation of Solid Particles*, Ghent, Belgium, pp. 277-287.

Wilson, K.C. and Thomas, A.D. (1985). A new analysis of the turbulent flow of non-Newtonian fluids, *Canad. J. Chem. Engrg.*, **63**, pp. 539-46.

Wilson, K.C., Clift, R. And Sellgren, A. (2002). Operating points for pipelines carrying concentrated heterogeneous slurries, *Powder Technology*, Vol. 123, pp. 19-24.

Wilson, K. C., Horsley, R. R, Kealy, T., Reizes, J. C. and Horsley, M. (2003). Direct prediction of fall velocities in non-Newtonian materials, *Int'l. J. Mineral Proc.*, **71**/1-4, pp 17-30.

Chapter 8

VERTICAL AND INCLINED SLURRY FLOW

8.1 Introduction

The previous chapters have dealt with horizontal flows, and it is now appropriate to consider other configurations. The case of vertical flow, which is reasonably straightforward, will be covered first, followed by a treatment of the effects of inclined pipes on deposition limit and friction loss.

Vertical flow is of particular importance in the mining industry, and mention should be made of the great depths from which coal-water mixtures have been pumped in Germany. Details are given in several papers by German engineers, many published in English in the Hydrotransport Conference series. In this application, the slurry is first loaded into chambers by low-head slurry pumps; using alternate connections, the material in successive chambers is then pushed into the vertical hoisting pipe by high-head water pumps. This type of lock-hopper feed arrangement is common for high-lift pumping of coarse-grained mixtures. The ultimate proposed application of hydraulic hoisting, however, is deep-ocean mining of manganese nodules, which are up to 50 mm in size and are found at depths as great as 5000 m.

The recent emphasis on back-filling mines with tailings or other waste material has also led to many instances of downward vertical flow. For downward flow there is usually more potential energy available than is required, and dissipation of excess energy can be bothersome. This is particularly so when coarse particles are involved, as they rapidly erode conventional chokes or valves. Full-scale backfilling tests carried out in a German coal mine (Geller 1978) involved dropping the solids in air for 900 m, at which point they hit the water surface in the stand- pipe, dissipated their energy and formed the slurry.

Short lengths of pipe on an upward incline are commonly encountered at the head end of transport systems; the ladder of a suction dredge is the most common example of this configuration. Lengths of pipe at both positive and negative inclination may be incorporated in pipelines running over rugged terrain (Venton & Cowper, 1986). Inclined lines will be considered later in this chapter, after the discussion of vertical flow.

8.2 Analysis of Vertical Flow

Consider a height z of a vertical riser, with pressure p_1 at the bottom and p_2 at the top, giving a net pressure force in the upward direction of $(p_1 - p_2)\pi D^2/4$. The weight of slurry is a downward force which can be written $(\rho_w g S_{mi} z)\pi D^2/4$, where ρ_w is the density of water and S_{mi} is the *in situ* relative density of the mixture, given by

$$S_{mi} = 1 + (S_s - 1)C_{vi} \tag{8.1}$$

Here S_s represents the relative density of the solids and C_{vi} the *in situ* volumetric solids fraction. The remaining force, directed downward as it resists motion, is based on the fluid shear stress at the pipe wall, τ_o, multiplied by the wall area πDz. The resulting force balance is given by

$$(p_1 - p_2)\pi D^2/4 = \rho_w g S_{mi} z \pi D^2/4 + \tau_o \pi Dz \tag{8.2}$$

or

$$(p_1 - p_2)/z = \rho_w g S_{mi} + 4\tau_o/D \tag{8.3}$$

The average upward velocity of the mixture is denoted V_m, and the particles are falling relative to V_m, with hindered settling velocity v_t', so that their net upward velocity is $V_m - v_t'$. If C_{vd} is the delivered volumetric solids fraction, then from continuity, C_{vi} equals $C_{vd} V_m /(V_m - v_t')$, or

$$C_{vi} = C_{vd} /(1 - v_t'/V_m) \tag{8.4}$$

The ratio v_t'/V_m (slip velocity/mean velocity) which appears in Eq. 8.4 can be very important in vertical hoisting.

Calculations for hindered settling velocity v_t' were mentioned in Chapter 2. For a typical sand-water slurry with average particle diameter 0.6 mm, the equivalent value of v_t is about 0.05 m/s. As the hindered values of settling velocity v_t' will be less than v_t, it can be seen that in many cases of practical interest (for which V_m will be 1.5 m/s or greater) the ratio v_t'/V_m will be significantly less than 0.1. This ratio can then be neglected in the term $(1 - v_t'/V_m)$, so that to a reasonable approximation $C_{vi} = C_{vd}$ and $S_{mi} = 1 + (S - 1)C_{vd}$.

In addition to their direct use in vertical conveying (which will be covered in section 8.3) the principles discussed above are applied in an instrument known as the U- tube, which comprises a combination of rising and descending vertical flows. This device, shown schematically on Fig. 8.1, was first proposed by Hagler (1956). Its behaviour has been studied by several authors, from Brook (1962) to Clift & Clift (1981). The latter analysis is more complete than those given previously, delineating conditions under which errors may be anticipated.

For the U-tube Eq. 8.1 to 8.4 apply directly to the riser, but require certain modifications for the downcomer. The *in situ* volumetric concentration of Eq. 8.1 has a different value for the two pipes, being smaller in the downcomer, in which the particles travel faster than the fluid. This effect also changes the sign of v_t'/V_m in Eq. 8.4. Moreover, the different values of C_{vi} in the two pipes affect the hindered settling velocity so that, strictly speaking, two values of v_t' must be employed. The force-balance equation for the downcomer is similar to Eq. 8.3, but with pressures p_3 and p_4 in place of p_1 and p_2, and a change of sign in the weight term. The mean velocity is the same in both upward and downward flows, and thus the wall shear stress should also be virtually equal, allowing the shear stress term to be eliminated by combining the equations for upward and downward flow. The result gives the following expression for $\overline{C}_{vi}$, the mean *in situ* solids concentration (i.e. the average for riser and downcomer).

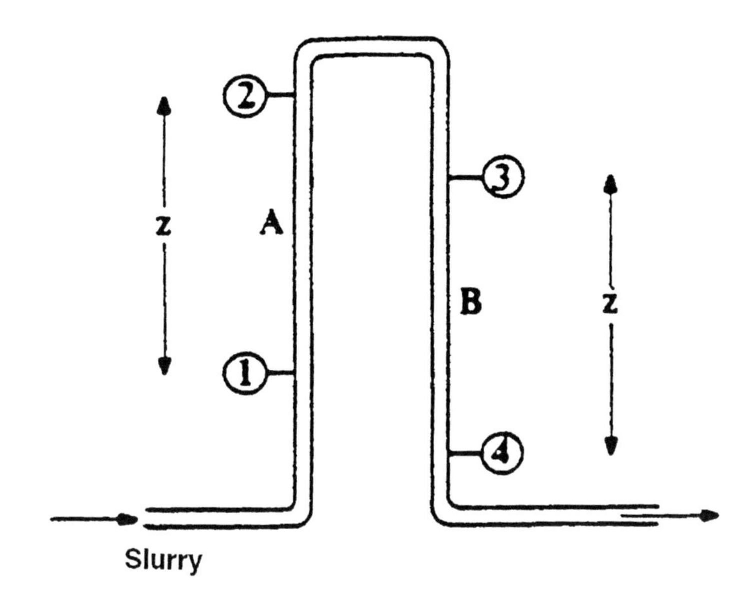

Figure 8.1. Profile of U-loop (schematic)

$$\overline{C}_{vi} = \frac{1}{(S_s - 1)} \left[\frac{(p_1 - p_2) + (p_4 - p_3)}{2 \rho_w g z} - 1 \right] \tag{8.5}$$

Although $\overline{C}_{vi}$ is usually very close to the delivered volumetric solids concentration C_{vd}, they are not identical. Clift & Clift (1981) derived an expression for the fractional difference using the hindered settling velocity correlation of Richardson and Zaki (1954). The magnitude of the fractional difference between C_{vd} and $\overline{C}_{vi}$ is generally less than $(v_t'/V_m)^2$, where v_t' is evaluated at the estimated delivered concentration. It follows that the difference is negligible in almost all cases of practical interest, implying that Eq. 8.5 can generally be used to calculate C_{vd}.

The combination of the force balance equations for the riser and downcomer can also be used to evaluate the wall shear stress τ_o (which, as noted above, is assumed to be the same for the two pipes). The equation for τ_o derived by Clift & Clift (1981) may be written as follows

$$\tau_o = \rho_w g \frac{D}{4}\left[\frac{(p_1 - p_2) - (p_4 - p_3)}{2\rho_w g z} - (S_s - 1) C_{vd} \frac{v_t'}{V_m}\right] \tag{8.6}$$

Here the first term within the brackets is the average manometric gradient for the ascending and descending flows, introduced in Section 4.6, where it was written $\bar{i}_v$. In the final term within the brackets the hindered particle settling velocity v_t' is to be evaluated at the discharge concentration C_{vd}. As this 'correction' or 'error' term involves v_t'/V_m to the first power, it is more likely to be significant than the concentration correction term which involved the square of this ratio. The fractional error in τ_o which would be caused by ignoring the final term is obtained by dividing it by i_w which is used as a first approximation to $\bar{i}_v$ (Clift & Clift, 1981). The value $(S_s - 1)C_{vd}v_t'/[V_m i_w]$ provides a convenient sieve, allowing the elimination of all data points where this fractional discrepancy exceeds, say, 0.05. On the other hand, since the sign of the correction term is known, the same ratio can be used to obtain corrected values of τ_o.

The final term in Eq. 8.6 can also be compared with the solids effect for horizontal flow, $i_m - i_f$. As indicated in Section 4.3, earlier evaluations of this solids effect, such as that proposed by Newitt et al. (1955), employed a proportionality between $i_m - i_f$ and $(S_s - 1)C_{vd}v_t/V_m$. On ignoring the difference between v_t and the hindered settling velocity v_t', it is reasonable to propose a proportional relationship such that the correction term in Eq. 8.6 (i.e. $(S_s - 1)C_{vd}\, v_t'/V_m$) is represented as a multiple, say λ, of $(i_m - i_f)$. As i_f equals $4\tau_o/(\rho_w g D)$, Eq. 8.6 now takes the form

$$i_f = \bar{i}_v - \lambda(i_m - i_f) \tag{8.7}$$

Rearrangement and elimination of higher order terms leads to the alternate version

$$i_f \approx \bar{i}_v - \lambda(i_m - \bar{i}_v) \tag{8.8}$$

Once λ is known, this form of the equation gives a method of evaluating i_f for cases of horizontal slurry transport where it is not appropriate to employ the usual simplification that $i_f = i_w$ in the expression for the stratification ratio. This condition typically occurs for slurries with a significant fines fraction which forms a non Newtonian carrier.

An estimate of the magnitude of λ can be obtained from the relation proposed by Newitt *et al.* (1955) for heterogeneous flow. Although it might no longer be considered sufficiently accurate for design calculations of the solids effect, this relation should be adequate for determining the coefficient of a correction term. On this basis $i_m - i_f$ equals $0.8\,(S_s-1)C_{vd}$ at $V_m = 17\,v_t$, allowing the coefficient λ to be evaluated as roughly 0.07. This value indicates that it was appropriate to dispense with terms involving λ to orders higher than the first, as done in arriving at Eq. 8.8. Moreover, it can give a reasonable indication as to whether it is also permissible to ignore the correction term in that equation.

8.3 Applications of Vertical Flow

As noted by Sellgren *et al.* (1989), in underground mines there is sometimes a considerable inflow of ground water that has to be removed. When the mine dewatering installations are integrated with a hydraulic hoisting system, the cost of power needed to pump out the ground water can be excluded from the cost of hoisting in making cost comparisons with other modes of transporting the solids to the surface.

The economic effectiveness of hydraulic hoisting, together with hydraulic design considerations, have been discussed by Kostuik (1965) and Sellgren (1985, 1987) for both small shallow mines and large deep underground mines. With hydraulic hoisting, only a shaft for ventilation, pipeline, and personnel are needed. On further development of the mine, the hydraulic hoisting installation can be moved to deeper levels. Conventional hoisting shafts are not always located close to the current exploitation of the ore body. A hydraulic hoisting system can replace, or partly replace, the shaft hoisting system and thus affect the load-carrying capacity of the main haulage system. The chief advantage of hydraulic hoisting is that the hoisting capacity can be increased at comparatively low total cost, because there is no need to sink new shafts. Alternative methods include deepening of the existing mechanical hoisting system, and transport in ramps to the present level, using trucks or conveyor belts. These methods often become very expensive.

For industrial application of vertical transportation of a solid-liquid mixture in a pipe, the operating velocity must be sufficient to maintain a continuous flow of solids at the discharge end. However, unnecessarily high velocity causes excessive pipe wear and energy losses. The appropriate operating velocity depends on the settling conditions of the solids, indicating that size, density, and concentration of particles are key parameters in the hydraulic design of a vertical particle-fluid transportation system.

The theoretical background for vertical conveying has been summarized in Section 8.2 above. The basic equation for rising vertical flow is Eq. 8.3. Substituting for S_{mi} and writing τ_o as $\rho f V_m^2/8$ gives

$$(p_1 - p_2)/z = \rho_w g(1+(S_s - 1)C_{vi}) + \rho g f\, V_m^2/(2gD) \tag{8.9}$$

The question immediately arises as to what to use for ρ and f in the final term of Eq. 8.9. If the solid particles are smaller than about 150 microns, then the equivalent-fluid approximation should be appropriate, and ρ will be the mixture density equal to $\rho_w(1+(S_s - 1)C_{vd})$. In this case, the friction factor f could be determined using the techniques for homogeneous flow, but in practice it is often appropriate simply to use the friction factor of a water flow with the same mean velocity, f_w.

Larger particles, in the size range from, say, 200 microns to 2 mm, tend to migrate away from the wall (discussion of the mechanics of this 'off-the-wall' effect is given in Chapter 6). When this occurs, the near-wall zone, in which most of the velocity is gained, can be considered as behaving like water. This gives $\rho = \rho_w$ in the final term of Eq. 8.9, together with an evaluation of f based on water flowing at the same mean velocity, i.e. f_w. Behaviour differing from that just described can occur at extremely high solids concentrations, which inhibit coring and force particles against the pipe wall. See Wilson *et al.* (1979) and Wilson (2004).

Many experimental studies have been made of vertical slurry transport. For example, Sellgren (1979) used a pilot-scale facility with a centrifugal pump to investigate important design parameters for ores and industrial minerals taken from in-plant crushing and milling. The results of these experiments are summarised in the following paragraphs.

It is suggested that the allowable minimum mixture velocity be based on the settling velocity of the largest particles in still water multiplied by a factor of 4 or 5. Provided the velocity exceeds this value, then in most industrial applications, with volumetric concentrations of 15-30%, the corresponding pressure requirement in the vertical system can be determined by the equivalent-fluid model. As noted in Section 4.2, this model is based on the density of the slurry and the friction factor for water. Applied to Eq. 8.9, it gives

$$p = \rho_w g\, S_m z \left(1 + \frac{f_w V_m^2}{2gD} \right) \quad (for\ V_m > V_{all}) \tag{8.10}$$

Here p is the pressure required, S_m is the relative density of the slurry, z is the length of vertical pipe, and V_{all} is the allowable minimum mixture velocity, approximately four times the settling velocity of the largest particle.

The settling velocity in still water of industrially-crushed mineral particles is normally reduced significantly compared to smooth spheres of corresponding size (see Section 2.5). Tests have shown that, on average the settling velocity for particles in the range of 1 mm to 30 mm is reduced approximately 50%. In other words ξ, defined in Section 2.5, is about 0.5. Therefore, the criterion given above for the minimum allowable velocity could alternatively be formulated as: V_{all} is twice the settling velocity of a smooth sphere of the same size as the largest particles.

Within the constraints discussed above, systems operate under conditions where the effect of relative velocity between the components appears to be negligible. The maximum particles sizes considered are in the range of 1 mm to 30 mm in pipe diameters of 0.1 m to 0.3 m. With larger particle sizes (up to 100 mm to 150 mm) and low concentrations the relative velocity between the components become significant. Particles larger than one-fifth the pipe diameter can promote slugging instability in vertical hoisting, and particles larger than one-third the pipe diameter may jam the pipe and should be avoided. Transitional effects associated with particle grading may also introduce certain instabilities which must be carefully evaluated, especially for long vertical risers (Shook 1988).

8.4 Effect of Pipe Inclination on Deposition Limit

Lengths of pipe with an adverse slope often form part of pipelines transporting solids. Compared to the horizontal case, flow up an incline tends to require higher throughput velocities in order to avoid deposition (Hashimoto et al., 1980). This is of greatest significance for coarse-particle flow.

In an experimental investigation carried out at Queen's University (Wilson & Tse, 1984), four particle sizes between 1 mm and 6 mm were tested in a pipe at angles of inclination up to 40 degrees from the horizontal. It was found that the velocity at the limit of deposition initially increases with the angle of upward inclination, θ, reaching a maximum when this angle is about 30 degrees. For the materials tested this maximum velocity was approximately 50 percent larger than that required to move a deposit in a horizontal pipe. This large difference is clearly a matter of importance for both design and operation of pipelines with inclined sections.

For each combination of particle size and pipe inclination angle, the experimental data for the limit of stationary deposition formed a locus which was consistent with the typical locus shape noted in Section 5.3, with a maximum velocity value V_{sm}. As indicated there, V_{sm} marks the lower end of the range of desirable operating velocities for a pipeline. For purposes of comparison, it is appropriate to represent the deposition limit discussed above by the dimensionless velocity, or Durand number, $V_{sm}/\sqrt{[2g(S_s - 1) D]}$. The analysis of the inclined-flow data is based on the mechanistic force-balance model which has been described in Chapter 5. In essence, this force-balance model states that a deposit of solids must move when the forces in the direction of flow exceed those resisting motion. The principal effect of pipe inclination is to introduce an axial component of submerged particle weight. This increases the force resisting motion, accounting for the increase in the velocity at the limit of deposition. The revised force-balance model incorporates this feature, and has been shown to give adequate predictions of the change of deposition velocity found for different angles of inclination (Wilson & Tse, 1984).

In applying the extension of the force-balance analysis to inclined pipes it is useful to consider the difference between the Durand number for inclined flow and that for horizontal flow. Figure 8.2 shows this difference, Δ_D, plotted against θ. The designer or operator concerned with coarse-particle transport in inclined pipes can take the conservative approach of using the envelope curve for the experimental points, shown dashed on the figure. In applying this method, the deposition limit velocity V_{sm} is first estimated for the horizontal case, then Δ_D is obtained from Fig. 8.2 for the required angle of inclination, θ, and the change of deposition velocity is calculated as $\Delta_D\{[2g(S_s\text{-}1)D]^{1/2}\}$. Adding this quantity to the value for the horizontal pipe gives the upper limit of stationary deposition when the pipe is inclined at angle θ. An example of correcting V_{sm} for inclination angle will be shown in the case study presented in Section 8.6.

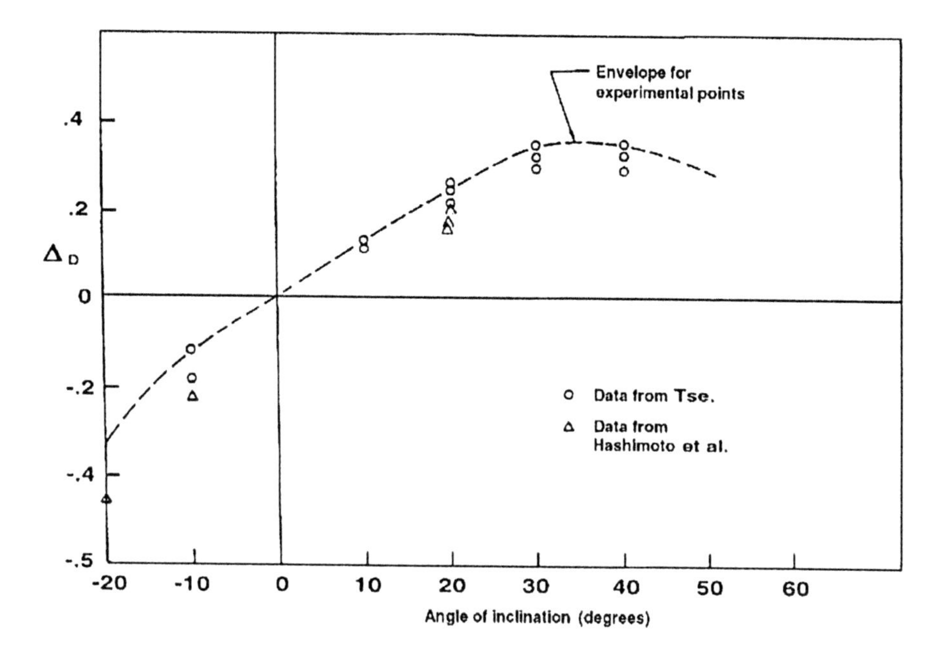

Figure 8.2. Effect of angle of inclination on Durand deposition parameter, after Wilson and Tse (1984)

8.5 Effect of Pipe Inclination on Friction Loss

It has been described in earlier chapters how the solids effect (i.e. the frictional gradient in excess of that for fluid alone) varies with the contact load, which transmits its submerged weight to the lower part of the pipe *via* granular contact. When the pipe is horizontal, all of the submerged weight of the contact load solids is transferred in this fashion, but when the pipe is inclined at angle θ to the horizontal, only the cross-pipe gravitational component $g \cos\theta$ is involved. Thus, in the absence of other changes the solids-effect component of frictional gradient will be multiplied by $\cos\theta$, whereas the fluid component will be unaffected by pipe inclination.

In horizontal flow it was not necessary to distinguish between pressure changes due to friction and those associated with differences in elevation, but this distinction is vital in inclined flow. Consider the two cases sketched on Fig. 8.3 , with the initial assumption that water alone is being pumped. In the first sketch the pressure difference between A and C reflects friction only (plus any losses at the elbows), and this difference would be unchanged if the pipe had been rotated to be flat on the ground (so that the profile became a plan view). The pressure at point B behaves differently. If the pipe were lying flat, the pressure at B would be the average of A and C, but the elevation difference Δz lowers the observed pressure at B by the amount $\rho_w g \Delta z$. However this pressure difference is balanced by a gain in potential energy, and thus is, in principle, recoverable. This recovery occurs in sketch (a) of Fig. 8.3 (provided of course that the pressure at B does not fall so far below atmospheric as to induce vapour pockets). For sketch (b) this pressure difference is not recovered, but the gain in potential energy remains.

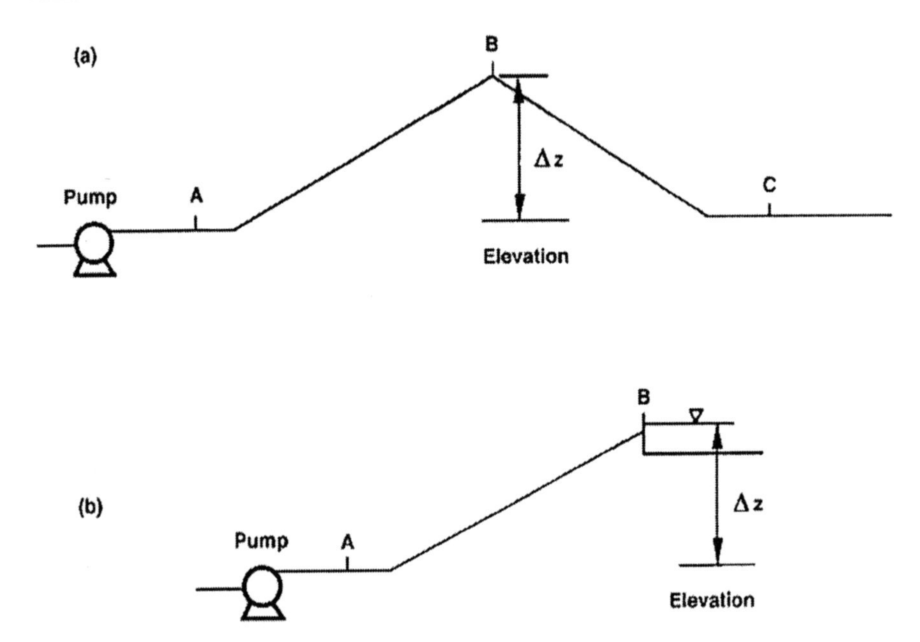

Figure 8.3. Inclined-flow systems (schematic)

If the water is replaced by a slurry, the same arguments apply, but now the gain of potential energy exceeds that of an equal flow of water by the submerged weight of the solids which are raised, i.e. by $\rho_w g(S_s\text{-}1)C_{vd}\Delta z$. It should be noted that it is the delivered solids concentration which is reflected in the potential energy gain. It is applied to the height difference Δz, and for a pipe at an upward inclination θ, the height difference per unit length of pipe equals $\sin\theta$. (The vertical flow dealt with in Section 8.2 is simply a special case for $\sin\theta = 1.00$.) Although the use of delivered concentration is required for calculating the gain in potential energy, it should be noted that the pressure needed to lift the particles depends on the resident (*in situ*) concentration. This point can be seen most clearly in vertical flow, where at small mean velocities the riser may be heavily loaded with particles (requiring a large pressure differential simply to hold them in place) but at the same time may be delivering very little in the way of solids throughput. This particle-hold-up effect can be important for large, heavy particles (those which give fully stratified flow in horizontal pipes), but it is not usually significant in heterogeneous flow, and is completely negligible in homogeneous flow (where the *in situ* concentration is virtually the same as that delivered).

It is interesting to compare the behaviour of the three flow types with an old and widely-used formula for the effect of inclination, that of Worster and Denny (1955). Their approach, based on water as the carrier fluid, deals with the extra pressure gradient ($\Delta i(\theta)$, expressed in m of water per m of pipe) beyond that for pumping water alone. For horizontal flow, this extra gradient is imply the solids effect (i_m - i_w), which may be written $\Delta i(0)$. Worster and Denny's formula states that

$$\Delta i(\theta) = \Delta i(0)\cos\theta + (S_s - 1)\,C_{vd}\sin\theta \tag{8.11}$$

For homogeneous flow $\Delta i(0)$ is zero, and $C_{vd} = C_{vi}$, so that the equation can be considered to hold.

For fully-stratified flow only the cross-pipe component of gravity acts to press the solids against the lower part of the pipe, and hence the solids-effect term $\Delta i(0)\cos\theta$ is applicable. However, for this type of flow hold-up is significant (i.e. $C_{vi} > C_{vd}$) so it can be expected that the final term, and thus Eq. 8.11 as a whole, will give an under-estimate. For short lengths of inclined pipe the equation may still be useful for initial estimation, but in the case of fully stratified flow it is not recommended in the case of suction piping, where an under-estimate of losses can produce NPSH problems (see Chapter 9, and Case Study 8.1. presented below). Improved accuracy can be obtained by using a computer algorithm designed to take particle hold-up into account (Wilson 1988), and this feature has been incorporated into the the GIW SLYSEL program. For heterogeneous flow the difference between C_{vd} and C_{vi} is usually not a major concern, but the solids-effect term $\Delta i(0)$ cos θ requires further consideration. This term was correct for fully-stratified flow where all the solids travel as contact load, but for heterogeneous or partly-stratified flow the stratification ratio (i.e. the fraction of solids travelling as contact load) changes with inclination angle. This change was shown by Wilson & Byberg (1987), using an earlier version of the model for heterogeneous flow. For the present version of the model, as described in Chapter 6, it is seen (Eq. 6.3 to 6.5) that the stratification ratio depends on V_{50}, which itself varies with the particle-associated velocity w, with w having, as its major component, the particle fall velocity v_t. If the flow is inclined, it is the cross-pipe component of the settling velocity that is required, so that the value of v_t, and thus, to a first approximation, of V_{50} must be multiplied by cosθ.

The stratification ratio depends on V_{50}^M, giving the effect of inclination on stratification ratio as a factor $(\cos\theta)^M$. The result is that the term $\Delta i(0)\cos\theta$ in Eq. 8.11 becomes $\Delta i(0)(\cos\theta)^M$ for heterogeneous flow. For narrow-graded particles with M equal to 1.7, the revised term is based on $(\cos\theta)^{1.7}$. This is less than cosθ, illustrating that the equation of Worster and Denny tends to give conservatively large estimates of the inclination effect on heterogeneous flow.

Recent experimental work by Matoušek (1996) with a 150 mm diameter pipe and inclinations angles between -35 degrees and +35 degrees showed a strong influence of pipe inclination on slurry flow stratification. He used four slurries — a medium sand, two coarse sands and a fine gravel – and measured concentration profiles with a collimated beam of γ rays. As expected from the mechanics of partially-stratified inclined flow, the observed concentration profiles were blunter (less stratified) for rising inclinations and more stratified for descending inclinations.

Now that the various features of slurry flow have been discussed, it is appropriate to broaden the focus to include other components of slurry transport systems. We will begin with centrifugal pumps, the subject of the following chapter.

8.6 Worked Examples and Case Study

Worked Examples of Vertical Hoisting

1. In an iron-ore mine the ore is ground to -100 μm in a sub-surface facility, and then pumped vertically 800 m to the surface. The pipe has a diameter of 0.20 m. The concentration by volume, C_{vd}, is 20% and S_s is 4.9. Determine the pressure requirement to pump the slurry to the surface at a velocity of 2 m/s.

Solution: According to the discussion leading to Eq. 8.9 it follows that:

$$(p_1 - p_2)/800 = \rho_w g\,(1 + (S_s - 1)\,C_{vd})(1 + f_w V_m^2/2gD)$$

The friction factor for water flow, for a smooth pipe, f_w, is 0.014.

$$(p_1 - p_2)/800 = 10^3 \times g\,(1 + (4.9 - 1) \times 0.20)(1 + 0.0014 \times 2^2 / 2g \times 0.2)$$

$$(p_1 - p_2) = 800 \times 10^3 \times g \times 1.78 \times (1 + 0.0143) = 14.2\ MPa$$

2. Centrifugal slurry pumps are used to pump a sand slurry (d_{50} = 1.5 mm) out of a quarry. The pipe is vertical with length 100 m and diameter 0.10 m. Tests have shown that the settling velocity of the largest particles is approximately 0.45 m/s. Select the operating velocity and calculate the head requirement in m of slurry.

Solution: Following the guidelines given in connection with Eq. 8.10, the velocity V_{all} is 4 times 0.45, i.e. 1.8 m/s. At this velocity, $V_m^2/2g$ is 0.165 m. With the friction factor f_w taken as 0.016 for smooth-pipe conditions, the head is obtained from Eq. 8.10 as

$$100\,[1 + \frac{0.016(0.165)}{0.1}] = 102.6\ m\ of\ slurry$$

or 102.6 S_m in m of water

3. Consider the upward vertical transport of coarse ore particles (diameter 75 mm) in a pipe flow with velocity V_m = 2.5 m/s. A delivered volumetric solids fraction of 0.05 (5%) is required. Estimate the resident volumetric solids fraction if $v'_t \approx v_t$ = 0.9 m/s.

Solution: Eq. 7.4 gives

$$C_{vi} = 0.05/(1 - 0.9/2.5) = 0.078 \qquad (7.8\%)$$

Case Study 8.1 - Inclined Flow in a Dredge Ladder

Case Studies 5.1 and 6.1 dealt with flow of sand slurries in a horizontal pipe of D = 0.65 m. This pipe will now be considered as the pressure line carrying the slurry from a suction dredge to a disposal area. However, in this example the friction factor will be not be re-evaluated to reflect dredge-line conditions. What will be considered here is the analysis of conditions in the inclined suction pipe located on the dredge ladder. This pipe (which has length 18 m in the present case) conveys the slurry from the cutter head to the main pump on board the dredge. The angle of the ladder can be adjusted according to the depth to be dredged. For simplicity we will consider only an angle of 30° to the horizontal. The dredging is taking place in fresh water, and it is assumed initially that the size of the suction pipe is the same as that of the discharge pipe and that the sand is as specified in Case Study 6.1, with d_{50} = 0.70 mm. As in that study the required transport rate of 4000 tonnes/hour (on a dry-weight basis) is satisfied by V_m = 6.30 m/s and C_{vd} = 0.20.

(a) The first point is to investigate the effect of the incline on the deposition velocity in the suction pipe. Figure 8.2 shows that an inclination angle of 30° gives the largest increase in deposition velocity. At this angle the upper limit of the envelope for experimental points has Δ_D = 0.33, equivalent to an increase in V_{sm} of $0.33\sqrt{(19.62)(1.65)(0.65)} \cong 1.51\ m/s$. As shown previously in Case Study 5.1, Eq. 5.10 gives a value of V_{sm} for the horizontal pipe of 4.84 m/s. If the maximum increase of 1.51 m/s is added, the resulting deposition limit velocity on the slope could be as large as 6.35 m/s. However this is rather pessimistic, as the typical experimental data (rather than the envelope) show $\Delta_D \cong$ 0.30, for a V_{sm} of about 6.2 m/s on the slope. Thus the operating velocity of 6.3 m/s, although well clear of deposition in the horizontal discharge line, is very close to the deposition point on the inclined suction pipe.

This deposition-limit condition is especially severe for the 0.7 mm particle, which is in the 'Murphian' size range for V_{sm}. If, say, the particle diameter were only 0.20 mm, V_{sm} would be 3.2 m/s for the horizontal pipe, and roughly 4.7 m/s on the incline, well below the operating velocity of 6.3 m/s.

(b) The next step is to consider the pressure drop in the suction pipe. For fully- stratified flow, detailed calculations would be required. However as noted in Section 8.5, for heterogeneous flow a reasonable approximation can be obtained from the formula of Worster and Denny (Eq. 8.11). In the present case, for the 0.70 mm sand slurry at V_m = 6.3 m/s, this formula gives the excess gradient, Δi (30°), as Δi (30°) = 0.0239 cos 30° + (1.65)(0.20) sin 30° where 0.0239 represents the solids effect in the horizontal pipe, as obtained in Case Study 6.1. Thus Δi (30°) equals 0.0207 + 0.1650 or 0.1857. The clear water gradient i_w was found in Case Study 6.1 to be 0.0373, giving a total value of 0.223, and on multiplying this by the suction-pipe length of 18 m, the drop is found to be 4.0 m of water. (The greater part of this arises from lifting the submerged particles from the bottom to the water level, this is represented by the term in $\sin\theta$ in Eq. 8.11). When the velocity head is added in, together with minor losses, it is estimated that the pressure at the suction of the pump (which is located close to the water line) will be subatmospheric to the extent of about 7 m of water.

In Chapter 9, it will be seen that only a limited sub-atmospheric pressure can be tolerated at the suction of a pump, otherwise cavitation will occur. In the present instance, the low pressure at the pump suction may well induce cavitation.

(c) As shown above, although the design conditions are well suited for operation in the discharge pipe, conditions are very near the deposition limit in the suction pipe, and the pressure drop in this pipe is probably enough to cause the pump to cavitate. This difficulty would not be alleviated by a change of particle size alone; for example if the vacuum at the suction side of the pump is 7.0 m of water for the 0.70 mm particles, a particle size of 0.20 mm would reduce only very slightly; to about 6.8 m, and cavitation would probably still occur in this case.

For water, and for any slurry that behaves as an equivalent fluid (i.e. $i_m = S_m i_w$, see Eq. 2.47), a normal method of eliminating cavitation problems is to have suction piping larger than that on the pressure side of the pump. Thus a larger suction pipe, say D = 0.70 m, might well have been installed along with the pressure-pipe diameter of 0.65 m. In this case the slurry of 0.2 mm particles (which, as seen in Case Study 6.1, behaves much like an equivalent fluid) will show satisfactory performance. The vacuum at pump entry is reduced from the figure of 6.8 m of water given above for the 0.65 m pipe to about 5.8 m for the 0.70 m pipe, which should be sufficient to eliminate cavitation. Despite the success of this strategy for a slurry of fine sand, the results of applying it to a coarse-sand slurry would be disastrous. For the 0.7 mm particles in the 0.7 m pipe, Eq. 5.11 gives V_{sm} = 5.02 m/s for a horizontal pipe, and the equivalent value on a 30° incline will be 6.59 m/s. However, for the required discharge the mean velocity in the 0.70 m suction pipe is much less than that in the 0.65 m discharge line (5.43 m/s *versus* 6.30 m/s). Thus deposition will take place immediately in the 0.70 m suction line. This deposit implies that the flow will become stratified. In the lower portion of the pipe the velocity will be zero, or even negative (at large angles of inclination reverse flow may actually be observed near the bottom of a pipe) and the required total flow rate Q_m can only be maintained if the velocity in the upper portion of the pipe is much larger than V_m, which in turn will cause extremely high velocity heads and friction losses. Moreover, the stratification in the suction line indicates that the *in situ* concentration of solids here will be much larger than that in the discharge line. Hence the term in sinθ which shows the effect of submerged weight of solids will rise abruptly when deposition occurs.

It follows from the points outlined above that for slurries of particles of coarse-sand size and above, use of an inclined suction pipe larger than the pressure line is highly likely to produce *both* deposition and cavitation. Such enlarged suction pipes are manifestly not to be recommended for coarse-particle slurries.

REFERENCES

Brook, N. (1962). Flow measurement of solid-liquid mixtures using Venturi and other meters. *Proc. Instit. Mech. Engrg.* **176**, 127-140.

Clift, R. & Clift, D.H.M. (1981). Continuous Measurement of the density of flowing slurries. *Int. J. Multiphase Flow,* Vol. 7, No. 5, pp. 555-561.

Geller, F.J. (1978). The hydraulic downward transportation of solids of highly heterogeneous composition. *Proc. Hydrotransport 5,* BHRA Fluid Engineering, Cranfield, UK, paper G3.

Hagler, T.W. Sr. (1956). Means for determining specific gravity of fluids and slurries in motion. *U. S. Pat.* 2 678 529.

Hashimoto, H., Noda, L., Masuyama, T. & Kawashima, T. (1980). Influence of pipe inclination on deposit velocity. *Proc. Hydrotransport 7*, BHRA Fluid Engineering, Cranfield, UK, 231-244.

Kostuik, S.P. (1966). Hydraulic Hoisting and Pilot-Plant Investigation of the Pipeline Transport of Crushed Magnetite, *The Canadian Mining and Metallurgical Bulletin*, January, pp. 25-38.

Matoušek, V. (1996). Internal structure of slurry flow in inclined pipe, experiments and mechanistic modelling. *Proc. Hydrotransport 13*, BHR Group, Cranfield, U.K., 187-210.

Newitt, D.M., Richardson, J.F., Abbot, M. & Turtle, R.B. (1955). Hydraulic conveying of solids in horizontal pipes. *Trans. Inst. of Chem Engrs.*, Vol. 33, London, U.K.

Richardson, J.F. & Zaki, W.N. (1954). Sedimentation and fluidisation - *I. Trans. Instit. Chem. Engrs.* **35**, 35-52.

Sellgren, A. (1979). Slurry transportation of Ores and Industrial Minerals in a Vertical Pipe by Centrifugal Pumps - A Pilot-Plant Investigation of Hydraulic Hoisting. Dissertation, Dept. of Hydraulics, Chalmers University of Technology, Gothenburg, Sweden.

Sellgren, A. (1985), Mine Water - A Resource for Transportation of Ores from Underground Mines. *Proceedings*, International Mine Water Association, Granada, Spain, Vol. 2, pp. 1027-1037.

Sellgren, A. (1987). *Slurry Transport - A Feasible Systems Solution to the Integrated Underground Mine of Tomorrow: Improvement of Mine Productivity and Overall Economy by Modern Technology,* Almgren, Berge, Matikainen, eds, A.A. Balkema: Rotterdam/Boston. Reprint 13th World Mining Congress, Stockholm, Sweden, pp. 751-757.

Sellgren, A., Jedborn, A., & Hansson, K. (1989). Hydraulic Hoisting: An Economic Alternative in the Deepening of Underground Mines, *Mining Engineering,* August.

Shook, C.A. (1988). Segregation and plug formation in hydraulic hoisting of solids. *Proc. Hydrotransport 11*, BHRA Fluid Engineering, Cranfield, UK, pp. 359-379.

Venton, P.B. & Cowper, N.T. (1986). The New Zealand Steel Ironsand Slurry Pipeline. *Proc., Hydrotransport 10,* Innsbruck, Austria, BHRA Fluid Engineering, Cranfield, Bedford, UK, pp. 237-247.

Wilson, K.C. (1988). Algorithm for coarse-particle transport in horizontal and inclined pipes. *Proc. Intern'l Symp. on Hydraulic Transp. of Coal and Other Minerals* (ISHT.88) CSIR and Indian Institute of metals. Bhubaneswar, India, 103-26.

Wilson, K.C. (2004). Energy consumption for highly-concentrated particulate slurries. *Proc. 12th Intn'l Conf. on Transport & Sedimentation of Solid Particles,* Prague, Czech Rep., 20-24 Sept.

Wilson, K.C. & Byberg, S.P. (1987). Stratification-Ratio Scaling Technique for Inclined Slurry Pipelines. *Proc. Twelfth International Conference on Slurry Technology,* Slurry Technology Association.

Wilson, K.C., & Tse, J.K.P. (1984). Deposition limit for coarse-particle transport in inclined pipes. *Proc. Hydrotransport 9,* BHRA Fluid Engineering, Cranfield, UK. pp. 149-169.

Wilson, K.C., Brown, N.P. & Streat, M. (1979). Hydraulic hoisting at high concentration: a new study of friction mechanisms. *Proc. Hydrotransport 6,* BHRA Fluid Engineering, Cranfield, UK, pp. 269-282.

Worster, R.C. & Denny, D.F. (1955). Hydraulic transport of solid materials in pipelines. *Proc. Institution of Mechanical Engineers,* London, Vol. 169, 1955, pp. 563-86.

Chapter 9

CENTRIFUGAL PUMPS

9.1 Introduction: Basic Concepts and Affinity Laws

The previous chapters, forming the first half of this book, concentrated on the behaviour of various kinds of slurries in pipes. The second half of the book deals with other components of a pipeline transport system, and with how the components interact. This portion of the work begins in the present chapter, with a discussion of centrifugal pumps. The importance of pumps of this type as prime movers for high-tonnage slurry operations was shown in Chapter 1; other types of prime movers, such as positive-displacement pumps and lock-hopper arrangements, are referred to where appropriate but their operation is not analysed in detail.

The present chapter deals with the behaviour, performance, selection and testing of centrifugal pumps. Major features of pump design are dealt with in a general sense, but the emphasis is on the information required by the pump user rather than the pump designer. The analysis is based on the classical case of the flow a simple liquid. Various differences between slurry flow and liquid flow are mentioned, but a complete treatment of the effect of solids on pump head and efficiency is reserved for Chapter 10, while Chapter 11 deals with the wear produced by solid particles. Some of the general geometric requirements for wear, such as thick sections and non-volute collectors are discussed in section 9.2. The variation of wear with design geometry is covered in Chapter 11. Other components of pipeline systems are described in Chapter 12, and the interaction of all components is analysed and described in Chapters 13 and 15. Pump selection is covered in Chapter 14.

We begin the analysis of pumps by considering the flow of a simple liquid of density ρ_f, which moves through the pump at a flow rate, or discharge, denoted Q and measured in volume per unit time. The pump adds energy to the fluid, enabling it to move through the piping, overcoming friction and the effect of any changes in elevation. It was seen in Chapter 2 that this energy represents the sum of pressure energy, potential energy and kinetic energy. Bernoulli's equation expresses each of these three energy terms as heights of the flowing liquid, as shown in Section 2.2. This concept of height or head is used in measuring the energy added by the pump as the total dynamic head, which is given the symbol TDH, or simply H. This quantity, measured in height of the liquid flowing, was defined by Eq. 2.10, in terms of pressures, velocities and elevations at pump entry (A) and at pump exit (B), thus

$$H = TDH = \frac{V_B^2 - V_A^2}{2g} + \frac{p_B - p_A}{\rho_f g} + (z_B - z_A) \tag{9.1}$$

The power output of the pump is determined by the product of Q and H, and is given by

$$(Power)_{out} = \rho_f g Q H \tag{9.2}$$

This relation applies in any consistent system of units. Thus for S.I. units Eq. 9.2 gives the power out in Watts, which is usually divided by 1000 to obtain kW. (In US customary units, Q is expressed in US gallons per minute, H in feet and power in horsepower; and a numerical coefficient is required in the equation.)

The input power to the pump is $T\omega$ where T is the torque on the shaft of the pump and ω is the angular velocity. With seconds as the unit of time, ω is equal to $2\pi n$ where n is measured in revolutions per second. In practice, the speed of rotation is usually measured in revolutions per minute, even where S.I. units are generally employed. The symbol N will be reserved for revolutions per minute, thus

$$\omega = 2\pi n = 2\pi N/60 \tag{9.3}$$

The input power is given by

$$(Power)_{in} = 2\pi nT = 2\pi NT/60 \tag{9.4}$$

With torque in Newton-metres, this equation gives power in Watts.

A final point concerning the pump's input and output is the efficiency, denoted η and equal to the ratio of output power to input power, i.e.

$$\eta = \frac{(Power)_{out}}{(Power)_{in}} \tag{9.5}$$

For an ideal pump, η is 1.00 or 100%. Although real pumps necessarily have lower values, efficiencies of over 90 percent can be achieved for large water pumps (efficiencies tend to be somewhat less in slurry applications, for reasons to be discussed in Chapter 10). In the past the term 'efficiency' was sometimes used to represent the ratio of actual to ideal values of quantities other than power. This usage can lead to confusion, and will be avoided here. In this book efficiency refers only to the power ratio defined by Eq. 9.5.

Efficiencies can be obtained from pump tests, which will be described in Section 9.5. If a given pump is driven at a constant shaft speed (i.e. fixed N), a series of readings of Q, H and T can be obtained at various openings of a throttling valve located downstream of the pump. The head is plotted directly against discharge, as shown on Fig. 9.1. This curve is known as the head-discharge characteristic, or the head-quantity (or head-capacity) relation, or simply the H-Q curve. The required power (Eq. 9.4) and the efficiency (Eq. 9.5) are also plotted against Q, as shown on the figure. The limiting suction performance curves (to be discussed in Sections 9.3 and 9.6) can also be plotted.

With n constant the efficiency η varies only with the ratio HQ/T, where T is always greater than zero. Thus η will be zero at the no-flow condition (Q=0) and again when the H-Q curve intercepts the discharge axis (here H = 0). Between these extremes the efficiency curve displays a maximum, as shown on the figure. This maximum defines the 'best efficiency point' or BEP, and the associated discharge and head are often identified as Q_{BEP} and H_{BEP}.

The curves shown on Fig. 9.1 refer to a single angular velocity, but if the tests were repeated with a different value of n, all the points shift. This behaviour can be plotted as a series of H-Q curves for various angular speeds, with contours of efficiency and power added, as shown on Fig. 9.2.

Test data are not required for each curve; instead, the various constant-speed curves are constructed on the basis of the following simple scaling relations. All discharges (including both Q_{BEP} and the discharge at H = 0) shift in direct proportion to N, while all heads (including both the no-flow head and H_{BEP}) shift in proportion to N^2.

Similar shifts are found if the shaft speed is kept constant but the scale of the pump is changed, while maintaining geometric similarity. The criterion of geometric similarity requires that all dimensions scale in the same ratio; thus the scale of each machine in the series can be specified on the basis of any one of its dimensions. The dimension which is customarily used is the external diameter of the impeller, denoted for this purpose by D. It is known that when the size of the unit as a whole is altered, heads scale in proportion to D^2, and discharges in proportion to D^3 (the theory underlying the scaling relations will be presented in Section 9.2, these scaling relations for geometrically-similar units differ somewhat from the relations appropriate for impeller trimming, which are discussed at the end of section 9.2).

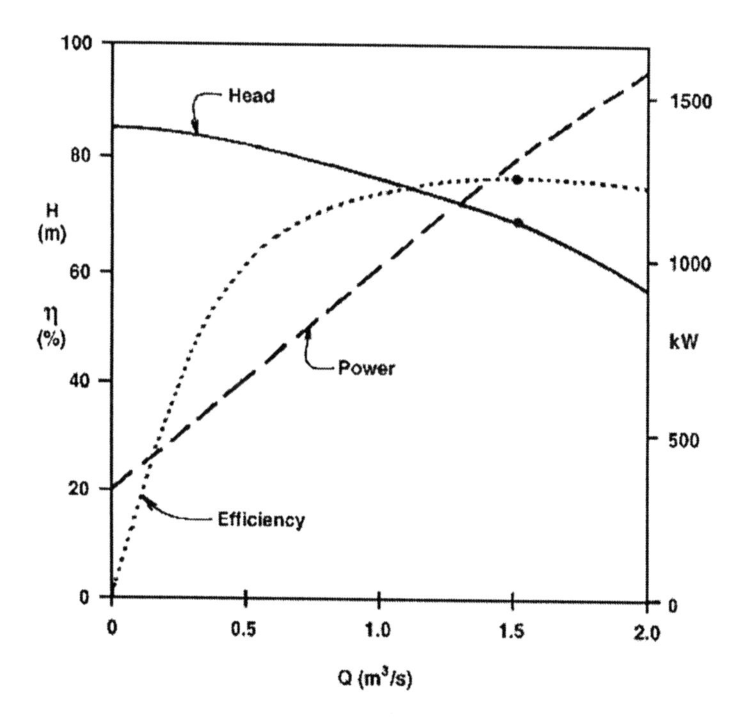

Figure 9.1. Representative pump characteristic curves

It follows from the scaling relations that the curves for all members of a set of geometrically similar (homologous) pumps operating at various n-values can be collapsed into a single entity by using an appropriate system of dimensionless axes. One system of this type (White 1986, p. 655) uses as abscissa the dimensionless discharge Q/nD^3, with n in revolutions per second. (The quantity is dimensionless only in consistent units e.g. Q in m^3/s and D in m, or Q in ft^3/s and D in feet). The efficiency η is already dimensionless, and the dimensionless head takes the form gH/n^2D^2 (again in consistent units).

The head and efficiency characteristics of a representative pump are plotted on Fig. 9.3, using this dimensionless axis system. Despite differences in pump size and angular velocity, all the points for geometrically similar pumps will lie on a single line of gH/n^2D^2 *versus* Q/nD^3. When the pump efficiency is plotted against Q/nD^3, the same basic uniformity occurs, but it is found that smaller pumps have somewhat reduced efficiency. This reduction is caused by the greater relative significance of friction (both mechanical and hydraulic) in small machines. Empirical formulas are available for estimating this effect (see, for example, White, 1986, p. 659).

The dimensionless characteristic curves shown on Fig. 9.3 display representative centrifugal-pump behaviour. The peak of the efficiency curve is gently rounded; operation slightly off the BEP does not carry a large penalty, but operation far from the BEP produces gross losses in efficiency. The dimensionless head-discharge curve is rather flat near the no-flow condition (i.e. at low Q/nD^3) but droops downward in the region near the best efficiency point.

The well-known affinity laws for pumps represent the scaling relations mentioned above, and are closely associated with dimensionless plots such as Fig. 9.3. The affinity laws are used for point-to-point transfer of test data for one pump size and shaft speed in order to plot characteristic curves for some other size or speed. (These laws apply for all members of any given set of geometrically similar or 'homologous' machines.) Since data points for any member of the set plot onto the same curve on the dimensionless axis system, it follows that they can also be plotted directly onto the curve for another member of the set. Thus each point from a test of a pump with impeller diameter D_1, run at angular speed n_1 can be transferred to the curve of a different pump with diameter D_2 and speed n_2. Although the affinity laws can be expressed in various ways, the system used here is consistent with that of the McGraw-Hill Pump Handbook (Cooper, 2001, Eq. 27, 29 and 32) and also with the affinity laws for slurry pipelines mentioned in earlier chapters.

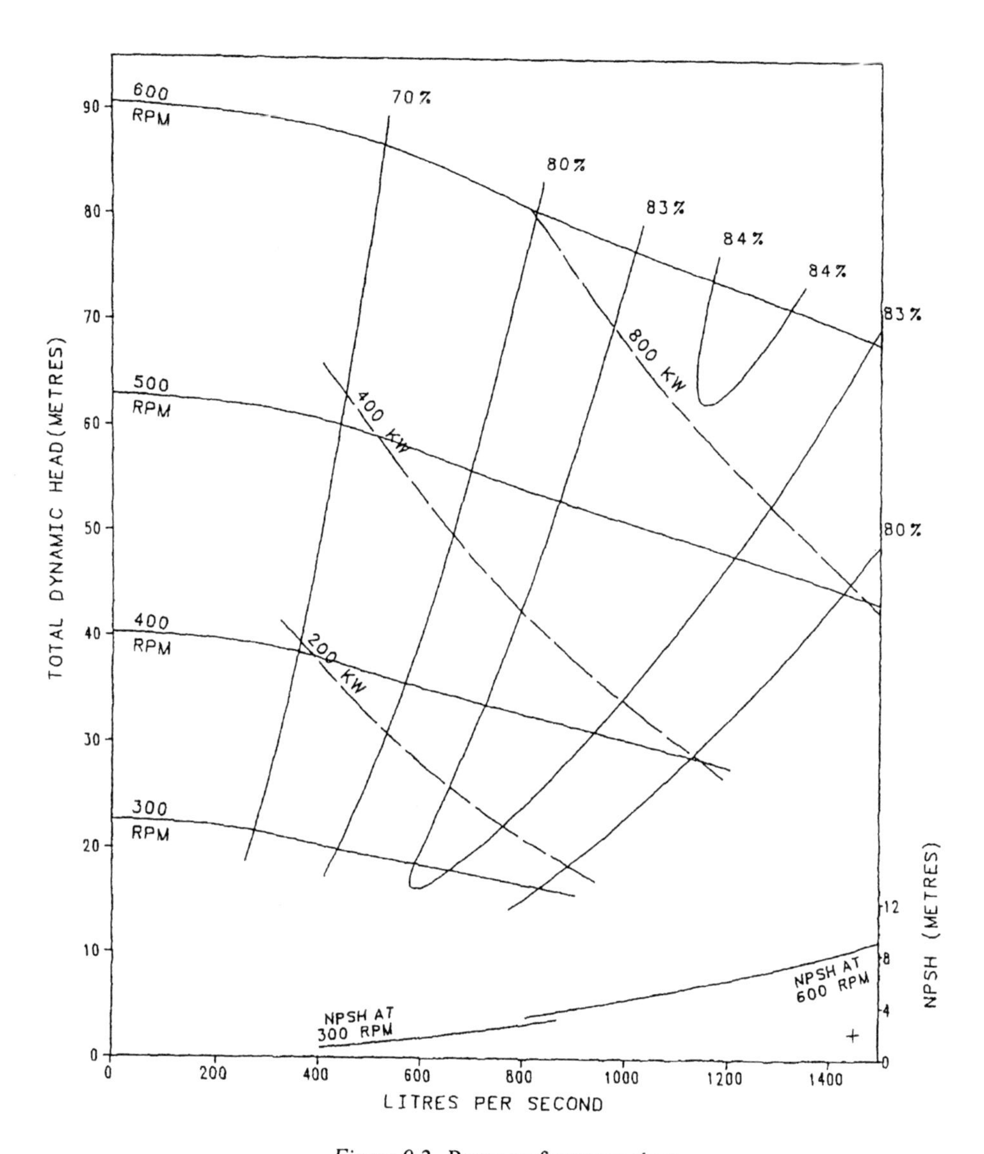

Figure 9.2. Pump performance chart

The affinity law for discharge ensures that the scaled point has the same value of Q/nD^3 as the test point, i.e.

$$Q_2 = Q_1\left(\frac{n_2}{n_1}\right)\left(\frac{D_2}{D_1}\right)^3 \tag{9.6}$$

Simultaneously, the scaled head H_2 is obtained from the test-point value H_1, by means of

$$H_2 = H_1 \left(\frac{n_2}{n_1}\right)^2 \left(\frac{D_2}{D_1}\right)^2 \tag{9.7}$$

This affinity law for head ensures that the value of gH/n^2D^2 is maintained. The gravitational acceleration has been cancelled from the equation. Also, the problem of units is side-stepped, as the affinity laws apply in any system of units. In practice N_1 and N_2 in revolutions per minute usually are used in place of n_1 and n_2 in revolutions per second. In the scaling process the variation of efficiency with pump size can often be ignored for large pumps (the drop-off of efficiency for small pumps was mentioned previously). The required power depends simply on the product of H and Q, which scales according to the relation

$$(Power)_2 = (Power)_1 \left(\frac{n_2}{n_1}\right)^3 \left(\frac{D_2}{D_1}\right)^5 \tag{9.8}$$

It is particularly important to note that discharge and head (and also power) must be scaled together. For example, a change of size affects discharge, head and power, but all in different ways in accord to the three affinity laws. To obtain, say, the H-Q curve for a scaled-up pump, each of the test data points must have both H and Q scaled by the affinity laws. The new characteristic curve is then drawn through the series of scaled-up points.

The scaling concept has been used in several places in previous chapters for the scale-up of friction loss from one pipe size to another on a point-by-point basis. The simplest example was for the laminar flow of non-Newtonian slurries, for which the affinity laws were given by Eq. 3.11 and 3.12. In a more general sense, the use of scaling relations forms a recurring theme in this book.

9.2 Hydraulic Design and Specific Speed

The previous section treated the centrifugal pump as a system component or 'black box' without examining the shape of the flow passages or the fluid mechanics of the pumping process. These physical considerations will receive more attention in the present section.

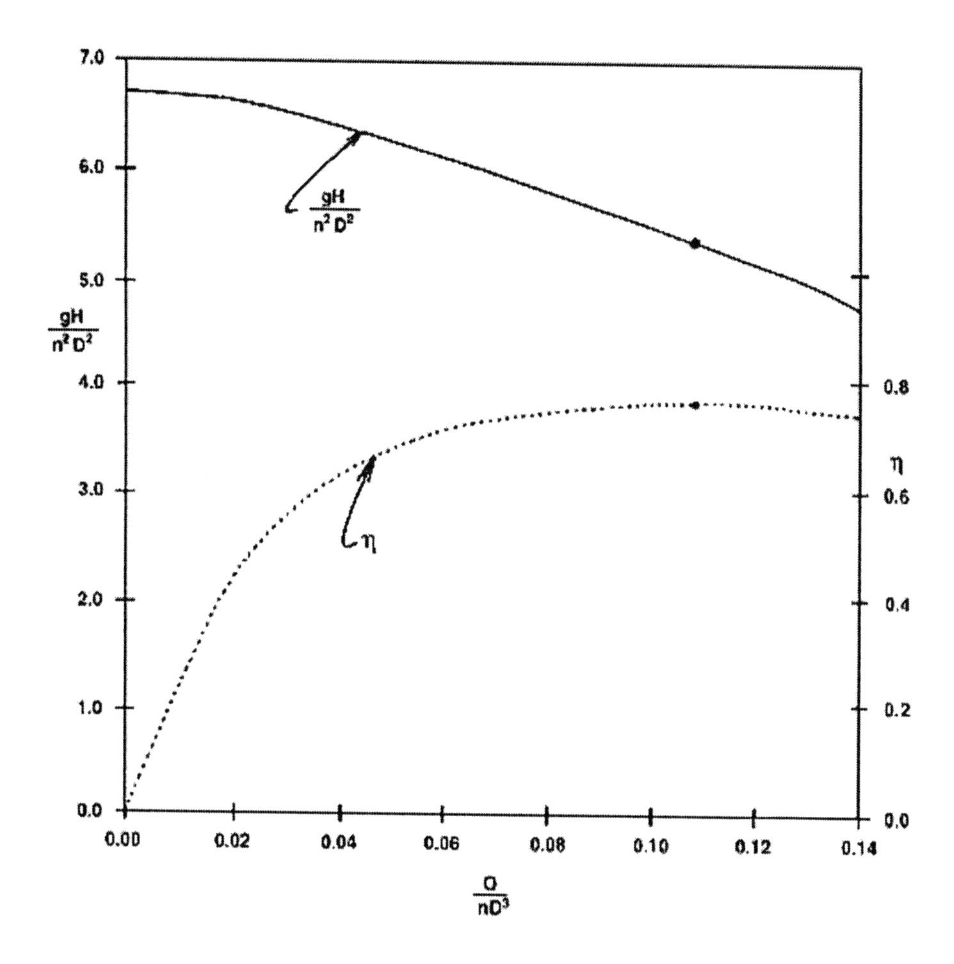

Figure 9.3. Dimensionless pump characteristics

A centrifugal pump has two main components: the first is the rotating element comprising the shaft and impeller, including the vanes which act on the fluid; the second is the stationary element made up of the casing or shell which encloses the impeller, together with the associated stuffing boxes and bearings. The hydraulic design of a centrifugal pump is concerned with dimensioning the impeller and the shell to provide the required performance characteristics. In any pump design there is usually more than one combination of component dimensions that can be arranged to give a specified performance. The combination selected will depend on the intended application, and on any hydraulic or mechanical limitations. In slurry pumps, a number of limitations are imposed. These include the need to pass large solids, the requirement for a robust rotating assembly because the slurry density exceeds that of water, and the desirability of thicker sections in order to minimise the effects of wear.

Centrifugal pumps exist in a wide variety of arrangements to suit different applications, and may comprise a number of stages of impeller and collector. Slurry pumps show less variety. They are normally single-stage end-suction type, and are usually of radial or mixed-flow configuration. They commonly have volute-type collectors, but these are often modified to concentric or semi-concentric form in order to reduce the effect of wear on the shell. Sections of a representative end-suction single-stage volute-casing pump are shown on Fig. 9.4.

Flow through pumps can be quite complicated, and in order to aid description, certain directions or co-ordinates must be specified. The axial direction is parallel to the shaft of the pump, and positive in the direction of the axial component of the inflow, and the radial direction is directly outward from the centre-line of the shaft. The tangential direction is perpendicular to both axial and radial directions, representing the tangent to the circular path of a rotating point. Points on the impeller have only tangential velocity, given by ωr or $2\pi nr$ where ω is angular velocity in radians per second, n is in revolutions per second, and r is the radius from the shaft centreline. A further direction, needed for mixed-flow pumps, is the meridional direction. This direction lies within a plane passing through the shaft centre-line, and follows the projection of the fluid streamlines onto this plane. Thus the meridional direction has both radial and axial components in the case of mixed flow, but coincides with the radial direction for a pump with radial flow.

The meridional and tangential velocities are used to plot the velocity triangles at the entry and exit of the impeller. The 'absolute' velocity of the fluid (i.e. its velocity measured relative to the ground) is denoted by c, with subscripts m and t denoting components in the meridional and tangential directions. The absolute velocity of a point on the rotating impeller, denoted by u, is necessarily in the tangential direction, so a directional subscript is not required in this case. Further, the subscripts 1 and 2 are used to distinguish conditions at the entry and exit of the impeller, respectively.

The velocity triangles are shown on Fig. 9.5. It should be noted that the velocity w, which closes the vector triangle, represents the velocity of the fluid relative to the impeller. As this velocity will follow the inclination of the blades, the angle β shown in the vector triangles will represent the blade angle, i.e. the angle between the impeller blade and a plane tangent to the impeller. The exit blade angle β_2 is an important design parameter, and the entry blade angle β_1 is set to minimize energy loss as the fluid enters the

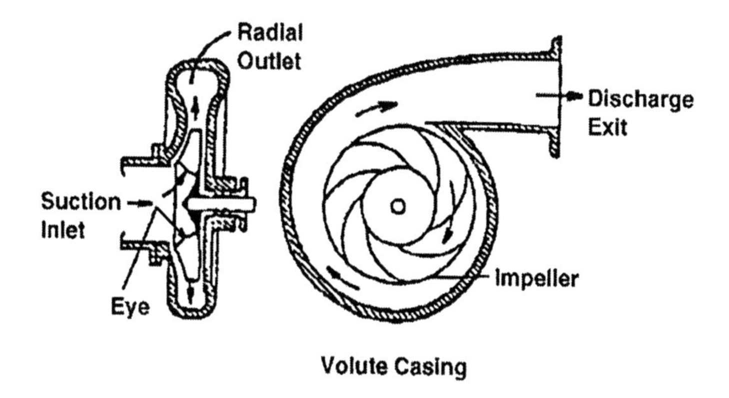

Figure 9.4. Sections of a representative slurry pump

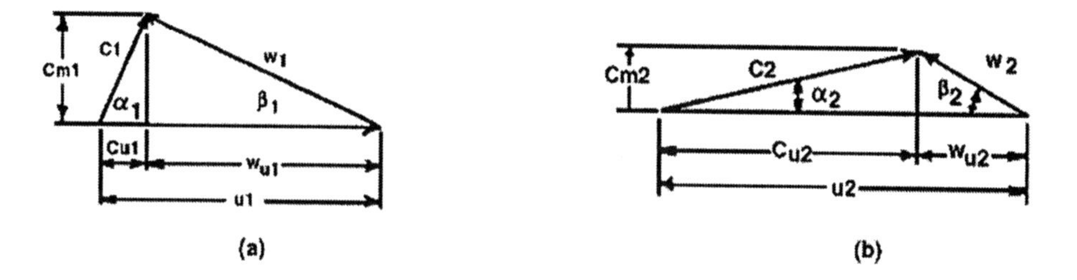

Figure 9.5. Velocity triangles at impeller (a) entry and (b) exit

impeller.

The vector triangles provide the information required for solving the moment-of-momentum equation. In its simplest form, applicable when conditions do not vary with time, this equation states the applied torque T must equal the moment of the net momentum flux passing through a stationary control volume. As the control volume is stationary, i.e. based on the ground not the impeller, the velocities used in calculating the momentum flux must also be 'absolute' or ground-based ones (for this reason absolute velocities were used for the vector triangles).

Referring to variables displayed on Fig 9.5, it can be shown that

$$T = \rho_f Q \left[c_{t2} r_2 - c_{t1} r_1 \right] \tag{9.9}$$

Multiplying Eq. 9.9 by ω gives the power, and on dividing both sides of the resulting equation by Q and g, one obtains

$$H_i = [u_2 c_{t2} - u_1 c_{t1}] / g \tag{9.10}$$

As losses have been disregarded, H_i is a theoretical head. Equation 9.10 is often called the Euler equation, after its originator (Euler, 1756). The term u_1c_{t1} refers to the flow entering the eye of the impeller, and at the best efficiency point this term effectively reduces to zero. Thus it is ignored when considering the idealised machine with efficiency of 100%.

The vector diagram at the exit of the impeller shows that

$$c_{2t} = u_2 - c_{m2} \cot \beta_2 \tag{9.11}$$

where c_{m2} is the meridional component of outlet velocity (directed radially outward for most slurry pumps), which in turn is given by the discharge Q divided by the exit area of the impeller, i.e.

$$c_{m2} = \frac{Q}{\pi D_2 b_2} \tag{9.12}$$

where b_2 is the breadth between the shrouds at the outlet of the impeller.

Equation 9.12, together with the evaluation of u_2 as $\pi n D_2$, can then be substituted into Eq. 9.11 and the result combined with Eq. 9.10, with the final term of that equation ignored for the ideal case. The result forms the basic head relation for the ideal pump, written

$$\frac{gH_i}{n^2 D_2^2} = \pi^2 - \frac{Q}{n D_2^3}\left(\frac{D_2}{b_2} \cot \beta_2\right) \tag{9.13}$$

The ratio D_2/b_2 and the blade angle β_2 will both be constant for all members of a set of geometrically similar pumps. Thus the head-capacity curve for any set of idealised pumps will give a single straight line when plotted on the dimensionless axis system of Fig. 9.3. The analysis given here is central to the theoretical demonstration of the validity of the affinity laws (Eq. 9.6 to 9.8). Real H-Q characteristics lie below the theoretical straight line, approaching it only near the best efficiency point. However, conditions near this point are of greatest practical interest. For all pumps of a given geometry, the BEP will lie at a single location on the dimensionless axis system of Fig. 9.3, but for different geometries this location shifts. As it happens, for most pump geometries of practical significance the best-efficiency points lie within a relatively narrow band on the dimensionless plot. Although values of the two dimensionless co-ordinates are required to define a BEP precisely, a single parameter which gives the position along the efficiency-point band can do almost as well. This parameter is the dimensionless specific speed, n_s, obtained by manipulating the ratios which form the axes of Fig. 9.3 in order to eliminate the impeller diameter. The definition of n_s used by White (1986) is

$$n_s = \frac{n\sqrt{Q}}{(gH)^{3/4}} \tag{9.14}$$

where n is in revolutions/second, Q in m^3/s and H in m. Both Q and H are determined at the best efficiency point. A similar dimensionless variable, based on angular speed ω rather than n, is used by Cooper (2001, p. 222).

In engineering practice, gravitational acceleration is often omitted, giving the customary specific speed denoted N_s and defined by

$$N_s = \frac{N\sqrt{Q}}{H^{3/4}} \tag{9.15}$$

In the United States N is measured in revolutions/minute, H in feet and Q in US gallons/minute. In other parts of the world, customary N_s is based on Q in m^3/s and H in m, but still uses N in revolutions/minute. In both cases the units are inconsistent. Numerically, N_s in US customary units = 51.64 times metric N_s.

The volute or casing of a pump has the task of converting the kinetic energy of the fluid leaving the impeller into pressure energy. In an idealized pump, it is considered that there are no losses in either the casing or the impeller. In practice, hydraulic losses occur in all the wetted passages of the pump. The head-capacity curve of an actual pump results from the subtraction of losses from the idealised pump characteristic, on the basis that there is only a single discharge for which the shock loss at the impeller inlet is zero. An example is shown on Fig. 9.6.

It can be expected that curves of this type will be geometrically similar for all members of a set of pumps with given specific speed. Thus one machine can be used as a model for all members of the set, with the affinity laws providing the basis for transferring the results from the model to other machines. If model testing is carried out accurately on pumps which are geometrically similar, performance can generally be predicted within 2%.

When model tests are not available, empirical methods must be used instead. In this case pump design is usually carried out in two stages: the first uses dimensionless performance coefficients to obtain initial approximations for the major dimensions, and to lay out the meridional cross section. The second uses vector triangles for the impeller blade angles, and slip estimates for the assumed dimensions, allowing the head and capacity performance to be checked against requirements.

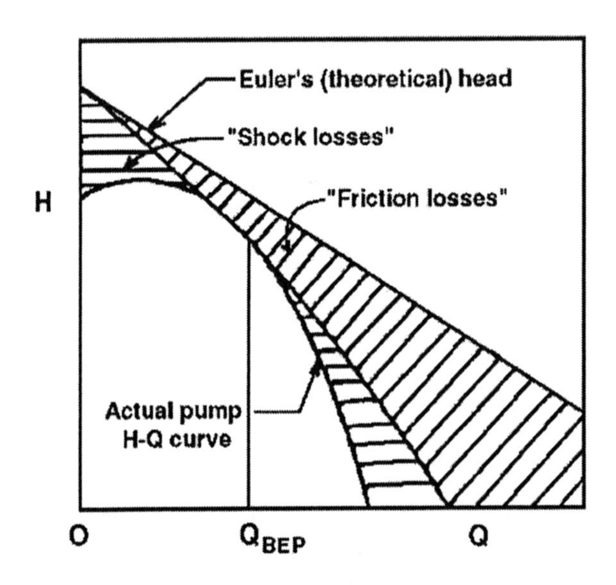

Figure 9.6. Head-discharge curve obtained by subtraction of hydraulic losses from ideal line

Stepanoff (1957) has provided empirical relations and coefficients for the calculation of impeller diameter and outlet width, shell throat areas, and other key dimensions for water pumps based on known (or assumed) values of specific speed and impeller outlet angle β_2. The head coefficient ψ, which is in common use, is defined by $\psi = gH_{BEP}/u_2^2$, and thus is proportional to the BEP value of the dimensionless ratio gH/n^2D^2 which was used in plotting Fig. 9.3. Given the shaft speed and the impeller diameter, the head coefficient can be used to calculate H_{BEP}. It should be noted that as the outlet blade angle increases so does the head produced. Values of β_2 usually range from 15° to 35°. Slurry pumps commonly have values around 30°.

Slurry pumps require thick sections and flow passages capable of passing large spheres, and as a result they have coefficient values which differ from those of water pumps. It is likely, for example, that a slurry pump will require larger impeller outlet widths than a water pump. Some manufacturers, in addition to large impeller outlet widths, also make the meridional section outlet concave as shown for Impeller 3 in Fig. 9.7. The intent of this modification is to create a wear-reducing flow pattern in the form of a double inward spiral inside the casing.

Hergt *et al.* (1994) investigated the impeller outlet velocity and its effect on wear of the three different types of impellers shown on Fig. 9.7. Impeller 1 is that of a conventional water pump, 2 represents a normal slurry pump and 3 is the special slurry type noted above. This research demonstrated that the concave outlet impeller provided no improvement in wear, and that Impeller 2 gave best efficiency and most favourable wear characteristics.

The combined action of the impeller and casing determine the location of the pump BEP. The larger-than-normal impeller outlet widths encountered in slurry pumps tend to shift the discharge at BEP to larger values, but this shift can be compensated by changes in the geometry of the casing, as shown by Worster (1968).

Wear in the casing must also be considered here. Depending on the operating circumstances and the best economic compromise between efficiency and wear, casings can have various sections -- volute, semi-volute or even annular. This point is discussed in greater detail in Chapter 11.

The head coefficients for twelve modern slurry pumps are shown on Fig. 9.8, a plot of ψ *versus* specific speed. This figure is based on Addie & Helmly (1989), where a more detailed description can be found. The table within the figure indicates impeller diameter D_2 (in inches) and outlet vane angle β_2. The use of Fig. 9.8 is shown in Case Study 9.1. Data reported by Visintainer & Addie (1993) provide details of efficiency, NPSH and wear performance for a large modern dredge pump.

In the idealised case, an impeller with a large number of infinitely-thin frictionless vanes would produce the highest efficiency in a pump. In practice, for a water pump this number ranges from five to nine. In a slurry or sewage pump it may be reduced to three or four in order to pass coarse solids and accommodate extra vane thickness. Fewer vanes result in a steeper H-Q curve and some reduction in efficiency. The reduction in efficiency can be held to 1 or 2% in most cases. The number of vanes to be used must be decided after considering the size of solids to be passed, the vane thickness, the location of the inlet and the overall design of the vane shape. To hold the efficiency loss to a minimum, the vanes should have the correct inlet angle for shock-free entry of the fluid at the design point, the outlet angle should set to give the desired performance, and the shape between the inlet and outlet should minimise the rate of change of velocity.

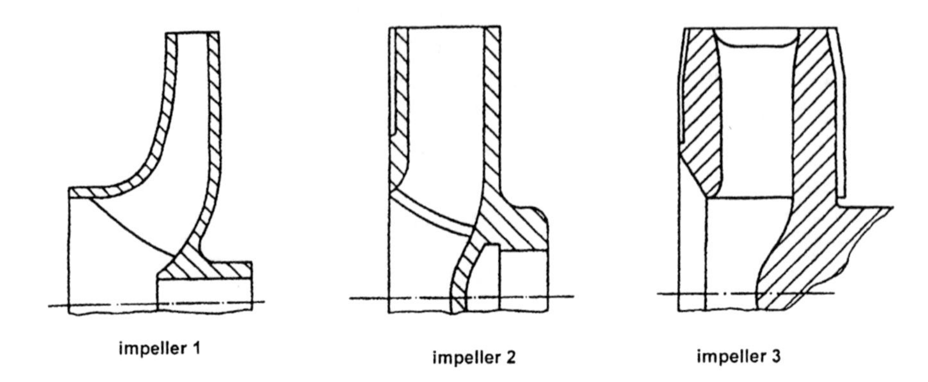

Figure 9.7. Meridional sections of three impellers

In reality the meridional section of the impeller, and the location of the inlet edge, almost always impose a flow across this edge that is a combination of axial and radial motion. As the tangential velocity of the inlet edge of the vane varies, the inlet angle that gives shock-free entry will also vary. This implies that twisted vanes are required for highest efficiency, and that radial vanes are necessarily a less efficient compromise solution. The suction performance of a pump is also strongly influenced by the location of the impeller-vane inlet edge in the cross-meridional section and by the inlet area. Placing the vane inlet at a small radius can give good performance, but if the radius is too small, the vane thickness can choke the eye. Conversely, location of the vane inlet at a large radius increases the tangential velocity and thus the chance of cavitation.

In large pumps the reductions in performance which were mentioned above can be significant, but where the pump size is small and the pump specific speed is low the disadvantage of having a radial vane may be acceptable, and the higher manufacturing costs of twisted vanes could outweigh their advantages. Slurry pump designs vary as a result of the intended service, the manufacturer's design expertise and even the vintage of the design. In very general terms it could be said that older impeller designs usually have simpler 'square' meridional sections with radial vanes that are relatively easy to mould. Meridional sections are shown schematically on Fig. 9.9 where the symbol RV refers to radial vanes. More modern low-to-medium specific speed designs, of the type labelled ME on the figure, often retain simple near-square meridional sections but employ modern twisted vanes, while higher specific speed larger and/or lighter service impellers are usually of the HE type.

Slurry pump casings vary from the true volute water pump T type shown on Fig. 9.9 through the semi-volute C type to the essentially annular A type. There are also a few examples of the OB type with recessed tongue and special extended necks intended for limiting wear in the tongue area due to excessive recirculation (which itself is symptomatic of misapplication).

Each combination of the types illustrated on Fig. 9.9 has its own hydraulic performance and wear characteristics. The HE/T combination generally has the highest performance but is not necessarily the most forgiving for wear, whereas the ME/C combination is capable of respectable efficiency while at the same time having more predictable wear performance. The wear of various types of casings and impellers is discussed further in Chapter 11. For additional information on the selection of different hydraulic types see Addie *et al.* (1995).

The maximum efficiency obtainable from properly-designed water pumps is given in the literature, for example by Karassik *et al.* (1986). Slurry pumps rarely achieve these values because of their larger mechanical losses (due to relatively larger shafts and bearings) and the hydraulic compromises necessary to improve wear life. With the use of the latest design technology, including numerical methods of hydraulic design (see Chapter 11), it is now possible to get to within a few percent of those values and still have a slurry pump which gives good wear performance.

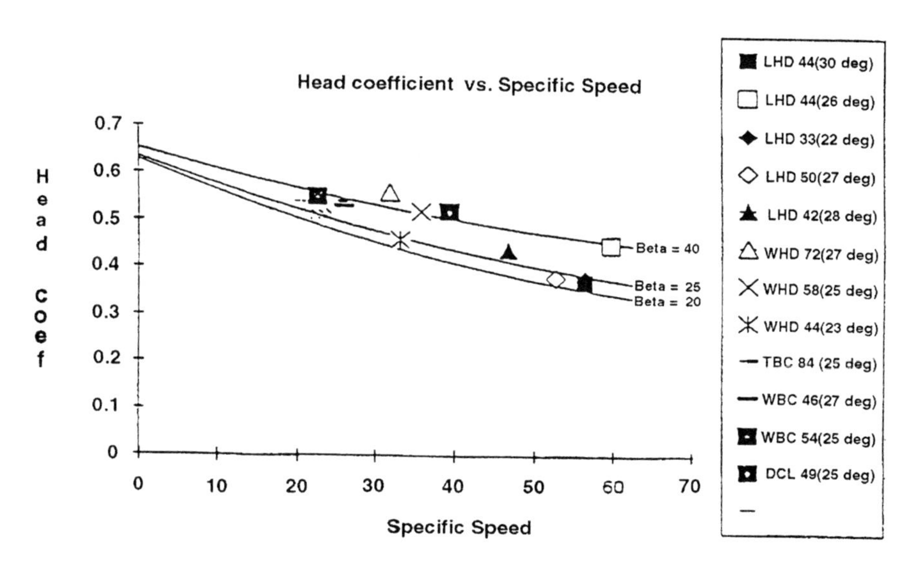

Figure 9.8. Head coefficient as a function of specific speed (based on m^3/s, m and revs/min), after Addie & Helmly (1989)

Figure 9.10 shows the efficiencies of the twelve slurry pumps mentioned in connection with Fig. 9.8 (Addie & Helmly, 1989). The curves on the figure represent the efficiency expected of water pumps of similar size, from Karassik *et al.* (1986). The abscissa of the figure is customary specific speed, based on m^3/s, m and revolutions/minute.

As indicated in earlier chapters - see, for example, Section 2.3 - pumps must be selected by matching their head-discharge performance to the requirements of the piping system. Figure 2.3 shows that the intercept of the pump characteristic with the system characteristic defines the operating head and discharge. For settling slurries in particular, selection of appropriate operating conditions raises special considerations, to be discussed in Chapter 13. Pump selection and cost considerations will be covered in Chapter 14. In the case of long lines, the total system head will be more than can be handled by a single pump. It is then necessary to use several pumps in series. This case, and the less common case of several pumps in parallel, is considered in Chapter 15.

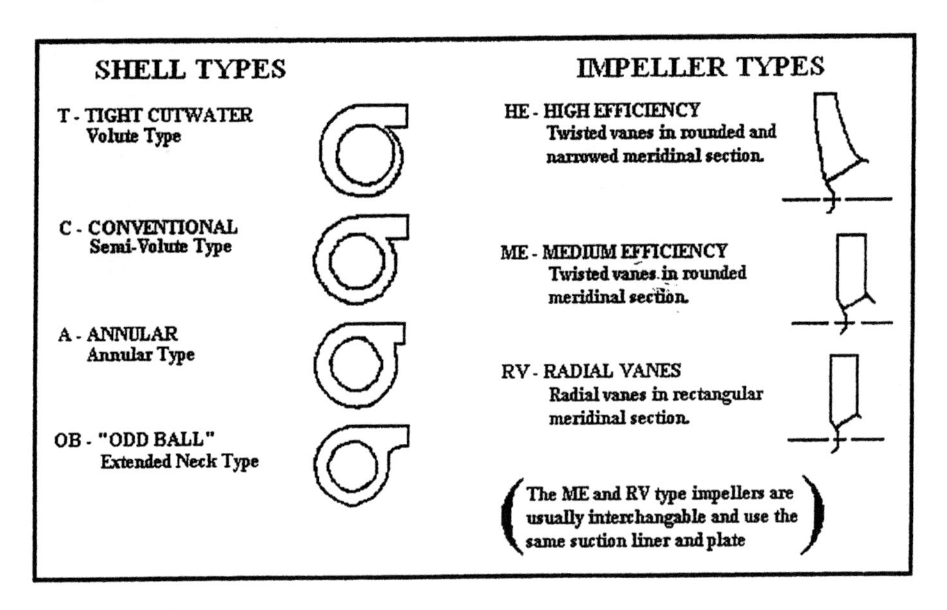

Figure 9.9. Types of shells and impellers.

Both the hydraulic designer and the pump user may be interested in modifying a pump's behaviour by 'trimming' the impeller, i.e. removing metal to reduce its diameter rather than using a scaled-down version which maintains geometric similarity. The effect of trimming can be obtained from Eq. 9.13. As the velocity u_2 varies in direct proportion to impeller diameter, the ratio gH/n^2D^2 retains its significance, so that the scaling law for head (Eq. 9.7) applies in the usual fashion.

For discharge a different scaling law is required. Assuming that the trimming operation involves a relatively minor change in diameter, the impeller outlet angle β_2 and the impeller outlet width b_2 remain fixed (Lazarkiewicz & Troskolanski, 1965), and the discharge ratio in Eq. 9.13 becomes Q/nD^2 rather than Q/nD^3 (as would be the case for geometrically similar impellers). On this basis Q for a trimmed impeller is proportional to D^2, equivalent to reducing the exponent of the diameter from 3 to 2 in Eq. 9.6. An alternative relation for discharge that is in customary use, particularly in the United States, employs an empirical proportionality between Q and D (Stepanoff, 1957, Kittredge & Cooper, 2001). Secondary adjustments which are applied to both relations ensure that the results are essentially equivalent from a practical viewpoint.

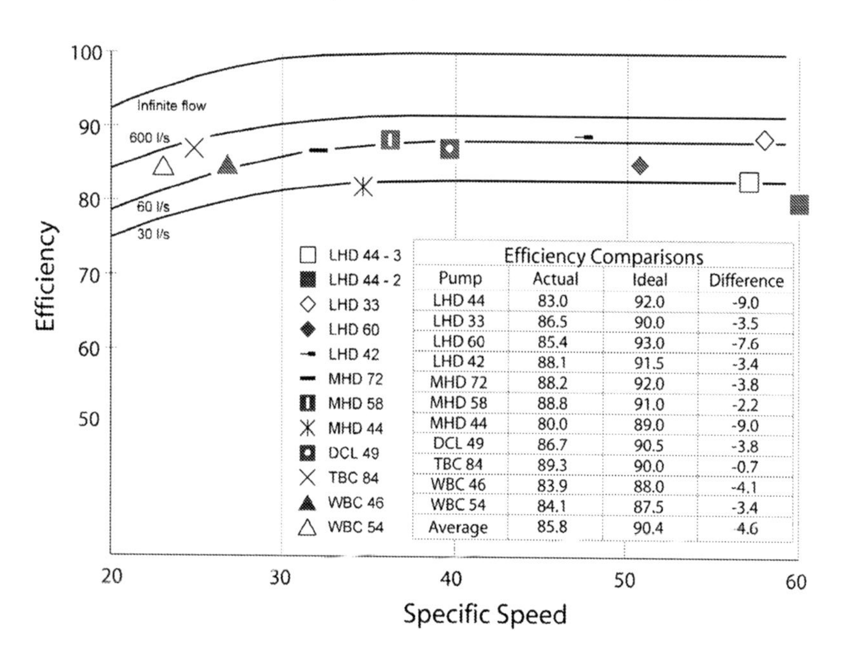

Efficiency Comparisons

Pump	Actual	Ideal	Difference
LHD 44	83.0	92.0	-9.0
LHD 33	86.5	90.0	-3.5
LHD 60	85.4	93.0	-7.6
LHD 42	88.1	91.5	-3.4
MHD 72	88.2	92.0	-3.8
MHD 58	88.8	91.0	-2.2
MHD 44	80.0	89.0	-9.0
DCL 49	86.7	90.5	-3.8
TBC 84	89.3	90.0	-0.7
WBC 46	83.9	88.0	-4.1
WBC 54	84.1	87.5	-3.4
Average	85.8	90.4	4.6

Figure 9.10. Pump efficiency as a function of specific speed (based on m^3/s, m and revs/min), after Addie & Helmly (1989).

9.3 Cavitation and Net Positive Suction Head

In most hydraulic applications the gauge pressure is used, i.e. the pressure above local atmospheric pressure. Negative gauge pressure (suction) is often found at the entry of pumps, and in some circumstances can produce the condition known as cavitation. Cavitation occurs when the local pressure falls below the vapour pressure of the liquid, i.e. the pressure at which it boils, p_v. Cavitation in a pump gives rise to vapour bubbles or 'cavities' which form and collapse. Slight cavitation leads to a reduction in pump efficiency and developed head. Severe cavitation is often audible - the pump 'sounds rough' and can damage the impeller or, more rarely, the casing. Eventually, the pump may cease pumping altogether. Severe surging and water hammer may then result (Carstens & Hagler, 1964). For all liquids, the vapour pressure increases with temperature, so that cavitation is more likely to occur when the liquid is hot. When the vapour pressure is expressed in terms of head of the liquid in question, evaluated using the liquid density at the operating temperature, it is known as the vapour-pressure head, h_v. Values h_v for water are shown as a function of temperature on Fig. 9.11, based on the data in Table 2.1.

Local pressure at some points in the impeller is even lower than the pressure at the 'eye'. Therefore, to ensure cavitation-free operation, every centrifugal pump requires that the absolute pressure at the impeller eye exceeds p_v by a certain margin. By convention, it is expressed in terms of head of the liquid being pumped, using the term 'Net Positive Suction Head' or NPSH to denote the difference between the total absolute head at the pump suction and the vapour-pressure head. If H_s is the suction head (i.e. the pressure head *below* atmospheric measured at the pump suction) and h_a is the atmospheric head (i.e. atmospheric pressure expressed as head of liquid being pumped), then the NPSH available at the pump suction is given by

$$NPSH = \left[\left(h_a - H_s\right) + \frac{V_A^2}{2g} \right] - h_v \qquad (9.16)$$

where $V_A^2/2g$ is the velocity head at the pump suction. If the suction head H_s has been measured, Eq. 9.16 gives the available NPSH directly. Otherwise H_s must be calculated by applying Bernoulli's law to the suction piping system, allowing for head losses due to friction and entry effects. The available NPSH varies with the flow, due to losses in the suction piping which may be calculated directly from the piping geometry and known or assumed friction coefficients. Note that the suction head H_s is not the same as the difference in level from the liquid surface in the sump to the eye of the pump.

If the available NPSH falls below the minimum NPSH required by the pump, then the pump will cavitate. Obviously, the minimum NPSH required to avoid cavitation, i.e. the value at the incipient cavitation point, is an important quantity in selecting a pump or designing a piping system. Although a well-designed pump has good suction performance, the minimum NPSH cannot be predicted with precision (see below). It must therefore be measured for a pump prototype. In determining the incipient cavitation point experimentally it is usual to carry out a series of tests, maintaining the shaft speed and discharge constant for each test but varying the suction head H_S. The test procedures will be discussed in detail in Section 9.5. As the magnitude of H_s is increased from a low initial value, i.e. as the suction pressure is reduced further below atmospheric, there is at first no change in the total dynamic head developed by the pump. However, when cavitation begins, the TDH diminishes. The incipient cavitation point is usually defined arbitrarily by some specified reduction in TDH. A reduction of 3% is usually used to define incipient cavitation, but the figure of 1% is also sometimes used. The difference is normally small, because once cavitation starts it usually increases rapidly. At constant shaft speed, the value of NPSH at incipient cavitation varies significantly with discharge. It is normal to carry out a series of constant-speed tests at different flows, and from these to draw a curve relating minimum NPSH to discharge flow. As the vapour-pressure head is temperature-dependent, the temperature of the water used in the test is monitored, and the appropriate value of h_v is used in calculating NPSH for each test run.

The suction performance of a pump can also be described using σ, a dimensionless cavitation parameter defined as the ratio of NPSH to the total head produced by the pump at the same discharge.

$$\sigma = \frac{NPSH}{TDH} \qquad (9.17)$$

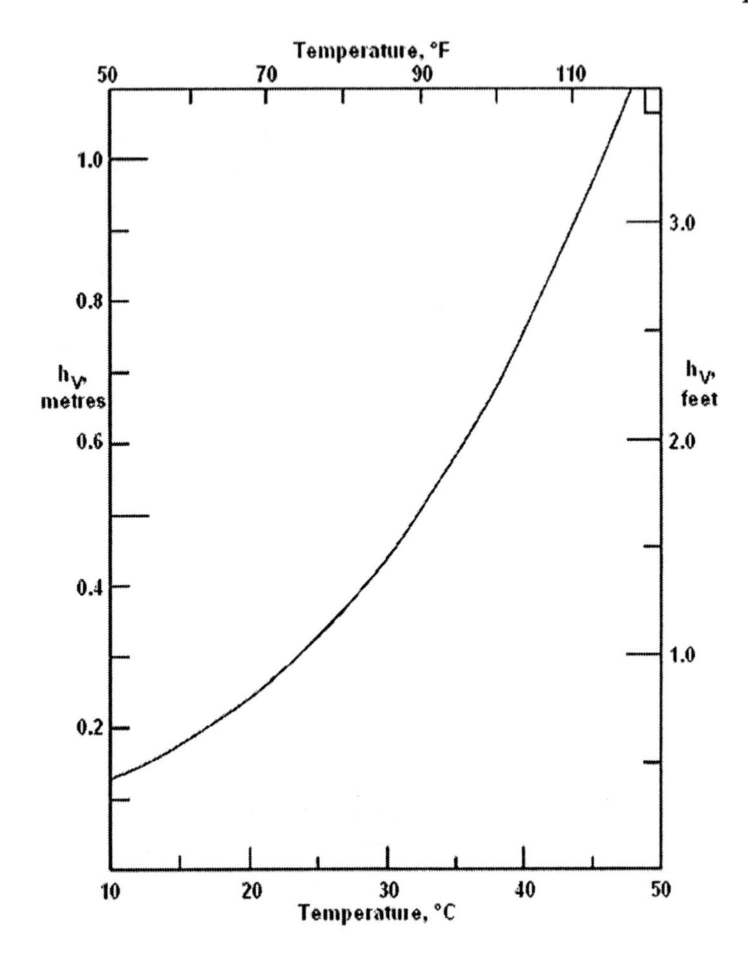

Figure 9.11. Vapour pressure head of water as a function of temperature

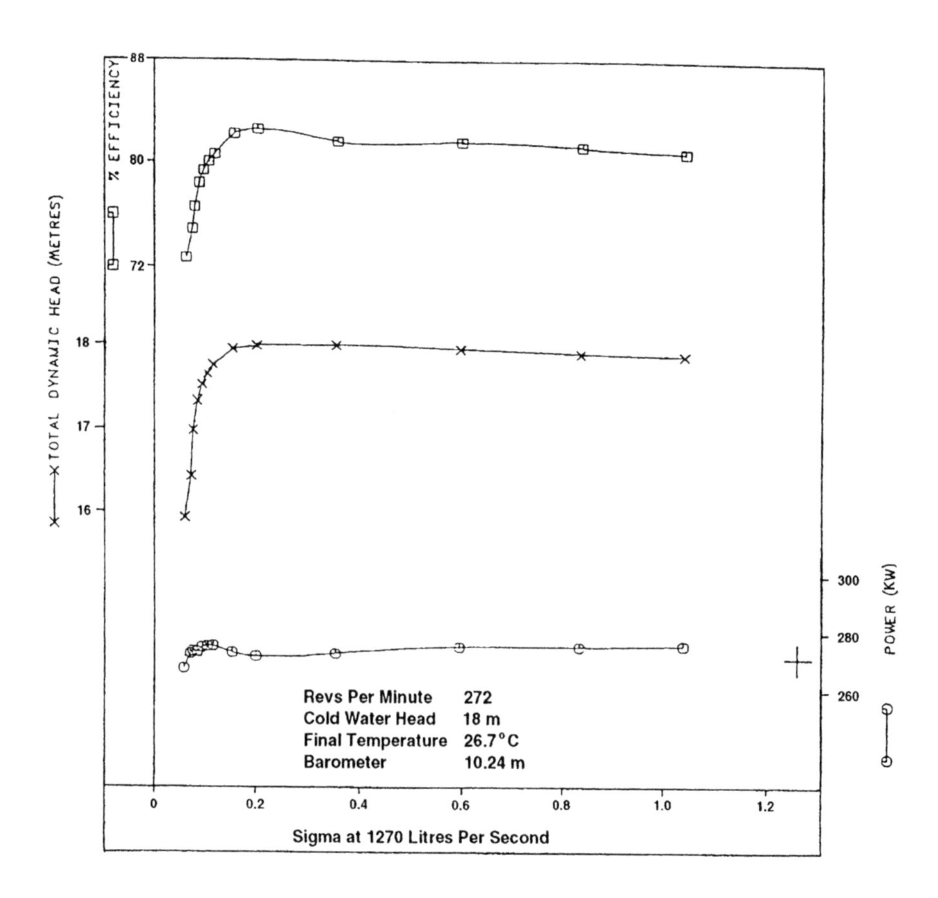

Figure 9.12. Experimental curves for onset of cavitation

Experimental cavitation curves based on this parameter are shown on Fig. 9.12. A representative NPSH curve was included in the pump performance plots shown on Fig. 9.2. Here the variation of NPSH at incipient cavitation was plotted against discharge, and it is worth noting that the curve shows NPSH to increase with increasing flow. This tendency is usually encountered in centrifugal pumps, indicating that discharge must be considered in evaluating NPSH. Angular speed also influences NPSH, but for a pump of a given design σ values depend only on the ratio Q/Q_{BEP}, irrespective of angular speed. An example curve of this type is plotted on Fig. 9.13. Suction performance can also be presented in other formats. For example, the dimensionless axes of Fig. 9.3 can be employed, using NPSH in place of H. Likewise, the concept of specific speed can be adapted for this purpose, defining a 'suction specific speed' in which the total dynamic head in Eq. 9.14 or 9.15 is replaced by the NPSH corresponding to incipient cavitation at the best efficiency point. This formulation should be supplemented by information on suction performance away from the BEP, such as that shown on Fig. 9.13.

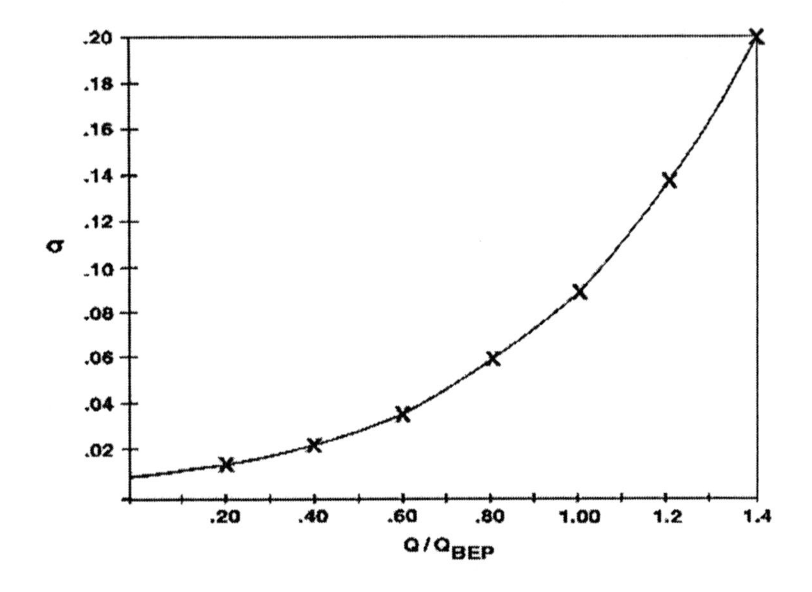

Figure 9.13. Dimensionless suction plot

The NPSH required (or the required value of σ) is usually specified by the pump manufacturer on the basis of tests or experience with similar machines, and depends on pump type and design. The required NPSH depends on a range of factors including eye diameter, suction area of the impeller, shape and number of impeller vanes, area between these vanes, shaft and impeller hub diameter, specific speed, and shape of the suction passages. Data for well-designed water pumps of a number of types and range of specific speeds are published by the Hydraulic Institute (1977). Results for the twelve slurry pumps tested by Addie & Helmly (1989) are shown here on Fig. 9.14, in the form of σ (at best efficiency point) as a function of the specific speed of the pump.

The impellers of slurry pumps have traditionally incorporated square meridional sections at inlet and relatively thick radial vanes, and both these conditions tend to affect suction performance adversely. Slurry pumps, however, are generally designed to run at relatively low speed in order to reduce wear. In pipeline applications with sumps to provide positive suction pressure, the NPSH available is commonly adequate for the pump regardless of the design. In pit and dredging service this is less likely to be the case, and for these applications the suction performance of the pump may have a great effect on the reliability and production rate of the system. This point was shown in the case study of a dredge ladder in Chapter 8, and will be discussed further in Chapter 12. It is recommended that every slurry pump application should be checked to confirm that the NPSH available in the system is greater than the NPSH required by the pump. If tests for the required NPSH have not been made with the slurry of interest, it will be necessary to adapt the results of water tests by expressing all heads, including the atmospheric and vapour pressure heads, as height of slurry rather than water. Further effects of slurries on NPSH will be mentioned in Chapter 10.

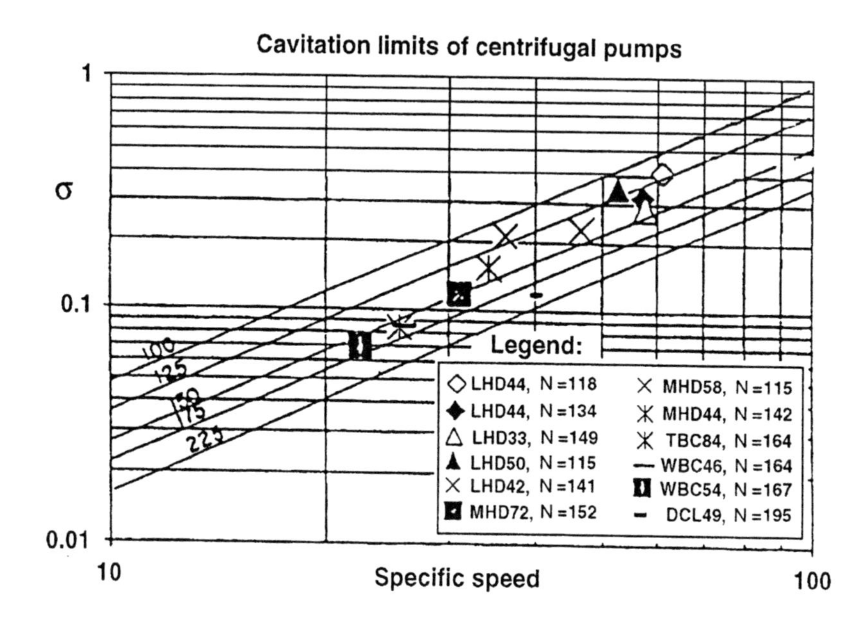

Figure 9.14. Cavitation parameter as a function of specific speed (based on m^3/s, m and revs/min), after Addie & Helmly (1989)

9.4 Mechanical Design

The mechanical design of a centrifugal pump must deal with the loads generated by hydraulic and dynamic forces, so that the pump will give trouble-free service for a wide range of hydraulic duties and operating conditions. The mechanical loads on a pump assembly may be divided into two main types. The first acts on the rotating assembly, giving rise to shaft and bearing loads. The second, caused by the pressure within the pump, acts on the casing assembly.

The loads on the rotating assembly comprise the torque load on the shaft, the hydraulic radial load on the impeller, the radial load due to any imbalance, the hydraulic axial-thrust load, the load due to the impeller weight, and any load due to a coupling or V-belt drive. The torque load during operation can be calculated from the input power and shaft speed using Eq. 9.4. However, the shaft must usually be designed for a torque larger than the operating value, reflecting the fact that the motor is usually oversized to some extent so that higher torques can occur during transients such as start-up.

The shaft load next in importance after torque is normally the radial hydraulic load, caused by the pressure distribution on the impeller. This load depends on the hydraulic design, rotational speed and size of each pump. The design value and the associated safety factors must be based primarily on experience, but various empirical tools are available to assist the designer (see, for example, Hydraulic Institute, 1977; Biheller, 1965; Agnostinelli *et al.*, 1960; and Addie & Sellgren, 2001).

The hydraulic thrust load depends on the pressure on the impeller and the effect of seal rings and clearing vanes. It can be estimated using simple empirically-based calculations, and is found to be most significant for multistage pumps. In most cases, the thrust load is applied directly to the pump thrust bearing, determining its size and life. Most slurry pumps are single stage, and for this configuration the main significance of the thrust load is in determining the life of any thrust bearing. Radial sealing rings are almost never used in slurry pumps because of wear considerations, but clearing vanes behind the back shroud are not unusual and can reduce the thrust load significantly. The thrust load is much greater for open impellers, i.e. impellers with no front shroud, which are commonly used for light duties or ‘stringy' materials.

Impeller front-shroud clearing vanes are common and are useful in that they generate head that reduces the leakage flow (and resulting wear) through the gap. Large front clearing vanes can reduce wear by as much as 50%, albeit at the cost of 2-5% in efficiency.

It has been shown that the eroding particles travel down the stationary face through the nose gap. A special impeller nose face shape and/or diverter that pushes particles into the rotating impeller surface (from where they will be thrown out), can reduce wear in the nose face area, and has been shown to reduce wear by as much as 50% or more.

The radial load due to the impeller imbalance is estimated from the pump speed and the balancing limits allowed on the impeller and shaft. In slurry pumps, which are usually of low specific speed and overhung configuration, static balancing is generally sufficient provided that the impeller is not excessively wide in relation to its diameter. The load due to misalignment of a coupling or a V-belt drive depends on the size and type of installation, and is usually best estimated from recommendations of the supplier of the drive.

Impeller attachment by Acme-type screw has generally been found best capable of carrying the heavy loads required, and the connections can be manufactured economically in the hard materials most commonly employed in impellers. The large loads associated with heavy-duty service require roller bearings with separate roller thrust bearings. Designs with single-row deep-groove ball bearings are suited only for light-service pumps.

The absence of radial sealing in slurry pumps allows shaft deflections larger than those found in water pumps, and these may limit the life of stuffing boxes and mechanical seals. Newer and better designs, however, tend to shorter shafts and smaller deflections, extending the seal life. A conventional packed stuffing box is still the simplest and most common rotating-assembly seal. Configurations are similar to those used for water pumps, with the lantern-ring supplying a clear-water flush to the centre for minimum dilution, or to the product side for maximum life. An expeller-type seal is popular where dilution of the product is unacceptable. These seals are limited to one stage and involve additional efficiency losses which usually range to 3% or more.

Mechanical seals are now available that take their coolant from the product and operate with no clear-water flush; they are mostly of the single partially-balanced type. In some cases, where a pump may run dry, a double type of mechanical seal must be used. At present, mechanical seals are the preferred shaft sealing method for fine-particle light-duty service such as pumps for flue gas desulpherisation. These seals are also fairly widely used in the aluminium industry for red-mud pumping service. A mechanical seal described by Maciejewski *et al.* (1993) is used for pumps handling oil-sand tailings at pump discharge pressures up to 2400 kPa, with d_{50} about 120 microns and average slurry density near 1600 kg/m^3. In this case the process fluid is at an average temperature of about 55°C and some 0.01 L/s of water is used for external cooling. In this application, representative of the limits of mechanical seal technology at the present, the average seal life is about 3000 hours.

We turn next to the materials used for wetted surfaces. For the coarsest slurries, hard metal is common, principally the so called NiHard, high-chrome or chrome-molybdenum alloys of 600-700 Brinell hardness. Impellers and shells using elastomers such as rubber, neoprene, and urethane tend to be limited to impeller tip speeds less than about 23 m/s (75 ft/sec), although this can rise with stiffer elastomers at some expense to wear life. Impellers employing elastomers require higher available NPSH because of the thicker impeller vane sections needed, and this condition may limit their use, for example in pumping flue-gas desulphurisation slurries. Pump casings made entirely of elastomeric materials have insufficient strength to withstand the pressure loads, so it is necessary to have an outer casing of some sort, a configuration which is commonly called a double-wall design. For versatility, some manufacturers make these designs for interchangeable wetted internal shell components of hard metal, rubber or urethane as the service warrants. For very large slurry pumps the double-wall configuration is heavy and costly. As a cost-effective design for this case, it is worth considering tie-bolt construction using high-tensile outer plates over a hard-metal shell. On the other hand, single-wall designs with the main shell casing in hard metal are simple to construct and maintain and thus are very popular. These can be used unlined or provided with liners of bonded rubber or other material. Ceramic materials are now available that are capable of several times the life of both elastomer and hard-metal components. At the moment, the cost effectiveness of these materials is such that their use tends to be limited to areas of high wear and other selected areas of wetted surfaces. Further information on wear and wear-resistant materials is given in Chapter 11.

In a shell of single-wall metal design, the wearing component must carry the water load. Wear tends to progress until leakage of the shell occurs, but the wear is almost always localised so that leakage occurs before the structural integrity of the assembly is affected. Single-wall designs may easily be ribbed or thickened locally to increase strength and wear life in a particular area. In small slurry pumps, minimum casing thicknesses usually produce safe designs of adequate pressure rating. In larger size units, it is necessary to carry out a full stiffness analysis and, in some cases, to verify stress levels by strain-gauge testing.

Pump drives are considered in Section 12.3. It may sometimes occur that an existing pump or drive must be used, with no possibility of free selection. Remedial action may then be needed. For example, if a drive of given rating is to be used with an existing pump, it may be necessary to trim the impeller in order to reduce the power required by the pump to match that available from the drive. The revised head-discharge performance of the pump must then be calculated, as explained in Section 9.2, to find the resulting operating point as defined by the intersection of the pump and system characteristics.

The size of the pump branches must also be checked: although they need not be exactly the same diameter as the pipeline, the difference should not be too great and allowance must be made for the head loss associated with flow through the expansion or contraction. If the branches are too large, solids deposition can occur (see Case Study 8.1). If they are too small, wear may be excessive. It is not unusual for the suction branch of a centrifugal pump to be one nominal size larger than the discharge branch. This configuration can cause difficulties when pumping settling slurries (see Section 12.2).

9.5 Performance Testing

The hydraulic design of centrifugal pumps is an extremely complex endeavour, based on a combination of scientific principles, and technological 'know how' and experience. In order to verify design methods, and to provide the data necessary to produce new and improved designs, performance testing is necessary. Accurate testing requires careful control of piping layout, instrumentation and test procedures. Representative testing standards are published by various organisations including the International Standards Organisation (1977), the Hydraulic Institute (1977), the British Standards Institution (1968) and the Standards Association of Australia (1968). The test procedures used for slurries are basically no different from those used for water, although precautions are necessary in taking some measurements.

The temperature of the fluid being pumped must be monitored throughout all tests. Standard tests are usually based on water, but investigations for specific applications are also carried out using the fluid or slurry of interest to give results of specific but limited applicability. Testing should also include a mechanical check of internal clearances, coupling alignments and gland operation. Laboratory or works tests can be more closely controlled, and thus are preferred to site tests which almost always give a less accurate indication of pump performance.

Because any pump may be used in a variety of systems, a test should include points at several different values of Q, allowing the shape of the characteristic curves to be fully defined. Figure 9.1 showed a representative pump head-discharge characteristic for constant-speed operation. The pump suction performance characteristics may also be required, usually in the form of curves such as those shown on Fig. 9.12 and Fig. 9.13.

As test points need to be taken under steady flow conditions, a closed loop is necessary for laboratory testing. Piping arrangements for the loop depend on the type and configuration of the pumps to be tested. Both pressurised loops and open loops are in common use. A representative pressurised system is shown schematically on Fig. 9.15, and Fig. 1.1 illustrates some features of a system of this type which is located at the GIW Hydraulic Laboratory. The advantage of the pressurised arrangement is that the absolute pressure in the system can be varied. Testing under positive pressure eliminates air leaks in any of the gauge and manometer lines. Changes in the system pressure do not affect the total developed head of the pump, but lowering the pressure in the system lowers the available net positive suction head, facilitating suction testing. The tank is necessary to trap any air in the system. Some method of cooling may be required if a test is lengthy, or if high head and power are involved.

If the test has to be carried out by drawing water from an open tank, care must be taken that the water does not become unduly aerated on returning to this tank. With an arrangement of this type, NPSH testing necessitates lowering the level in the tank, which is not desirable as it increases the time for a test.

The diameter of the piping should preferably not be less than that of the suction branch of the pump being tested. Bends or tapers directly on the discharge and suction flanges of the pump are not permissible. Variation in system resistance is best achieved by a valve near the discharge of the pump. Throttling on the suction should be avoided, as it might lower the suction pressure, promoting cavitation. Air-bleed valves are necessary at several locations around the loop, for use in priming and after start-up.

Instrumentation is required for measuring the flow through the system, the pressures on both sides of the pump, and the power supplied to the pump. Instrumentation for slurry systems will be discussed in detail in Chapter 12. Flow measurement may be carried out in several ways, but many procedures used for water testing are poorly suited for slurry tests. An example is direct volumetric measurement using a tank for which the volume of the contents is calibrated as a function of liquid depth. This method is potentially very accurate, but the surface area of the tank must be large to minimise head changes during the time required for an accurate measurement. Thus the tank volumes become very large, and the large quantity of slurry needed renders this method impractical for slurry testing. Other traditional methods of flow measurement which are appropriate for water but not for slurries include V-notch weirs and orifice plates. The square-edged orifice is accepted as a standard technique for water flow measurement, but is not generally applicable to slurries because of wear. Venturi meters and nozzles can be used more readily for slurries, but have the disadvantage of limited range, while the Venturi also suffers from relatively high cost. For slurry flows accuracy is reduced, and may be influenced by wear in the instrument.

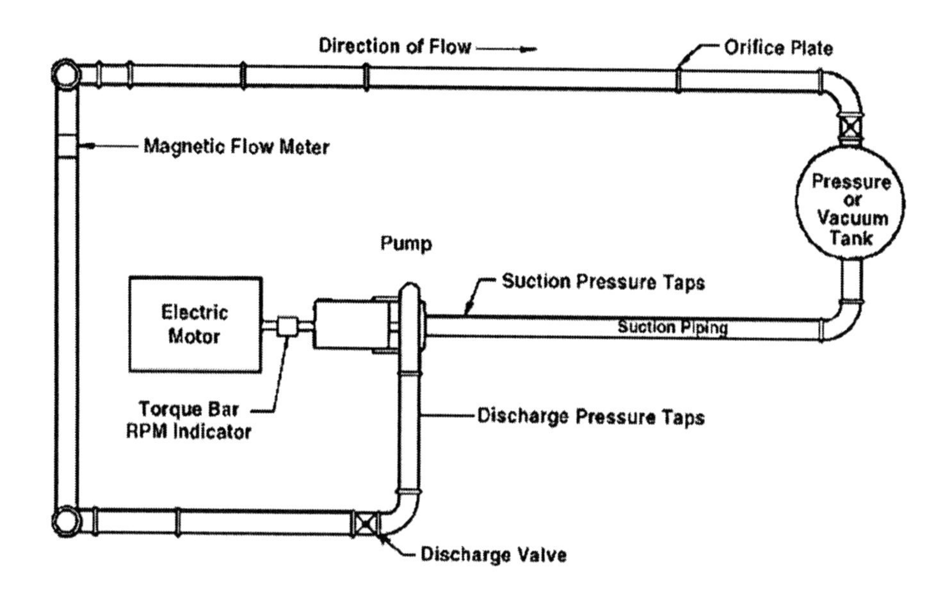

Figure 9.15. Layout of pressurised test loop (schematic)

Measuring the differential head across a bend gives a method of determining discharge which is cheap and does not impede the flow. As with the Venturi meter, the flow rate is proportional to the square root of the head difference, but this difference is small and must be measured accurately. The calibration coefficient for the bend depends on both the radius and the pipe roughness, and is best determined from comparison with independent flow measurements. This method is useful for slurry flow, but loses accuracy when the slurry particles are large.

The instrument of preference for measuring slurry discharge is the magnetic-flux flow meter, because this type of device provides no interference with the flow. It is claimed also that these meters may be located close to bends and in any position. Accuracy varies between models, but is often claimed as 0.5% of full-scale reading. In practice, magnetic flowmeters are susceptible to stray electrical currents and earthing problems, and may have their accuracy affected by rust or dirt on the probes. Wherever possible they should not carry significant pipe loads, as this can also affect accuracy. It is recommended that some means of calibration be provided in addition to calibration relations supplied by the manufacturer. Magnetic meters are probably the most common and accurate method of measuring slurry flows. For settling slurries, the meter should ideally be located in a vertical pipe section, to avoid possible problems associated with deposition of stationary solids.

In present practice, an electronic pressure transducer is used to measure the pressure at the pump discharge. The instrument should be valved off when not in use. The tapping points should be four-tap ring types located 1 to 2 pipe diameters from the discharge of the pump. Suction pressure was traditionally measured using a mercury U-tube, but now pressure transducers are also used here. Tapping points, of ring type, should be located two to three pipe diameters from the suction flange. The various transducers used in pressure measurement should be selected and arranged so that an accuracy of ±1% may be obtained. The lines connecting them to the pump system should be kept full of water, with provision for draining prior to testing. For slurry duty, it is advisable to provide settling pots to collect any particles which enter the lines, and a clean-water flush to avoid clogging.

The total dynamic head developed by a pump is equal to the pump discharge pressure less the suction pressure plus any velocity-head difference given to the water, expressed in units of height of liquid (see Eq. 2.10). If the inlet and outlet branches are the same diameter, the velocity-head difference will be zero. It the suction and discharge gauges are at different levels, then a correction must be applied to bring all readings to the datum of the pump centreline. The output power provided by the pump is based on the product of flow rate and head which occurs in the expression $\rho_f gQH$ (Eq. 9.2).

Measurement of power input to the pump may be carried out by three main methods. The first involves using a precision torque bar and multiplying by pump speed to determine input power to the pump. This is the preferred method because electrical calibration is generally accurate and reliable, but has the disadvantage of requiring a special drive mounting for the test. The second method of measuring power input involves a DC motor dynamometer drive (or a similar unit) set up to measure torque and rpm at the motor. Because of the expense of such units and possible difficulties with availability, this method is usually limited to pumps with small power requirements. It needs a separate dedicated drive mounting for the test. The third method involves calculation from the power input to an electric motor using certified or other reliable motor efficiencies. This method requires the use of high-accuracy potential and current measuring transformers in conjunction with a two-element type wattmeter. It has the advantage of providing the overall efficiency for the entire pump and motor unit, but gives less accurate measurements of the efficiency of the pump itself.

In other circumstances, drive losses must be estimated. If a V-belt drive exists with no means of measuring the associated loss, the drive train loss can be estimated as 3 to 10%. In the case of fluid or magnetic coupling, a similar allowance should be made.

In addition to these measurements, slurry density must be monitored throughout a test. This can be accomplished using a nuclear density gauge, or the inverted U-tube device described in Chapter 8 and discussed further in Chapter 12. The output of an H-Q test should consist of readings of suction pressure, discharge pressure, flow, input power and shaft speed. Total developed head, pump efficiency and input power are then calculated and plotted against flow. Pump speed should be kept essentially constant for each series of readings, but if necessary values of Q, H and power can be adjusted to different speeds using the affinity laws (Eqs. 9.6, 9.7, 9.8). The readings at each flow should be averaged from a series of observations. The water in the test loop should, if possible, be kept below 30°C, and the time of testing should be kept short but must be sufficient to allow steady flow for each reading. The number of points required will vary, but at least eight will be needed for a full curve.

Simple data-acquisition systems make use of programmable controllers to collect and display the results of pump performance tests. The more sophisticated systems now used by most pump manufacturers include analogue-to-digital conversion units in conjunction with electronic transducers, and also employ computers to process, store, display, print and plot the results of performance tests. For suction performance testing, the most satisfactory equipment is a closed-loop system with provision for deaeration and pressure control. A series of tests should be performed; for each test the angular speed and flow are maintained constant, while the suction pressure is reduced until cavitation occurs. The net positive suction head is calculated from the pump suction reading and the velocity in the suction branch using Eq. 9.16, and is employed to obtain the cavitation ratio σ (Eq. 9.17). As noted in Section 9.3, incipient cavitation is conventionally considered to occur at the σ value corresponding to a 3% loss of developed head. Any gas dissolved in a liquid will come out of solution when the pressure is reduced, and any entrained gas will expand. This effect is significant when pumping at low available NPSH, causing a reduction in liquid capacity pumped. Thus, testing for NPSH should generally be carried out on degassed liquids.

The effect of solids on pump performance is discussed in the following Chapter. Further aspects of instrumentation for slurry flows will be discussed in Chapter 12.

9.6 Case Studies

Case Study 9.1 - Sizing a Water Pump

A pump is required for a clear-water flow of 2.5 m^3/s (2500 L/s) and 50 metres of head. It is to be operated at 400 rpm because of an existing driver.

Estimate the impeller diameter, the expected efficiency (assuming operation at best efficiency point), and the required net positive suction head.

(a) From the speed, head and flow, the design specific speed of the pump (at the best efficiency point) will be

$$N_s = 400 \times \frac{\sqrt{2.5}}{50^{0.75}} = 33.64$$

An impeller vane outlet angle of 25° is an appropriate selection. This is used in Fig. 9.8, together with the specific speed, to obtain a value of 0.463 for the head coefficient, which is defined as

$$\psi = \frac{gH_{BEP}}{u_2^2}$$

Here H_{BEP} is the head at the best efficiency point, and u_2 is the tangential velocity of the impeller outlet, equal to $\frac{\pi DN}{60}$. Rearranging gives

$$D = \frac{60}{\pi N} \times \sqrt{\frac{gH}{\psi}} = \frac{1}{\pi} \times \frac{60}{400} \times \sqrt{\frac{9.8 \times 50}{0.463}} = 1.55\ m$$

(b) From Fig. 9.10, we can interpolate from the solid lines that a water pump of this specific speed and of this size should achieve an (ideal) efficiency of about 92%, but a well designed slurry pump will have efficiency approximately 4.5% less, i.e. 87.5%.

(c) From Fig. 9.14, the sample slurry pumps shown indicate that it is reasonable to expect a σ value of 0.15.

Therefore

$$NPSH = 0.15\,(50) = 7.5\ m\ of\ water$$

To operate properly, the pump should therefore be provided with at least 7.5 metres of NPSH.

Case Study 9.2 - 'Trimming' a Pump Impeller

A pump with characteristics shown on Fig. 9.2 is to be operated at 500 rpm on water at 800 litres/s against a system head of 45 m. Calculate the reduced impeller diameter necessary, neglecting any slip losses. Determine the power needed, and the NPSH required.

(a) For impeller turn-down, as noted at the end of section 9.2, the head equation is

$$\left(\frac{H_1}{H_2}\right) = \left(\frac{D_1}{D_2}\right)^2$$

where D is the impeller diameter and 2 and 1 refer to initial and final values. For the discharge relation, the equation in customary use in the United States (Kittredge & Cooper, 2001) will be employed, i.e.

$$\frac{Q_2}{Q_1} = \frac{D_2}{D_1}$$

Using Fig. 9.2 as a guide we may find (by trial and error), the ratio of D_1/D_2 which transforms the design Q and H to a point on the 500 rpm head characteristic. This ratio was found to be 1.091, giving

$$H = 45 \times (1.091)^2 = 53.6\ m$$

and

$$Q = 800\,(1.091) = 873\ litres/s$$

where these values refer to the head and flow at the point on the 500 rpm head line.

Using the same ratio we can calculate the trimmed impeller diameter as

$$D_2 = \frac{1170}{1.091} = 1072\ mm$$

In this case, the turndown diameter ratio is 0.916 and is well above the value of 0.8 regarded as the limit where significant slip losses reduce head and efficiency.

(b) Thus negligible slip will occur and the efficiency can be taken to be the same as that found by interpolation at the 500 rpm take-off point on Fig. 9.2. This efficiency is 82.8%.

In this case, the pump input power for the water-only duty is

$$\frac{800\,(45)\,(9.81)}{1000\,(0.828)} = 426\ kW$$

(c) The pump suction geometry and speed are the prime factors governing the suction capability of the pump. Provided the turndown is kept small as in this case, the NPSH at the duty for all practical purposes will be that of the take-off point of 864 litres/s flow, 52.5 m head at 500 rpm.

Figure 9.2 only provides us with NPSH for 600 rpm. It is necessary therefore to scale this back to 500 rpm. Figure 9.13 shows a plot of σ versus Q/Q_{BEP} derived directly from Fig. 9.2. The take off point is at 82% of the BEP flow. This corresponds to a σ of 0.07 on Fig. 9.13. The NPSH at the duty therefore equals 0.07 (45) or 3.2 m.

REFERENCES

Addie, G.R. & Helmly, F.W. (1989). Recent Improvements in Dredge Pump Efficiencies and Suction Performances. *Proc. Europort Dredging Seminar*, Central Dredging Association.

Addie, G.A. & Sellgren, A. (2001). Application and construction of centrifugal solids-handling pumps. pp. 9.351-9.367 of Karassik et al. (2001).

Addie, G., Dunens, E. & Mosher, R. (1995). Performance of High Pressure Slurry Pumps in Oil Sand Tailing Application, *Proc. 8th International Symposium on Freight Pipelines.*

Agostinelli, A., Nobles, D. & Mockridge, C.R. (1960). An Experimental Investigation of

Radial Thrust in Centrifugal Pumps, *Transactions of the ASME Journal of Engineering for Power, **82**, 120-6.*

Biheller, H.J. (1965). Radial Force on the Impeller of Centrifugal Pumps with Volute, Semivolute, and Fully Concentric Casings. *Journal of Engineering for Power*, July, pp. 319-323.

British Standards Institution (1968). *British Standard BS 599: Methods of Testing Pumps.* British Standards Institution, London.

Carstens, M.R. & Hagler, T.W. (1964). Water hammer resulting from cavitating pumps. *Proc. Am. Soc. Civil Engrs.*, Vol. 90, No. HY6, pp. 161-184.

Cooper, P. (2001). Centrifugal pump theory. Section 2.1 of *Pump Handbook, Third Edition,* Eds. Karassik, I.J., Messina, J.P., Cooper, P. & Heald, C.C., McGraw-Hill, New York, NY, pp 2.19-2.22.

Euler, L. (1756). Théorie plus complette des machines qui sont mises en mouvement par la réaction de l'eau. Mémoires de l'academic des sciences de Berlin, Vol. 10, pp. 227-295. Reprinted (1957) in *Leonhardi Euleri Opera Omnia,* (Ed. J. Ackeret), Ser. 2, Vol. 15, pp. 157-218. Societas Scientiarum Naturalium Helveticæ, Lausanne.

Hergt, P., Pagalthivarthi, K., Brodersen, S. & Visintainer, R. (1994). A Study of the Outlet Velocity Characteristics of Slurry Pump Impellers. *Proc. 5th Intern'l Symposium on Liquid-Solid Flows.* ASME, Lake Tahoe, NV.

Hydraulic Institute (1977). *Hydraulic Institute Standards for Centrifugal, Rotary and Reciprocating Pumps, 14th Ed.* Hydraulic Institute, Cleveland, Ohio.

International Organization for Standardization (1977). *ISO-3555 Centrifugal, mixed flow and axial pumps - Code for Acceptance Tests - Class B.* ISO, Switzerland.

Karassik, I.J., Krutzsch, W., Fraser, W.H., & Messina, J.P. (1986). *Pump Handbook, 2nd Ed.,* McGraw Hill, New York. p. 2.13.

Kittredge, C.P. & Cooper, P. (2001). Centrifugal pumps: general performance characteristics, Section 2.3.1 of *PumpHandbook, Third Edition,* Eds. Karassik, I.J., Messina, J.P., Cooper, P. & Heald, C.C., McGraw-Hill, New York, NY, pp 2.338-2.339.

Lazarkiewicz, S. & Troskolanski, A.T. (1965). *Impeller Pumps.* Pergamon, Oxford.

Maciejewski, W.M., Oxenford, J. & Shook, C.A., (1993). Transport of Coarse Rock with Sand Clay Slurries, *Proc. Hydrotransport 12*, BHRA, Brugge, Belgium.

Standards Association of Australia (1968). *Australian Standard CB9-1968, SAA Pump Test Code,* Standards Association of Australia, North Sydney, N.W.S.

Stepanoff, A.J. (1957). *Centrifugal and Axial Flow Pumps, 2nd Ed.* John Wiley & Sons, New York.

Stepanoff, A.J. (1965). *Pumps and Blowers, Selected Advanced Topics: Two-Phase Flow*, John Wiley & Sons, New York.

Visintainer, R. & Addie, G. (1993). Development of a New Large Dredge Pump. *Proc. Central Dredging Association* (CEDA), Amsterdam.

White, F.M. (1986). *Fluid Mechanics, 2nd Ed.*, McGraw Hill, New York.

Worster, R.C. (1968). *The Flow in Volutes and its Effect on Centrifugal Pump Performance.* BHRA Hydraulic Plant and Machinery Group, Harlow, UK.

Chapter 10

EFFECT OF SOLIDS ON PUMP PERFORMANCE

10.1 Introduction and Previous Correlations

The presence of solid particles in the flow tends to produce adverse effects on pump performance, and detailed information on these effects is needed to achieve reliable and energy-efficient operation. Head-discharge curves for centrifugal pumps are often rather flat, as noted in the previous chapter. Also, the pipeline characteristic for slurry flow very often displays a minimum (as shown in Chapters 4, 5 and 6), followed by a slow rise with increasing discharge. As a result, the two characteristic curves often intercept at a rather shallow angle. The full implications of this typc of intercept will be examined in Chapter 13, but it is important to note here that even a small diminution in pump head can produce a disproportionately large drop in flow rate. The resulting difficulties in operation can cause large and expensive systems to run inefficiently or not run at all.

The analysis of solids effect on pump performance is not yet as well-developed as that detailed in earlier chapters for the effect of solids on resistance to flow in pipelines, but similar lines of reasoning can be employed, as will be shown subsequently. There are two related phenomena, however, which have been studied in detail in connection with pumping simple fluids: the effects of varying density and the effects of varying (Newtonian) viscosity.

If only the fluid density is changed, while the viscosity remains constant, the affinity laws (see Chapter 9) provide a direct method of evaluating the results. At a given discharge (for constant pump size and rotational speed) the internal flow characteristics within the pump are unchanged. Hence the efficiency is unaffected, and so is the head, when measured in terms of height of the fluid being pumped. The relative density of the fluid (fluid density/density of water) is S_f, and the resulting power requirement is S_f times that needed to pump water at an equal flow rate. For pumps of given size and rotational speed, it can readily be shown that the flow rate at the best efficiency point is not influenced by the density change. Likewise, the density-change effect does not depend on pump size, provided, of course, that the comparison is made within a set of geometrically similar (homologous) machines.

As a slurry is denser than water, it appears attractive to use the behaviour of a denser fluid as a model of slurry behaviour, though it may be an approximate one. This approach corresponds to the 'equivalent fluid' model, considered in Chapters 2 and 4 (see Eq. 4.3) as a first approximation for slurry flow in pipes. Figure 10.1 represents the results of applying the equivalent-fluid model to a centrifugal pump. Note that the head is expressed in height of the equivalent fluid which is numerically the same as the head of water for an equal volumetric flow rate of water alone.

As mentioned above, the effects of a change of fluid viscosity are also well known. If the density of the fluid being pumped is kept constant while its viscosity is increased, then the head and efficiency are lowered, and the power needed to maintain a constant capacity is increased. This behaviour is shown on Fig. 10.2. Note that the effects of viscosity are more pronounced at higher flow rates, and as a result the flow rate at best efficiency point tends to become smaller as viscosity is increased. This shift may not always be significant, but another influence, which cannot be represented on the curves for a given pump, is of greater importance. This is the scale effect; an increase in viscosity has a greater influence on small pumps than on large ones.

Although the presence of solid particles introduces more complicated effects than those accompanying a viscosity increase in a simple fluid, there is some rough qualitative similarity between slurry flow and the flow of a fluid having values of both density and viscosity greater than those for water. The effects on pump characteristics are shown schematically on Fig. 10.3, which is a definition sketch for illustrating the reduction in head and efficiency of a centrifugal pump operating at constant rotary speed and handling a solid-water mixture. In this sketch, and the discussion which follows, η_m represents the pump efficiency in slurry service, and η_w is the clear-water equivalent. Likewise, P_m and P_w are the power requirements for slurry service and water service, respectively. The head H_m is developed in slurry service, measured in height of slurry, while H_w represents the head developed in water service, in height of water. The head ratio H_r and the efficiency ratio η_r are defined as H_m/H_w and η_m/η_w, respectively. The fractional reduction in head (the head reduction factor) is denoted by R_H and defined as $1\text{-}H_r$; for efficiency the fractional reduction (efficiency reduction factor) is R_η, given by $1\text{-}\eta_r$.

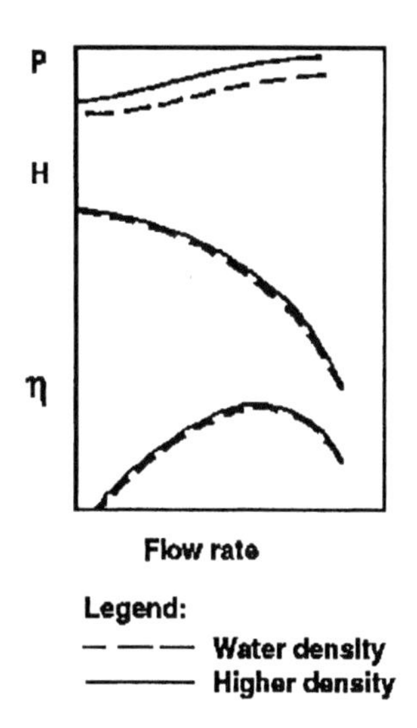

Figure 10.1. Effect of fluid density on pump characteristics (schematic)

Obviously, it is important to relate quantities such as R_H and R_η to the slurry properties which were defined in Chapter 2. These include the relative density of solids S_s, the delivered volumetric concentration C_{vd}, and the relative density of the slurry S_{md}, which equals $1 + (S_s\text{-}1)C_{vd}$.

Investigations reported in the literature have been reviewed by Holzenberger (1980) and Engin & Gur (2003) and others. In the course of comparisons of different proposed relations, they found that there was often a large scatter attributable to the semi-empirical nature of the equations. Nevertheless, most investigations which were based on performance tests of centrifugal pumps handling industrial slurries show certain points of agreement. In the present section, we will deal only with moderate delivered solids concentrations, say $C_{vd} \leq 0.20$ (higher concentrations will be considered in Section 10.3). For this condition, it has often been found that the power consumption at the same flow rate increases directly with the relative density of the slurry, i.e.

$$P_m = S_m P_w \qquad [C_{vd} \leq 0.20] \tag{10.1}$$

The efficiency tends to be lower for slurry flows than for water flows; and if Eq. 10.1 holds, the head ratio and the efficiency ratio must undergo an equivalent reduction giving

$$H_r = \eta_r \qquad [C_{vd} \leq 0.20] \tag{10.2}$$

McElvain (1974) proposed that for centrifugal pumps the head reduction factor R_H (and the efficiency reduction factor R_η) are directly proportional to the delivered concentration C_{vd}, giving

$$R_H = R_\eta = K\left(\frac{C_{vd}}{0.20}\right) \tag{10.3}$$

where 0.20 is simply a convenient reference value of concentration. McElvain related the factor K to a relative density of solids and to average particle size by means of reduction curves of which one is shown on Fig. 10.4 for S_s =2.65.For example, if a coarse sand with d = 1.5 mm is pumped at C_{vd} = 0.20, the reduction of head and efficiency is indicated to be nearly 30%.

A factor to consider is that slurry pump design has changed over the years. Prior to 1980, for example, most slurry pump impellers employed this inefficient radial-vane impeller design, whereas new types now employ vanes twisted so that the inlet angles are correct for the design duty, and the vane shape and sweep are optimized to minimize shock and other losses. Walker *et al.* (1992) investigated the influence of vane design with pumps with impeller diameters of 0.36-0.40 m. and 2-6 vanes. They used narrowly graded sands (d_{50} = 0.4 and 0.8 mm). The results showed that the vane shape had a pronounced effect, with a curved shape giving better performance. However, the inlet and outlet angles, vane numbers and outlet width did not change the performance.

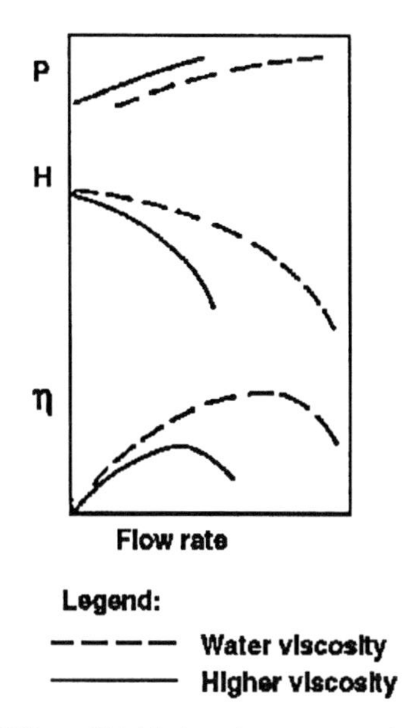

Figure 10.2. Effect of fluid viscosity on pump characteristics (schematic)

In this connection, it should be noted that the lower reduction curve in Fig. 10.4 were based on older designs, i.e. mostly radial-vane type. Tests were made by Sellgren & Addie (1993) of pumps with three twisted vanes of new design transporting sand slurries. The results are plotted on Fig.10.4 where they are compared with McElvain's reduction curve for sand-weight solids. The impeller diameters of the pumps were 0.3 m and 0.4 m and the d_{50} of the sands lay between 0.4 mm and 1.12 mm, with σ_g between 1.5 and 3.2,where $\sigma_g = d_{85}/d_{50}$ and d_{85} is the sieve diameter for which 85% of the particles are finer, and d_{50} is the mass-median diameter. Values of σ_g smaller than 1.8 are considered here to define narrow size distributions (see Section 6.6), while larger values indicate intermediate and broad gradings.

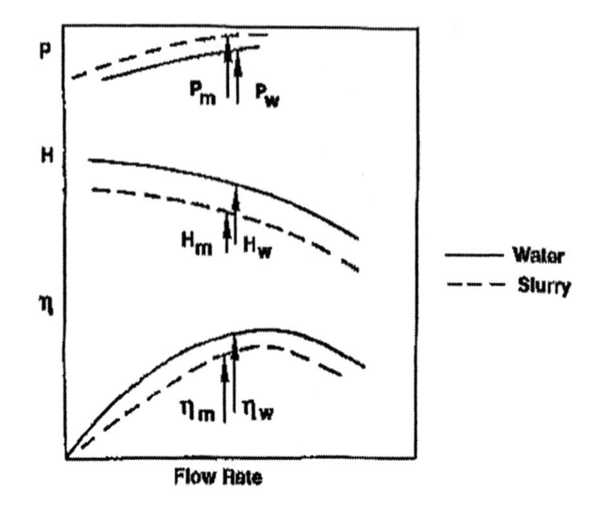

Figure 10.3. Effect of slurry on pump characteristics (schematic)

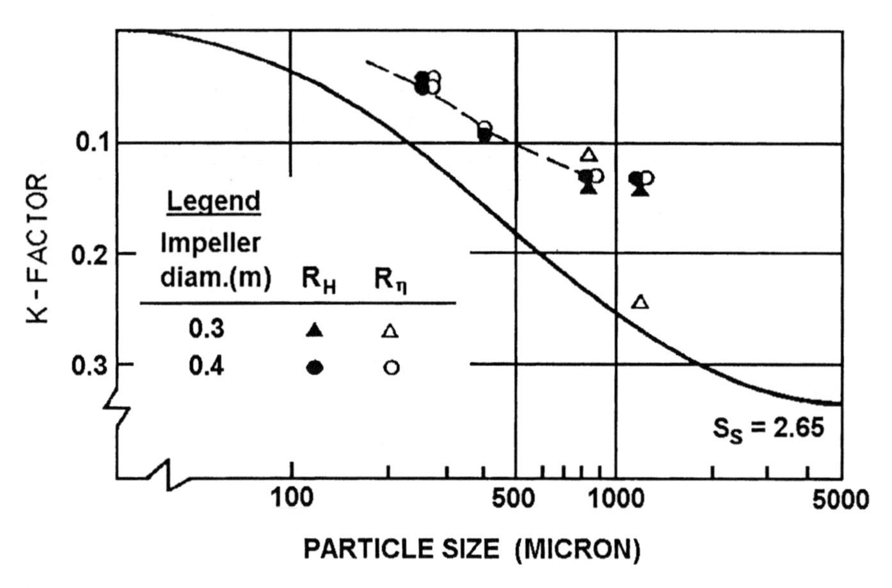

Figure 10.4. Reduction in head and efficiency for newier pumps of 3-vane twisted-vane design, from Sellgren & Addie (1993). Observed values reduction factor K for various sands (d_{50} between 0,30 and 1.12 mm) are compared with the reduction from McElvain (1974) ($S_S = 2.65$)

It is seen from Fig. 10.4 that the solids effect for the newer pumps is considerably smaller. The only exception is the depression in efficiency for the smallest pump when pumping the coarsest sand (d_{50} = 1.12 mm).

Experimental results of Sellgren & Addie (1989, 1993) for pumps of various sizes also confirmed that the solids effect could be considered to be independent of rotary speed. A slight dependence on flow rate was found, but it was concluded that this could normally be neglected in practical applications.

Figure 10.4 indicates that the quantities $(S_s\text{-}1)$, C_{vd} and d affect the fractional head reduction in pumps in the same fashion as they influence the solids effect in pipelines. Presumably there is only one set of physical mechanisms involved, and thus it is desirable to try to link the solids effect in pumps to that in pipelines. As shown in Chapters 4, 5 and 6, in heterogeneous pipeline flow the deleterious head change produced by the solids can be expressed as the product of $(S_s\text{-}1)C_{vd}$ and the stratification ratio. At given mean velocity the stratification ratio in a pipe increases with particle fall velocity (and hence with particle size) until a plateau is reached with large particles, for which the flow is fully stratified. It will be recalled that the immersed weight of solids per unit length of pipe was approximated by $\rho_w g\,(S_s - 1)\,C_{vd}\dfrac{\pi D^2}{4}$, while the stratification ratio indicates the fraction of this weight which is transferred to the pipe wall by granular contact, with resulting mechanical friction. The stratification ratio depends on the particle settling velocity, and thus shows a variation with particle diameter which is similar to that of McElvain's K factor. Hence the stratification ratio and the K factor can be set aside for the moment, focussing attention on the mechanical friction force. The volume of particles in the pump can be taken as proportional to $C_{vd}D^3$ where D signifies the pump impeller diameter. Also, in a pump it is not the gravitational acceleration which dominates, but rather the centrifugal acceleration, proportional to N^2D, where N represents rotational speed. Thus, for a given stratification ratio the force of granular contact can be taken as proportional to $(S_s - 1)C_{vd}N^2D^4$. Multiplication by a representative velocity, *i.e.* ND, gives a measure of the extra power required to overcome friction. If this is divided by the flowrate, a measure of the lost head is obtained. By the affinity laws, the flow rate is proportional to ND^3, giving the lost head as proportional to $(S_s\text{-}1)C_{vd}N^2D^2$. Since the affinity law for head shows it to be proportional to N^2D^2, the head reduction factor R_H (i.e. lost head/head) should vary directly with $(S_s\text{-}1)C_{vd}$.

This simple proportionality between H_r and $(S_s\text{-}1)C_{vd}$ is based on the assumption that the pumps under consideration are all geometrically similar.

The stratification ratio, which is also a factor in the proportionality for H_r, was assumed to be constant, or at least independent of pump size. The results are seen to be in rather good accord with Eq. 10.3 and Fig. 10.4, which indicate that H_r is directly proportional to C_{vd} and increases with (S_s-1). The change of K with particle diameter shown on Fig. 10.4 is similar to the variation of stratification ratio in pipes; near zero for small particles, then increasing with particle size up to a plateau for particles the size of coarse sand or gravel.

10.2 Effect of Pump Size

The data discussed in the previous section had, in general, been obtained from pumps of modest size, and it was felt that tests with large pumps were necessary to extend the data range. This was accomplished by experiments by Sellgren & Addie (1989) with pumps having impeller diameters 1.1 to 1.2 m, using slurries of narrowly graded fine sand, coarse sand and gravel.

The results clearly demonstrated that the effect of solids is much smaller in larger pumps. For example, Fig. 10.4 indicates a 20% reduction in head and efficiency when pumping a sand with d_{50} = 1.5 mm at C_{vd} = 0.15; but the value found for the large pump is only 7%, i.e. about one third as much.

Sellgren & Addie (1989) found that the reduction in head R_H depended on pump impeller diameter raised to the power -0.9. The equivalent scaling law can be based on a reference reduction factor R_{Ho} for a pump with an impeller diameter D_o larger than 0.35 m. The reduction factor R_H for an arbitrary diameter D larger than 0.35 m is then:

$$R_H = R_{Ho}\left(\frac{D}{D_o}\right)^{-0.9} \tag{10.4}$$

Extensive large–scale field measurements of pipeline friction losses and pump performances were presented by van den Berg *et al.* (1999) and Vercruijsse et al. (2002). Pump solids effects were evaluated for a 4-vane twisted design impeller (diameter ~2.5m) GIW centrifugal pump with inlet and discharge diameters of 1.2 and 1.1 m, respectively. Average pump solids effect results for sands with average particle sizes from 200 to 1100 microns. showed that reductions in head and efficiency were practically negligible. These large-scale results indicate that eq.10.4 gives slightly conservative estimations for very large pumps.

Equation 10.4 implies that the stratification ratio depends on pump size. This point can be investigated using the simple relation proposed by Newitt *et al.* (1955), in which the stratification ratio varies with the ratio of the particle settling velocity to the mean velocity of the flow. As noted in Section 7.2, although this relation has been superseded for detailed analysis, its simplicity makes it suitable for providing initial estimates. The affinity laws show that the mean velocity through the pump is proportional to ND. This product, in turn, is proportional to $\sqrt{(gH)}$, so that N is proportional to $H^{1/2}D^{-1}$. As noted previously, in pumps the gravitational acceleration must be replaced by the centrifugal acceleration, proportional to N^2D, i.e. H/D. This substitution affects the particle settling velocity, which for present purposes can be expressed as proportional to gravitational acceleration g raised to some power, say γ. This power can be obtained from the relations set out in Section 2.5; it varies from 1.0 in the Stokes settling range to 0.5 in the constant-C_D range. On above basis, it is expected that the stratification ratio will depend on head and impeller diameter according to $H^{\gamma-1/2}D^{-\gamma}$. With γ in the range 0.5 to 1.0, the power of the head is not great, and the experimental range of H may also have been rather limited, so that head did not appear as a parameter in the empirical correlation. On the other hand, the expected variation of stratification ratio with $D^{-\gamma}$ is in very encouraging accord with the empirical variation with $D^{-0.9}$ given by Eq. 10.4.

10.3 Effect of Solids Concentration

Centrifugal slurry pumps can be used for very high solids concentrations. Experimental results from the GIW Hydraulic Laboratory with a coarse sand in 0.25 and 0.30 m in diameter loop system showed that the upper limit of solids concentration by volume of about 49% was not set by the pump but by dramatically varying pipeline friction losses with unstable operating conditions, Addie & Hammer (1993). The average and maximum particle size was about 600 and 10 000 microns, respectively.

During the research with pumps with different impeller diameters Sellgren & Addie (1989, 1993) found that the reductions in head and efficiency could be related linearly to delivered volumetric concentrations C_{vd} of less value than about 20%. As remarked previously, Eqs. 10.1 and 10.2 apply for $C_{vd} \leq 0.20$. For significantly higher values of concentration, the equivalence of the head reduction factor and the efficiency reduction factor no longer hold, and the fractional reduction in efficiency is more than that in head (Sellgren & Vappling 1986). As a result, the power consumption must be larger than that given by Eq. 10.1, which now becomes

$$P_m > S_m P_w \quad [For\ C_{vd} > 0.20] \tag{10.5}$$

while Eq. 10.2 is changed to

$$H_r > \eta_r \qquad [For\ C_{vd} > 0.20] \tag{10.6}$$

The behaviour indicated by Eqs.10.5 and 10.6 is very clear for the data of Sellgren & Vappling (1986) which are plotted on Fig. 10.5.

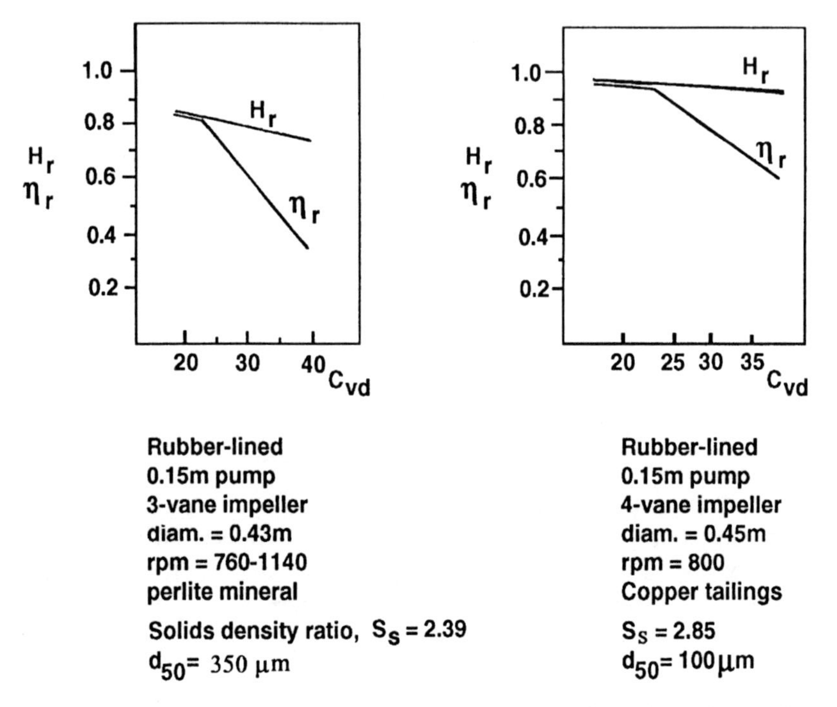

Figure 10.5. Head and efficiency curves for two products, after Sellgren & Vappling (1986)

Recent high-concentration results from the GIW Hydraulic Laboratory for a sand with d_{50} = 350 microns and a pump with impeller diameter 0.63 m are shown in Figure 10.6.

Comparison of Figures 10.5 and 10.6 indicates that the change in reductions moves to higher C in larger pumps. However, Ni *et al.* (1999) found that Eq. (10.1) held for C_{vd}-values up to approximately 35% when pumping a narrowly graded sand with a d_{50} of 370 microns in an all metal 3-vane pump with an impeller diameter of 0.4 m. R_H and R_η were about 15%.

The large-scale field data by van den Berg *et al.* (1999) and Vercruijsse *et al.* (2002) discussed earlier in connection with eq.10.4 with an 2.5 m diameter impeller pump showed that the average pump solids effect for fine sands had reductions in head and efficiency that were practically negligible up to C_{vd}-values as high as 48%.A similar trend was found for the efficiency with the coarsest sand. However, the head was influenced slightly with a lowering to about R_H=5% at C=48%.

In an effort to generalize the results discussed above, the parameter d_{50}/D may represent a scaling parameter when expressing the feasibility of Eq. (10.1), see Fig. 10.7.

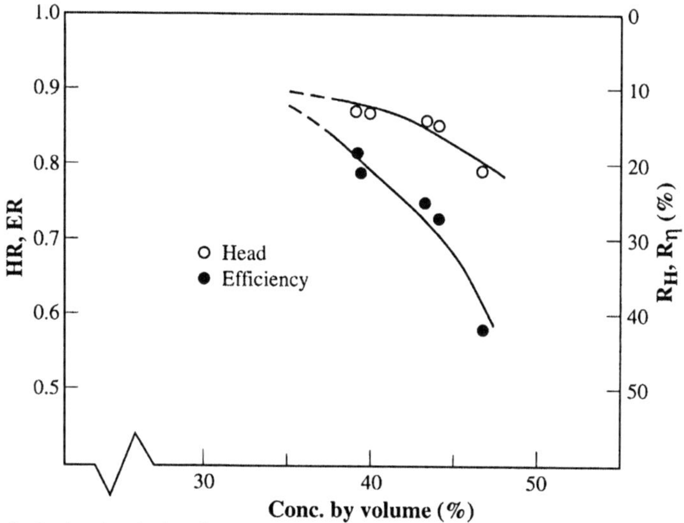

Figure 10.6. Reduction in head and efficiency at high solids concentrations for a sand with d_{50} = 350 μm and an all metal pump with impeller diameter 0.63. Sellgren *et al.*(2002)

The way the power requirement develops with increased C_{vd} in cases when $P_m > S_m \cdot P_w$, depends strongly on the individual properties of the solids. It is an unclear situation similar to the development of friction losses in pipelines for very high C_{vd}-values.

In opposition to the tendencies discussed above, there have also been investigations where the reductions remain small at very high concentrations, specifically with broad particle-size distributions.

10.4 Effect of Particle Properties

The influence of relative density of the solid particles, and their size grading, can be important in determining the reduction of pump head and efficiency. The influence of relative density of solids has been investigated in several studies, For example, data of Burgess & Reizes (1976) had been obtained for an ore product of $S_s = 4.35$. This material had a narrow particle grading, with $d_{50} = 290$ μm, and showed a head reduction ratio R_H of 0.15 at $C_{vd} = 0.15$. For a similar sand product ($S_s = 2.65$) the corresponding value of R_H was 0.10. This result, taken together with the findings of McElvain (1974) and Sellgren (1979), suggests that the effect of solids density in reducing efficiency can be approximated by a proportionality to $(S_s\text{-}1)^\lambda$, with λ having a value somewhat less than unity, say 0.6 or 0.7.

Sellgren & Addie (1993) found that the particle size distribution is very important in determining the reductions of pump head and efficiency. A broad size distribution and a large fraction of fine particles generally has a favourable effect on the reductions. As proposed in Section 6.7, dealing with pipeline flow, the fine particles combine with water to form a carrier fluid which has enhanced viscosity, or non-Newtonian viscous properties. The result is a reduction in the settling velocity of the larger particles, diminishing stratification and lessening the resistance to flow. If the mass fraction of particles less than 40 μm (X_h) combines with water to form a Newtonian medium, then the settling of larger particles with sizes of 500 - 1000 μm will be reduced by a factor of approximately $(1\text{-}X_h)^2$. If this reduction in settling velocity is related directly to the reduction in head then R_H should be multiplied by $(1\text{-}X_h)^2$. This preliminary quantification applies for X_h in the range 0.05-0.50 and has been found valid for various experimental data.

With coarser solids products, particle angularity increases the solids effect compared to normal particle roundness of sands and processed mineral products, as shown by Sellgren et al.(1997) and Sundqvist & Sellgren (2004) for iron ore and waste rock pumping, respectively, with crushed particles of sizes of up to 100 mm.

The effect of concentration, particle size distribution, particle angularity and fine particle content show great similarities to recent findings concerning the influence of these parameters on the friction loss in pipelines.

10.5 Non-settling Slurries

With very fine particles, or large X_h-values and high solids concentrations, most slurries are practically non-settling. In this case the solids effect is mainly related to the rheological behaviour. When pumping non-settling slurries in industrial applications the slurry normally behaves in a non-Newtonian way, giving widely varied pump performance effects. In general, however, small pumps are affected more than large units when pumping highly viscous or non-Newtonian media. Furthermore, the influence on the efficiency is normally larger than on the head. Some typical results are shown in Table 10.1.

Table 10.1. Typical pump solids effects for non-settling industrial slurries at flow rates near Q_{BEP}. Impeller size 0.63m.

Product	Bauxite	Red mud	Kaolin	Phosphate clay
Typical slurry density ratio: S_m	1.6	1.55	1.35	1.1
Head reduction R_H (%)	2	3	3	2
Efficiency reduction $R\eta$ (%)	5	6	6	3

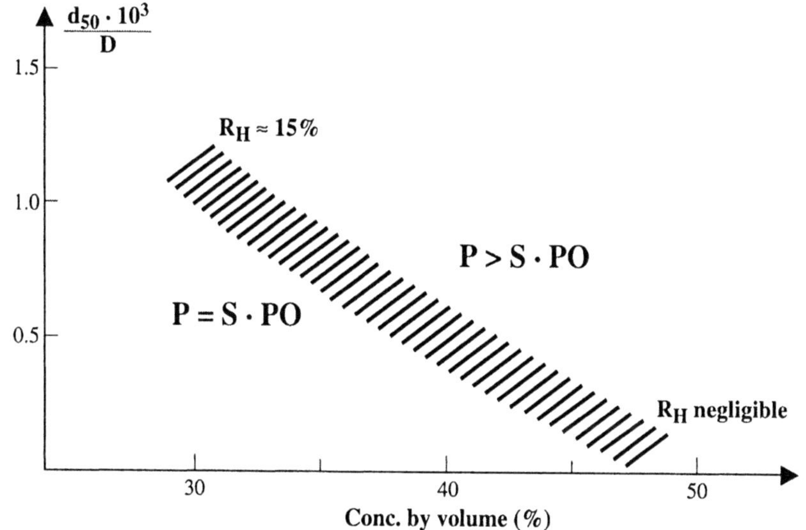

Figure 10.7. Rough generalisation of upper limit of volumetric solids concentration for which $P_m = S_m P_w$. Mainly applicable for narrowly graded sands with d_{50} in range of 150 to 400 μm and a small amount of fine particles< 40 microns. Sellgren *et al.* (2002)

Walker & Goulas (1984) found from experiments with small centrifugal pumps and non-settling slurries that the drop in head and efficiency near flows at BEP could be approximately correlated with the Hydraulic Institute (1983) correction charts for Newtonian viscous fluids. At large shear rates they used the 'tangent' viscosity (see Section 3.2). They also related the performance to an impeller Reynolds number:

$$Re = \rho_m u_2 D / \eta_t \qquad (10.7)$$

where D is impeller diameter and u_2 is peripheral velocity. Values of this Reynolds number less than about 10^6 give significant reductions in efficiency. Results for pumps with impeller diameters from 0.4 to 1m by Shook *et al.* (1995), Sery *et al.*(2002) and Xu et al.(2002) with Bingham type of slurries with yield-values of 100-200 Pa, support the use of these parameters and correlations. Slatter & Sery (2004) suggested a modification of the peripheral velocity in Eq.10.7 to include the blade angle.

Pumping experiments by Whitlock & Sellgren (2002) with a mixture of clay and fine sand for yield stresses of about 200 Pa resulted in reductions in head and efficiency of 10 and 15%, respectively with an 3-vane open shrouded all-metal centrifugal pump with an impeller diameter of 0.3 m arranged with an auger-like inducer. With yield stresses in excess of about 200 Pa, the head could not be maintained for lower flow rates.

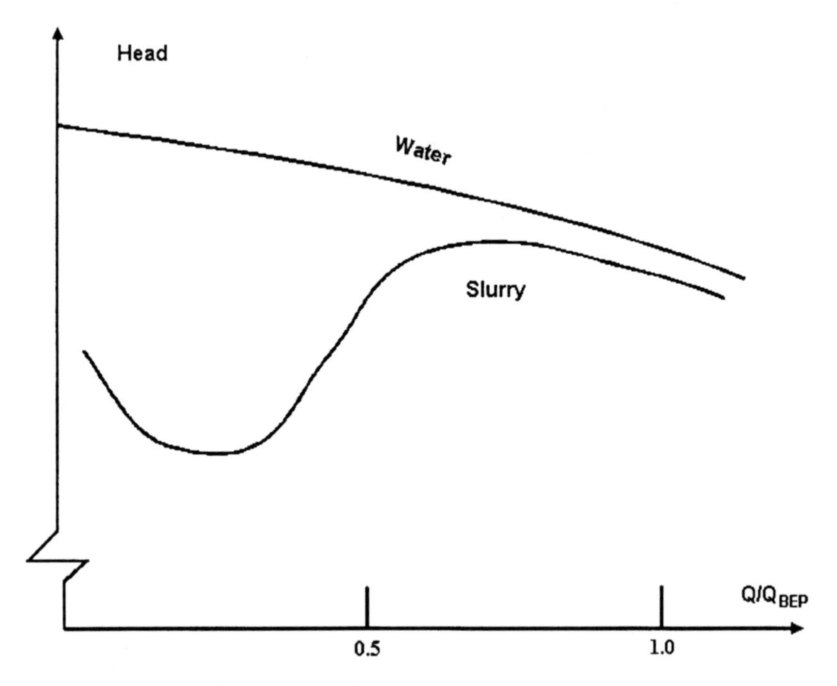

Figure 10.8. Effect of highly non-Newtonian slurries on head (schematic)

Pumping highly non-Newtonian slurries at flow rates much lower than Q_{BEP} may cause a dramatic drop in head which creates an unstable head curve (Sellgren *et al.* 1999) see Fig. 10.8. This drop in head has been found to correspond to a flow rate of about 50% of Q_{BEP}. Thus, the situation can become very complex when operating outside the BEP region. Guidelines on pump curve reductions for viscous *Newtonian* fluids show principally the same trends as in Fig.10.2 page 10.2, i.e. a larger pump should be chosen to pump the required flow rate with operation within the best efficiency region. Using these guidelines resulting in a larger pump for a highly *non-Newtonian* slurry with pronounced yield behaviour might introduce a risk of operation at low flow rates compared to Q_{BEP} where an unstable head curve can occur.

10.6 Suction Performance

Cavitation and related problems with pump suction performance when pumping water were dealt with in section 9.3.The NPSHR specified by the pump manufacturer based on controlled tests with water, is normally expressed in terms of head of slurry and considered to be equal with NPSHR for water, i.e. an equivalent fluid behaviour of the slurry. Standardized closed vacuum tank results by Herbich (1975) for sands at various slurry densities support this similarity. Roudnev (2004) reported similar sand and ore results. Based on reported results and correlations for Bingham-like fiber and ore suspensions he also pointed out that NPSHR may need to be corrected when pumping non-Newtonian slurries.

In an attempt to simulate field suction side conditions in a phosphate dredging pit, Whitlock *et al.* (2001) induced cavitation with an open tank-valve laboratory arrangement with a 1.4 m diameter impeller pump. The NPSHR in metres of slurry was about 50% higher than the corresponding specified water value when pumping various sands for C_{vd}-values of up to 24%.

10.7 Generalised Solids-Effect Diagram for Settling Slurries

Various correlations have been suggested relating the solids effect to solids concentration and density, a representative size of the particles and the pump size. Engin and Gur (2003) based their correlation on own experimental data and results from Kazim *et al.* (1997), Gandhi *et al.* (2001), Ni (2000) and reported older data. However, the data used and resulting correlation were limited to small pumps, mainly with impeller diameters less than 0.5 m.

The similarities of pump solids effect mechanisms to the influence of leading parameters on slurry friction losses in pipelines have been pointed out in the end of Sections 10.3 , 10.4 and 10,5 and by for example Shook (1995), Sellgren & Addie (2001) and Xu *et al.* (2002). However, energy losses related to relative motion between settling type of particles and water and to wall friction are strongly influenced by the comparatively large radial acceleration in the flow passages in a centrifugal pump.

Observed reductions in head and efficiency are sometimes unclear. The defined reductions H_r and η_r in the interrelated Eqs.(10.1) and (10.2) and their coupling to the power in Eq.(10.3) may reflect how imparted energy is transferred and balanced in a complex way into useful pump work and energy losses.

Concentrating on leading parameters, a generalised design diagram for estimation of the solids effect is given in Figure 10.9 It has been based mainly on tests at the GIW Hydraulic Laboratory during the last 15 years. The diagram gives R_H in terms of pump impeller diameter (D) and solid size (d_{50}), with corrections for concentration (C_{vd}), relative density of solids (S_s), and content of fine particles (X_h). The use of the diagram and corrections is explained in the following worked example.

Corrections for various values of volumetric concentrations, solids densities and fines will be given below together with estimations of the reduction in efficiency, $R_\eta \cdot R_H$-values obtained from Fig.10.9 may be multiplied by the following factors when the solids density ratio and the fine particle content is different from 2.65 and zero, respectively.

Solids SG: $\left[\dfrac{S_s - 1}{1.65}\right]^{0.65}$

Fine particle content: $(1 - X_h)^2$

R_H can normally be related linearly to the volumetric solids concentrations when it is less than 20%, i.e. values from the diagram are multiplied by $C_{vd}/15$ where C_{vd} is in percent. With narrowly graded products, R_H may increase more than in direct proportion to C_{vd}. With higher concentrations and broad particle distributions the dependence on C_{vd} may be smaller.

For the larger pumps in the figure, the reduction in efficiency, R_η, is normally less than R_H. For the smaller pumps, R_η usually equals R_H; however, it may be sensitive to solids properties. Independent of the pump size R_η may exceed R_H if the volumetric concentration exceeds about 20%. In opposition to the tendencies discussed above, there have also been investigations where the reduction in efficiency remains small at very high concentrations, specifically with broad particle size distributions.

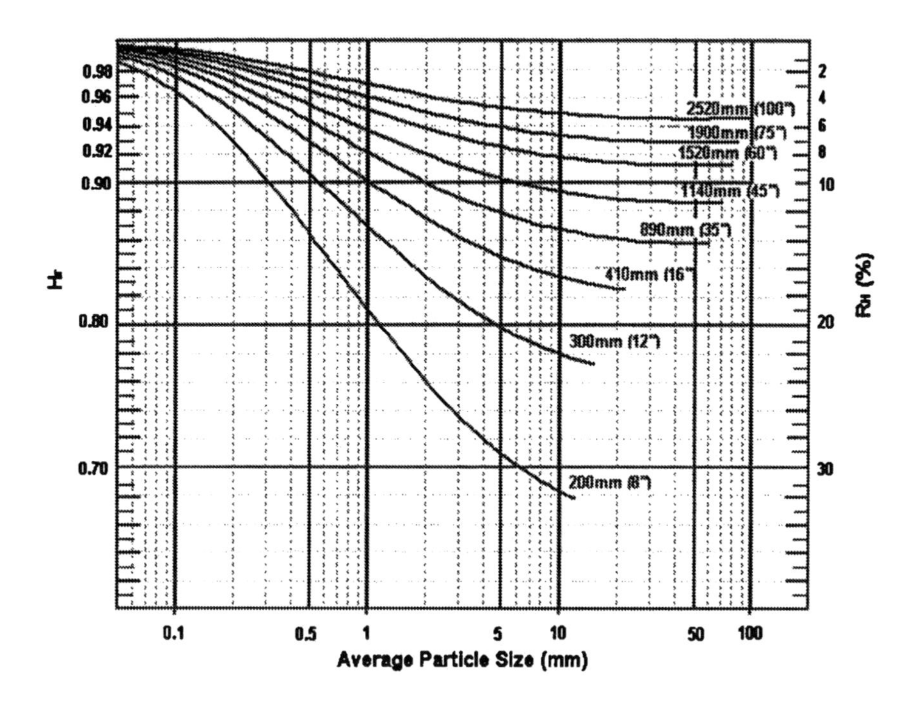

Figure 10.9. Generalised solids-effect diagram for pumps of various sizes (impeller diameters) based mainly on tests at The GIW Hydraulic Laboratory. For solids concentration by volume, C_{vd} = 15% with relative density of solids, S_s = 2.65 and a negligible amount of fine particles

Example 10.1

A sand slurry is to be pumped using a large pump with a 1320 mm impeller. Determine the reduction in head R_H if the solids concentration by volume C_{vd} = 0.20, the solids specific gravity S_s = 2.65 and the average particle size d_{50} = 400 μm.

It follows from the diagram that R_H = 3.5% for S_s = 2.65, d_{50} = 400 μm and C_{vd} = 15%. Correction for C_{vd} = 20% gives: 3.5 (20/15) = 4.7%.

Example 10.2

Determine R_H for a pump with an impeller diameter of 800 mm pumping an ore product (S_s = 4.0) with d_{50} = 500 μm, X_h = 0.28 and delivered volumetric concentration of 20%.

It follows from the diagram that R_H = 6% for d_{50} = 500 μm for S_s = 2.65 and X_H = 0. Correction factor for C_{vd} = 20%, S_s = 4 and X_h = 0.28 are; respectively:

$$\frac{20}{15} = 1.33$$

$$\left[\frac{4-1}{1.65}\right]^{0.65} = 1.475$$

$$(1 - 0.28)^2 = 0.52$$

R_H is then $6 \cdot 1.33 \cdot 1.47 \cdot 0.52 = 6.2\%$.

Example 10.3

A smaller pump is to be used to pump a slurry of the same solids as in Example 10.1, but at a volumetric concentration of 10%. The pump has an impeller diameter of 380 mm.

Following Example 10.1 with a pump impeller diameter 380 mm then the diagram gives R_H = 7%. Correction for C_{vd} = 10% means that R_H = $7 \cdot (10/15) = 4.7\%$, say 5%. Assuming that R_η equal to R_H, it is also 5%. The head required for this application is 45m of slurry at a flow rater of 0.0625 m^3/s. The pump has a 100mm discharge branch, 150mm suction branch and an impeller diameter of 380 mm. Its clear-water performance characteristics are shown on Figure 10.10. The running speed and power requirement for pumping this slurry are to be determined.

The head ratio H_r was defined in section 10.1 as 1 – R_H, and for the present example this equals 0.95, i.e.

$$H_r = \frac{H_m}{H_w} = 0.95.$$

In order to be able to produce the required 45m head of slurry, the pump must be capable of producing a head of water, H_w, given by

$$H_w = \frac{45.0}{0.95} = 47.4\text{m of water}$$

For this water head, and the discharge of 62.5 L/s, the pump characteristics shown on Fig. 10.10 indicate that the pump must run at 1500 rpm, and that the clear water efficiency η_w will be 75%.

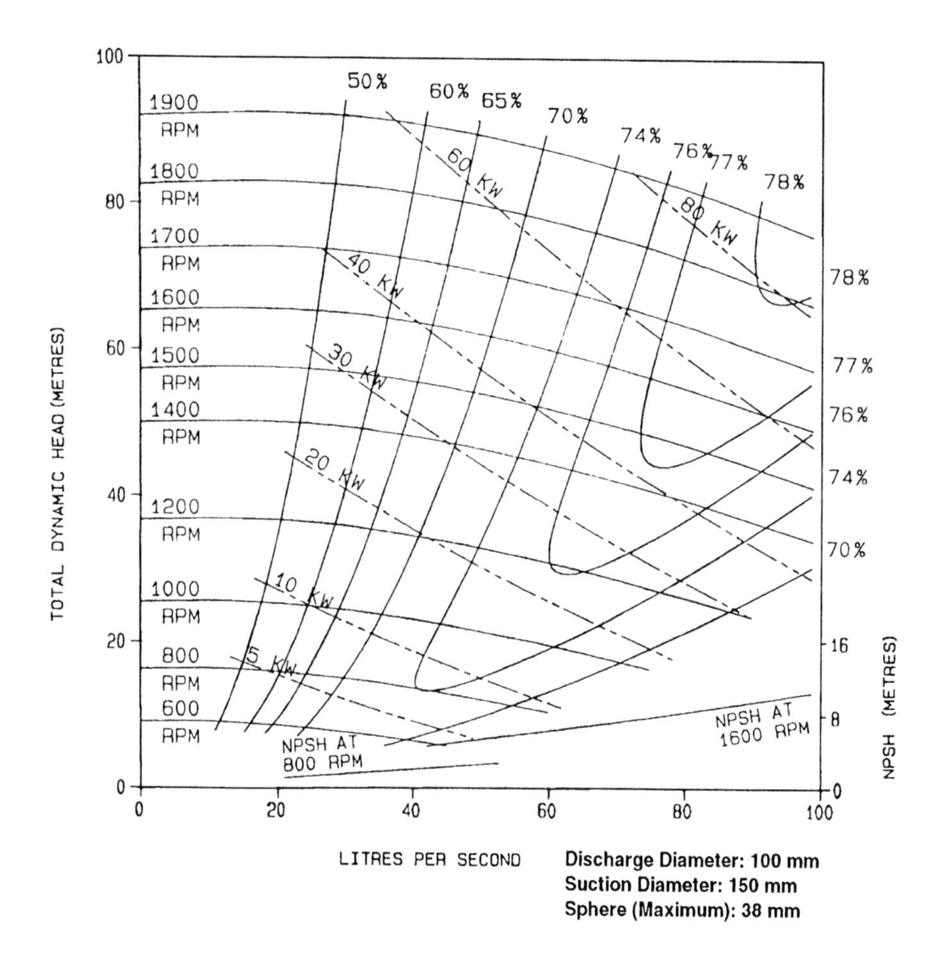

Figure 10.10. Pump characteristic curves for Example 10.4

With R_η assumed to be equal to R_H, then

$$\eta_r = 1 - R_\eta = 0.95 = \frac{\eta_m}{\eta_w}$$

Thus

$$\eta_m = \eta_r \, \eta w = 0.95\,(0.75) = 0.713 \text{ or } 71.3$$

The slurry specific gravity, S_m, is 1 + 0.10 (2.65 - 1) or 1.165, and by Eq. 10.1 P_m equals S_mP_w. For the calculated values of η_m and other quantities the power requirement is found to be

$$P_m = \frac{1.165(9.81)(45.0)(0.0625)}{0.713} = 45.1\,kW$$

REFERENCES

Addie, G. & Hammer J. (1993): Pipeline head loss of a settling slurry at concentrations up to 49 % by volume, *Proc.14th Int .Conference on the Hydraulic Transport of Solids in Pipe*, Brügge, Belgium, 28-30 September.

Addie G. & Sellgren A. (2001): Application and construction of centrifugal solids handling pumps. Subsection 9.16.1. In Karassik I. J. et al. (2001): Pump handbook, Third ed. McGraw–Hill.

Burgess, K.E., & Reizes, J.A. (1976). The effect of sizing, specific gravity and concentration on the performance of centrifugal slurry pumps. *Proc. Inst. Mech. Eng.*, UK, Vol. 190, No. 36, pp. 391-399.

Engin & Gur (2003) Comparative evaluation of some existing correlations to predict head degradation of centrifugal slurry pumps. ASME J. of Fluids Eng. 125,January,pp149-157

Gandhi B. K., Singh S. N. & Seshadri, V. (2001) Performance characteristics of centrifugal slurry pumps ,ASME J. Fluid Eng. 123,pp 271-280

Herbisch J.B. (1975) *Coastal and Deep Ocean Dredging*, Gulf Publishing., Houston TX,U.S.A.

Holzenberger, K. (1980). Betriebsverhalten von Kreiselpumpen beim hydraulischen Feststofftransport. *VDI-Berichte* Nr. 371, pp. 59-66.

Hydraulic Institute Standards for Centrifugal, Rotary and Reciprocating Pumps, (2004) Hydraulic Institute, Cleveland, Ohio, X?th Ed.

Kazim K.A., Maiti B. & Chand P. (1997) A correlation to predict the performance characteristics of centrifugal pumps handling slurries. *Proc. Instn. Mech. Engrs.*, 21A,pp 147-157

McElvain,R.E.(1974). High pressure pumping. *Skilings Mining Review*, Vol. 63 No. 4, pp. 1-14.

Newitt, D.M., Richardson, J.F., Abbott, M. & Turtle, R.B. (1955). Hydraulic conveying of solids in horizontal pipes. *Trans. Inst. of Chem. Engrs.*, **33**, London, U.K., 93-113.

Ni, F., Vlasblom, W.J., & Zwartbol, A. (1999) Effect of high solids concentration on characteristics of a slurry pump, *Proc.14th Int.Conference on the Hydraulic Transport of Solids in Pipes,* Maastricht, The Netherlands, 8-10 September.

Roudnev, A. S.(2004) Slurry pump suction performance considerations, *Proc.14th Int.Conference on Hydrotransports,BHR Group,* Santiago, Chile

Sellgren, A. (1979). Performance of centrifugal pump when pumping ores and industrial minerals. *Proc. 6th International Conference on the Hydraulic Transport of Solids in Pipes,* BHRA Fluid Engineering, Cranfield, Bedford, U.K., pp. 291-304.

Sellgren, A., & Addie, G.R. (1989). Effect of solids on large centrifugal pump head and efficiency. *Proc.The CEDA Dredging Day*, Amsterdam, The Netherlands.

Sellgren, A. & Addie, G.R. (1993). Solids effects on the characteristics of centrifugal slurry pumps, *Proc. 12th Int. Conference on the Hydraulic Transport of Solids in Pipes,* Brugge, Belgium.

Sellgren, A., & Vappling, L. (1986). Effects of highly concentrated slurries on the performance of centrifugal pumps. *Proc. International Symposium on Slurry Flows*, FED Vol. 38, ASME, USA, pp. 143-148

Sellgren, A., Addie, G. & Yu, W.C. (1999) Effects of non-Newtonian mineral suspensions on the performance of centrifugal pumps, *Journal of Mineral Processing and Extraction Metallurgy Review,* vol. 20, pp 239-249.

Sellgren, A., Addie, G. & Scott, S. (2001): The effect of sand-clay slurries on the performance of centrifugal pumps, *The Canadian Journal of Chemical Eng.* Vol 78, No 4, August, 764-769.

Sellgren, A., Addie, G. & Whitlock, L. (2002). Using centrifugal pumps for highly concentrated tailings slurries, *Proc. 15 th Int. Conference on Hydrotransport,* BHR Group U.K. June.

Sellgren, A. & Whitlock, L. (2002). Pumping paste with a modified centrifugal pump, *Proc. Tailings and Mine Waste 02,* Col. State Univ. Ft. Collins, Col., U.S.A. Jan. 27-30.

Sellgren, A., Luther, D. & Addie, G. (1997). Effect of Coarse and Heavy Solid Particles on Centrifugal Pumps Head and Efficiency. *Proc. 47th Canadian Chemical Engineering Conference*, Edmonton, AB,Canada, October 5-8.

Sery, G.A. & Slatter, P.T. (2002). Pump derating for non-Newtonian slurries.,*Proc.15th Int Conference on Hydrotransport* ,Banff Canada,pp 679-692

Shook, C.A., Gillies, R.G. & McKibben, M. (1995). Derating of a centrifugal pump by large solid particles, *Proc. 8th International Freight Pipeline Society Symposium*, Pittsburgh, PA, pp. 144-151.

Slatter, P.T. & Sery, G. (2004). Centrifugal pump performance Reynolds number for non-Newtonian Slurries. *Proc. 12th Intn'l Conf. on Transport and Sedimentation of Solid Particles,* Prague, Czech Republic, pp. 601-609.

Sundqvist, A. & Sellgren, A. (2004). Large-scale testing of coarse waste rock pumping, *Proc.16th Int.Conference on Hydrotransport*, Santiago, Chile

van den Berg, C.H., Vercruijsse, P.M. & van den Broek, M. (1999): The hydraulic transport of highly concentrated sand-water mixtures using large pumps and pipeline diameters. *Proc.14th Int.Conference on the Hydraulic Transport of Solids in Pipes,* Maastricht, The Netherlands, 8-10 September

Vercruijsse, P.M. & Corveleyn, F. (2002). The solids effect on pump and pipeline characteristics- Keeping up with present trends in the dredging industry.*Proc.15th Int conference on Hydrotransport,* Banff, Canada, 3-5 June,711-723

Walker, C.I. & Goulas, A. (1984). Performance characteristics of centrifugal pumps when handling non-Newtonian homogeneous slurries, *Proc. Instn. Mech. Engrs.* Vol. 198A, No. 1, pp. 41-49.

Walker, C.I., Wells, P.J. & Pomat, C. (1992). The effect of impeller geometry on the performance of centrifugal slurry pumps. *Proc. 4th Int. Conf. on Bulk Material Handling and Transportation,* IE Aust, Wollongong, Australia, pp. 97-101.

Whitlock, L., Sellgren, A. & Wilson, K.C. (2001). Net positive suction head requirement for centrifugal slurry pumps, Handbook of conveying and handling of particulate solids, Editors: A. Levy, H. Kalman, Elsevier Science B.V., 491-497.

Xu, J., Tipman, R., Gillies, R. & Shook, C. (2002). Centrifugal pump performance with newtonian and non-Newtonian slurries, *Proc.15th Int.Conference on Hydrotranport,* Banff, Canada, BHR Group,UK.

Chapter 11

WEAR AND ATTRITION

11.1 Introduction

The useful life of most slurry transport equipment is limited by erosive wear of the wetted passages. As a result, wear performance must often be evaluated in connection with the design or operation of slurry systems. Wear is a common industrial problem, leading to frequent maintenance and replacement of components, and possibly also to reduced operating efficiencies. In simplest terms, erosive wear amounts to the progressive removal of material from a solid surface. In practice, the mechanisms by which this erosion occurs are diverse.

As the factors affecting wear performance are manifold, and the gamut of slurry applications broad, a good deal of wear-performance evaluation has occurred *post facto*, when the system is already in operation. A body of experience and insight gathered by this method has accumulated over time, and much of the current design for wear performance of slurry systems is based on this experience.

Recent years have also seen the introduction of more rigorous approaches to wear-performance evaluation. These include standardised laboratory tests for ranking slurry abrasivities and material wear resistances, and electron microscopy for providing close examination of the micro-mechanisms of wear for both laboratory and field-collected samples. The approach of numerical modelling of slurry flow is also gaining popularity as more powerful computers become widely available, and as numerical techniques become more refined. This chapter presents an overview of some of the more important operational and design considerations affecting the wear performance of pumps and pipelines. It also provides an introduction to some of the recent developments in wear testing and computer modelling of slurry flows, and what they tell us about expected wear rates of wet-end pump components.

The major erosive mechanisms are sliding abrasion and particle impact. The sliding-abrasion mode of wear typically involves a bed of contact-load particles bearing against a surface and moving tangent to it, as illustrated on Fig. 11.1. In pipelines the stress normal to the surface is caused by gravity, as shown in the analysis of contact load which was presented earlier in this book. As described in Chapter 5, the submerged weight of particles which are not suspended by the fluid must be carried by intergranular contacts. The analysis of the motion of these contact-load (bed-load) solids in pipes has been developed over many years (Wilson *et al.* 1973 and other references cited in Chapter 5), with some of the basic concepts dating from Bagnold's (1956) work on the flow of cohesionless grains in fluids. For sliding abrasion the erosion rate depends on the properties of particles and wear surfaces, the normal stress and the relative velocity.

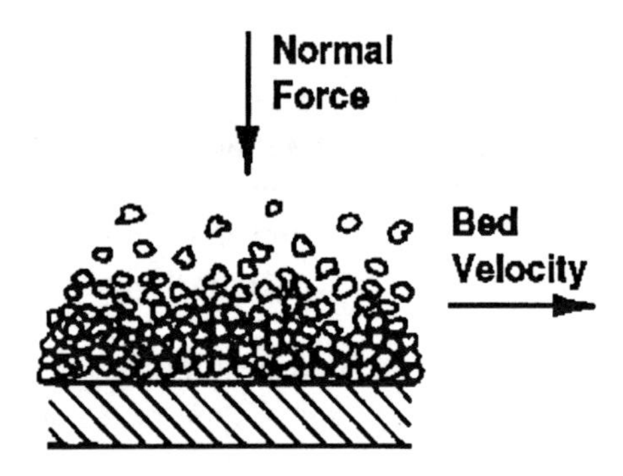

Figure 11.1. Erosion by sliding abrasion

The normal stress is enhanced when the flow streamlines are curved, as in an elbow. In this case there is a centrifugal acceleration, equal to u^2/r, where u is the local velocity and r is the radius of curvature of the stream lines. This acceleration can often be much greater than that of gravity, producing a commensurate increase in the normal stress between the moving contact-load solids and the wall material, and hence a greatly-increased rate of sliding-abrasion wear. This type of behaviour can cause elbows to wear through when adjacent straight-pipe sections of the same material have hardly been affected by erosion, and is also very important in pump casings where sliding abrasion tends to dominate in most areas. Near the tongue of the pump casing, however, the other major wear type is more significant.

This second type of wear is the particle-impact mode, which occurs where individual particles strike the wearing surface at an angle, despite the fact that the fluid component of the slurry is moving along the surface (see Fig. 11.2). Removal of material over time occurs through small-scale deformation, cutting, fatigue cracking or a combination of these, and thus depends on the properties of both the wearing surface and the particles. Ductile materials tend to exhibit erosion primarily by deformation and cutting, with the specific type depending on the angularity of the eroding particles. Brittle or hardened materials tend to exhibit fatigue-cracking erosion under repeated particle impacts. For a given slurry, the erosion rate depends on properties of the wearing surface: hardness, ductility, toughness and microstructure. The mean impact velocity and mean angle of impact of the solids are also important variables, as are particle characteristics such as size, shape and hardness, and the concentration of solids near the surface.

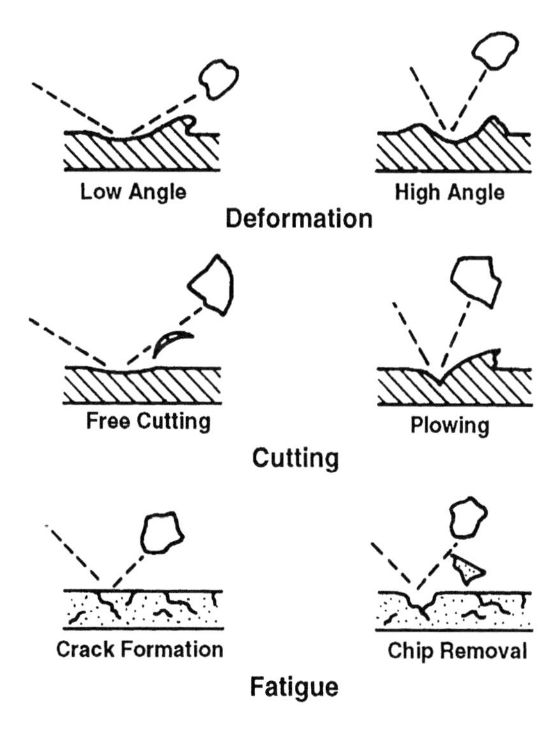

Figure 11.2. Mechanisms of particle-impact erosion

The particle-impact type of erosion occurs because the trajectories of the individual particles do not follow the streamlines of the average flow. This behaviour is important where the acceleration produced by strongly curving streamlines throws the particles towards a near-by surface, as occurs at the tongue region of pump casings. Moreover, pumps and pipelines transporting water-based settling slurries almost invariably operate in a highly-turbulent flow regime, and particles can also be driven toward the boundary of the flow by turbulent eddies. The impact erosion associated with turbulence is best exemplified by the conditions in the upper portion of a settling-slurry pipeline. Here removal of material occurs as in other types of impact erosion, although the impact velocities and angles are more random.

As noted in Section 6.4, gravity, turbulent eddies and other features of the flow may drive particles toward a boundary, but hydrodynamic lift forces caused by the velocity profile can provide a counterbalancing effect. There will be cases where hydrodynamic lift will be sufficient to remove particles from the walls of a straight pipe but will be inadequate to do so for the curvilinear flow in an elbow or a pump casing. The tendency for high wear rates in pump casings should follow trends similar to those noted in Chapter 10 in connection with the effect of solids on pump performance. In a general sense, it can be argued that the power lost for useful purposes by a reduction of pump efficiency may be diverted to erosive activity. Nevertheless, efficiency loss is a global phenomenon, whereas in wear analysis it is the maximum local erosion which is of major interest.

For example, the solids effect in pumps shows a simple increase with solids concentration, but the same does not necessarily apply to wear rates at any specified location within a pump casing. It is expected that erosion by particle impact will be more effective than sliding abrasion provided that an equal number of particles is involved in each mechanism. The required conditions apply for low solids concentrations, or cases where only a small fraction of the solids moves as contact load. Here the moving contact-load particles coming from upstream will be spaced sufficiently far apart to allow speedy incoming particles to erode the surface by impact. As a result, the local wear rate may be high. However, at larger solids concentrations the contact-load particles will be closer together. They can interact to reduce the mean impact velocity, and may even form a moving layer several particle diameters in thickness, enhancing sliding abrasion but protecting the boundary from the more damaging impact erosion. The result is a local wear rate which does not increase monotonically with concentration of solids, but instead peaks at some specific concentration.

A further complexity is introduced by secondary currents, which occur in curved flows such as elbows (Patanker *et al.,* 1975) and pump casings. Secondary currents cause high-velocity flow to migrate to the outer portion of any bend, setting up a return flow along the side walls. These currents can sweep contact load away from certain locations, exposing them to impact erosion. The general principle is well understood, but it is extremely difficult to predict the resulting behaviour in specific instances.

The different modes of particle erosion, although sometimes difficult to distinguish, and often occurring simultaneously, do exhibit typical worn surface forms or 'fingerprints'. The sliding contact-load type of wear often results in a scalloped or sand-dune-like wear surface such as that seen on Fig. 11.3. This configuration is especially common in cases where the flow as a whole is highly turbulent, as in centrifugal pumps, or in pipelines operating at high velocities. In such cases, very smooth surfaces are usually observed, with wear concentrations appearing at any discontinuities of material or geometry, however small. Well-defined sculpted lines are often seen at the dune peaks, or in the 'eddy' areas downstream. By contrast, surfaces worn by impact often have a frosted look with few well-defined lines or edges. Their appearance resembles that of sand-blasted surfaces.

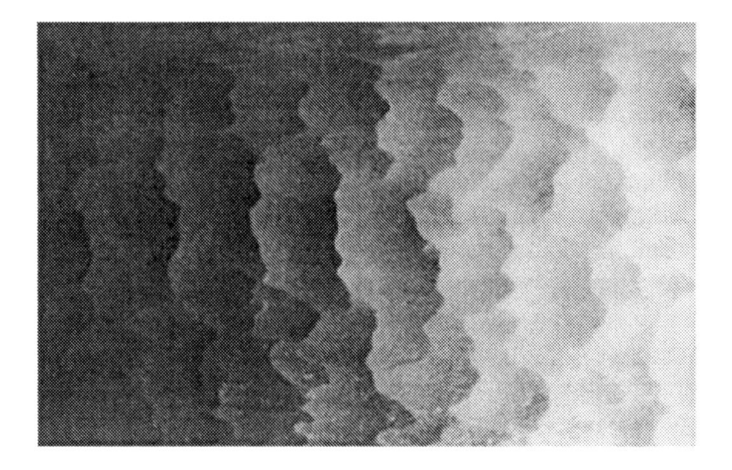

Figure 11.3. Appearance of surface subjected to sliding abrasion

11.2 Wear-Resistant Materials

As wear mechanisms vary considerably, so also do the engineering materials which resist them (Tian *et al.* 2003). A survey of all the available wear-resistant materials and their applications would provide subject matter for a volume in itself. Nevertheless, there are a few groups of materials which have found widespread successful application against erosion in slurry systems, and thus merit particular attention. These materials fall into the broad categories of hardened-metals, elastomers (rubbers and urethanes), and ceramics. Materials in each of these categories have their advantages and disadvantages; and the final choice of a material must account for wear resistance, strength, ease of maintenance, direct cost, and indirect costs such as plant downtime for maintenance or failure, safety of operation and effect on system efficiency.

Not included in the above list of materials, though widely used, is common carbon steel. Several times inferior to many other wear-resistant metals, carbon steel still finds wide application in slurry service, especially where its qualities of toughness, weldability and low cost are desirable, and where erosive wear is low, as in pipelines and tanks. The use of a better-wearing material can often be justified by the savings in the cost of downtime resulting from wear-induced failure of components. In cases where the high toughness of a steel is required to resist shock, or to carry loads, manganese steel, an alloy containing over 1% carbon and 11 to 14% manganese, provides an alternative. An austenitic steel which work-hardens and transforms to martensite in surface layers under heavy abrasion, this steel is found primarily in crushing, sanding and earth moving equipment, but is also used in slurry pumps.

Within the metals group, superior wear resistance is given by the high-alloy white cast irons, particularly nickel-chrome (*e.g.* NiHard), chrome-moly (*e.g.* 15% chromium and 2% molybdenum) and high chrome (*e.g.* 27% chromium) alloys. These materials exhibit higher hardness (600 plus Brinnel), but less toughness than steel alloys of similar tensile strength. The high carbon content and resultant massive carbides present in these metals gives them excellent resistance to all forms of particle erosion, especially in sliding or impact at low angles (cases often encountered in slurry applications). These materials provide an excellent compromise between wear resistance and toughness, and are widely employed in all manner of slurry system machinery including slurry pumps, grinding, crushing and conveying equipment, pipeline fittings and valves. Recent advances in white iron technology have produced grades with various specialized properties including fine and globular carbide morphology for increased toughness and strength, and ever higher carbide fractions for improved wear resistance.

For slurries without coarse particles, elastomers such as rubber, neoprene and urethane tend to wear better than hard metal, especially for liners and shells. This improvement applies only if there is no 'tramp' (i.e. extraneous material such as tools, bolts, or pieces of broken castings). Elastomer selection may also be determined by the corrosive characteristics of the slurry. Natural soft rubber remains an excellent and economical pump lining material to handle fine abrasive and corrosive slurries. Carbon black can be added for additional strength, hardness and tear resistance, so as better to withstand the impact of large particles in the slurry. Like all other materials, rubber has limitations, which should be considered in the selection process. Wear rates in the presence of coarse material or high tip speeds may make it uneconomical, and tramp material or adverse suction conditions can tear the rubber. High temperature or the presence of oils or chemicals may require the use of one of the higher-cost synthetic materials. For example neoprene has been found to tolerate higher temperatures and higher impeller tip speeds than can be used with natural rubber.

Urethanes show higher durometer-test values than either natural rubbers or neoprenes, and thus are more tolerant to the presence of some tramp material, and also can tolerate high impeller tip speeds (urethanes and high-chromium iron alloys are similar in this regard). For certain slurries, urethanes can provide wear performance superior to both metals and rubbers; these slurries typically have particles which are too small to create an adequate rebound from a resilient rubber surface but are large enough to erode the matrix of a white iron. Urethanes have the disadvantage of limited temperature range (similar to that for natural rubbers), and their unit cost exceeds that of rubber, making them expensive when used in large quantities. However, where small or medium quantities are required, urethane products enjoy considerable advantages, since they are based on a two-part resin with good bonding characteristics, and do not require vulcanisation, with its attendant high temperatures and expensive moulds and presses. Ceramics comprise the remaining group of engineering materials, they are the most wear-resistant, but are low in toughness and impact strength. If we consider the series of materials which begin with steels and proceed through white irons, it is seen that ceramics represent the extremity of the series, the culmination of the trend to lower toughness and higher wear resistance. The more successful types of ceramic materials are also more costly than their metallic counterparts. Current applications of ceramics tend to be limited to non-load-bearing liners strategically placed in high-wear locations such as pipe elbows or chutes, or in localized areas of high wear in centrifugal pumps. Some effort has been made towards producing fully-ceramic-lined slurry pumps, but to date these have found few successful applications. A ceramic's low tolerance to shock and vibration generally requires that, if included as part of a pump assembly, it be mounted in compression or somehow isolated from the load-carrying components.

The creation of improved ceramics appears to provide great promise for future development, and research is proceeding in this area with the view of fabricating stronger and tougher materials for load-bearing components. Metal-ceramic composites have also been developed, and these may some day provide the toughness of a metal with the wear resistance of a ceramic. Also, ceramics are being used with some success as filler material for various plastics and urethanes in the form of granules, or plates aligned with the surface. The resulting composite can provide the toughness and resilience of a plastic plus the hardness of a ceramic. Other sorts of composite include resin-type materials reinforced by high-strength graphite fibres which resist wear and corrosion and are machinable into impellers and liners.

11.3 Combined Erosion and Corrosion

Wear in slurry transport applications is caused primarily by the solids, but in addition to erosion, cavitation and corrosion also occur. Fortunately, cavitation may often be controlled through good system design, and corrosion is usually of significant concern only in chemical or seawater applications (as opposed to typical slurries based on fresh water). Still, it is important to understand these phenomena, since even small levels of cavitation or corrosion can influence wear performance significantly when coupled with the surface effects of particle erosion. Corrosion attack associated with erosion may appear in different forms such as uniform corrosion and localized pitting corrosion. In a corrosive slurry system, corrosion occurs when two or more electrochemical reactions take place on a wearing surface. On the attacked surface layer the corrosion products are usually loose, soft and easily removed by erosion from the slurry stream.

On the other hand, cavitation occurs on surfaces where pressure differences cause formation and collapse of vapour cavities. When the cavities collapse in a high-pressure zone, shocks will be generated, causing localized deformation or pitting damage on the surface. Cavitation damage is most often seen near the inlet of a centrifugal pump impeller, downstream of an obstruction in a pipeline such as a valve or orifice, or anywhere that a sudden drop in fluid pressure may occur. Its form is very distinctive and easily identified when occurring in isolation, having the appearance of cratering or pitting, or of a severe but localized attack of corrosion. Sharp edges are also characteristic of cavitation damage. This appearance is associated with the extremely high local pressures caused by cavitation (see Section 9.3).

As the presence of cavitation in slurry applications is often masked by particle erosion which is occurring simultaneously, cavitation is not always detectable by examination of the wear surfaces. When cavitation is suspected but not verifiable by examination, the equipment should be monitored when it is running. Cavitation produces characteristic broad-frequency popping or rumbling noises which may be detected by ear in some cases, although monitoring by a vibration-detection device can give better resolution. For a centrifugal pump, NPSH performance curves should be available from the manufacturer, and can be compared to the system conditions (see Chapter 9) to indicate whether cavitation may be expected.

In a combined erosion-corrosion situation, either erosion or corrosion can be the predominant factor (Tian *et al.* 1997). In most slurry transport applications, erosion is the control factor, and corrosion may be difficult to identify as it often occurs together with particle erosion, which may have a masking effect. Corrosion attack may become much more severe when acids are found in the slurry, particularly in the presence of impurities such as chlorides or fluorides. In cases where the suspect chemicals may be separated from the solids in the slurry flow, erosion and corrosion may be tested separately using a standardised wear test for comparison with the combined effect on the material. It is not uncommon for the combined corrosion-erosion effect to be more severe than the sum of the individual effects. This situation occurs when the particulate erosion accelerates the corrosion by removing the surface corrosion products, allowing the fluid access to a fresh surface. Where it is suspected that corrosion is a significant factor, and it is not possible to separate the solids from the corrosive component of the slurry, materials highly resistant to corrosion (or corrosion-erosion) should be tested as potential replacements for the existing material. The relative particle-erosion resistance of these materials, although important, should not be rigorously imposed to limit the materials chosen for testing, since the particle-erosion component may be greatly reduced once the corrosion-erosion cycle is broken.

11.4 Experimental Testing Methods

Wear testing is critical for the prediction and control of wear in slurry-handling equipment such as pipelines and pumps. Numerous studies have been conducted to quantify and characterise wear in a system, including those of Finnie (1960), Tuzson *et al.* (1984), Miller (1987), Roco *et al.* (1987), Pagalthivarthi & Helmly (1990), Sundararajan (1991), Hutchings (1992), Tian *et al.* (2005) and Li *et al.* (2005). In any wear test, the goal is either to rank the wear resistance of several materials or to obtain an absolute wear coefficient (energy consumed per unit volume of material removed) for some specified system or application. The latter is typically used for predicting wear rate and effective life. In selecting a wear test, attention must be paid to the types of wear to be encountered under the actual operating conditions, and how well these can be simulated. For instance, the operating environment might include long-term corrosive damage, which is hard to implement experimentally. Tests must also provide realistic and measurable wear rates. Often these requirements are contradictory, since actual wear rates may be too small to permit accurate measurement within the period of testing. Conversely, tests with artificially-increased wear rates may not represent prototype wear adequately, and scale-up of erosion measurements is not easy.

In planning a wear test, three factors must be taken into account: slurry characteristics, material characteristics, and operating characteristics of the equipment. The wear characteristic of the slurry, also known as slurry abrasivity, will be determined by the median particle size and the size distribution and by particle shape (whether angular or rounded), particle hardness and particle friability. Large particles tend to cause impact damage, and angular particles are instrumental in scratching and scouring the surface. The hardness of the particles in relation to that of the surface material is very important in determining crack propagation, especially when impact damage is of concern. Also, hard particles tend to retain their sharp edges after scratching the surface, but are likely to be friable upon impact. For this reason they may lose some of their ability to cause further damage, although the new sharp edges resulting from particle breakage can produce additional scratching. The erosive ability of the slurry is compounded if it has a non-neutral pH, since in this case the added mechanism of corrosion will be significant.

The most important material characteristics that affect wear are composition, hardness, microstructure, toughness, and surface finish. If the material is harder than the particles, the particles will be unable to scratch the wear surface effectively. Toughness of the material is significant when the wear mechanism is fatigue-controlled, and for many applications a good combination of hardness and toughness is necessary. Important microstructural considerations are the distribution of second-phase materials such as carbides in iron alloys, and also the grain size and the orientation and shape of the second phase. The chemical composition of the material is especially significant in corrosive media. Finally, surface finish can have a significant effect on crack propagation.

A detailed description of the various types of test apparatus used for quantifying wear and ranking wear resistance of materials cannot be undertaken here, but salient features of some of the most relevant tests deserve mention. The Miller machine is commonly employed for ranking materials in the order of their abrasive wear resistance, Miller (1987). In this test a wear sample is forced to reciprocate for a specified length of time in a slurry of 50% volumetric concentration of sand. A weight is used to impose a normal force between the slurry and the sample. Measurements of the rate at which the sample loses mass serve to determine the slurry abrasion resistance (SAR) number for each material of interest. A high SAR number indicates a relatively poor wear resistance of the material. Alternately, the Miller test may be performed with a standardized wear material, while changing the slurry from test to test. From this procedure a Miller number may be determined for each slurry of interest. A high Miller number indicates a relatively high abrasivity of the slurry. Other wear test rigs can also be employed for this two-way use, measuring both wear resistance of materials and slurry abrasivity.

The Miller test is easy to perform and yields useful results in wear situations where abrasion dominates. Even for coarse-slurry pipelines, where this condition is met, the Miller test can only give relative wear resistances for various materials used for pipes and linings. It cannot provide the absolute wear coefficients required to estimate the actual wear rate in a pipeline.

The ASTM G2 committee on wear testing, which oversees the Miller machine and many other wear standards, occasionally considers new standards for slurry wear tests more closely modelling the combination of wear mechanisms encountered in centrifugal pumps and other slurry system equipment. An ongoing review of committee publications can be an effective way of keeping abreast of trends in this area.

While the bulk of wear test methods have traditionally been designed for the purpose of relative ranking between materials, recent advances in numerical modelling of turbulent slurry flows have created a need for 'absolute' wear data which quantify wear rate in terms of the numerically predictable properties of the flow field, (*e.g.* particle velocity, concentration, shear stress at the wear surface). In the case of slurry pumps, this need has encouraged the development of wear tests which strive to simulate the actual wear mechanisms while allowing for easy calculation or direct measurement of the flow field properties.

The centrifugal sliding-wear machine was originally proposed by Tuzson *et al.* (1984). A modified version of this machine, developed at GIW Industries, is shown on Fig. 11.4. It simulates wear in pump casings, where, apart from the tongue area, sliding wear is often the predominant wear mechanism. In addition

Figure 11.4. Centrifugal machine for sliding abrasion

to being one of the most accurate tests for determining the relative ranking of various wear materials in slurry applications, it can also be used to obtain absolute sliding wear coefficients (Tian *et al.*, 2005). These are used for input to numerical analyses, based on a function of specific energy dissipated at the wear surface. In this test a rotating channel has a wear specimen embedded in it. Slurry is forced through the channel, and the motion of the channel and the slurry induces a Coriolis force on the particles, causing them to press against the wear specimen as they move. The relatively low velocities experienced by the slurry at all other locations within the machine's circulatory system serves to minimise particle degradation while the forces involved at the wear sample accelerate the wear mechanism and allow for reasonably short testing times. Tuzson *et al.* (1984) observed that the wear marks on the test specimen resemble those found in a pump, suggesting a basic similarity in the wear mechanisms. Further testing at GIW has shown that the exact pattern of wear on identical specimens can vary considerably while still maintaining repeatability in overall weight loss and allowing correlation to the energy expended in creating the wear. This indicates that the random or chaotic nature of wear parallels that of fluid turbulence, where significant real-time fluctuations are often superimposed on an apparently steady (time-averaged) flow.

The wedge test shown on Fig. 11.5 is useful for determining the impact wear coefficients. In practice there are some uncertainties concerning actual angles of impact at any location except near the tip, combined with the inevitable loss of the tip itself as a result of particles impacting directly on the stagnation point. However, by a combination of careful measurements of the actual wear depth at several evenly spaced lines downstream of the tip and appropriate extrapolations back for the tip (generally non-linear), useful coefficients can be obtained.

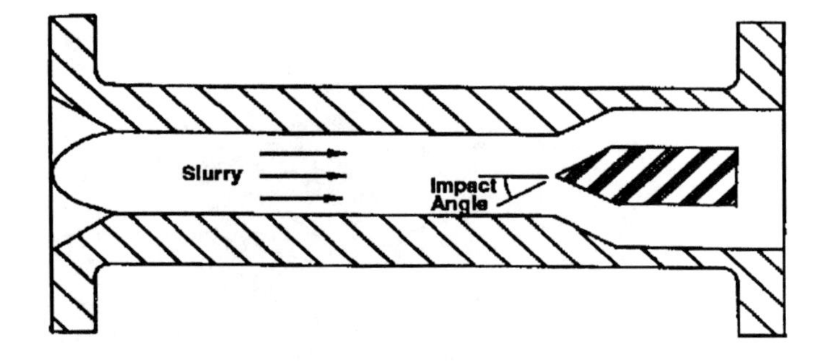

Figure 11.5. Wedge test for particle impact

The difficulties in determining absolute wear coefficients, noted above, show the need for wear measurements on full scale pumps in order to verify or 'back track' the application of wear coefficients. Laboratories which have carried out systematic pump wear measurements include the University of Paderborn in Germany (Wiedenroth 1984a, 1984b, 1988) and the GIW Hydraulic Laboratory (Roco & Addie 1983, Roco *et al.* 1984a). Full-scale testing has the great advantage that actual wear mechanisms and operating conditions are duplicated, removing the 'speculation factor' from the test results. Another advantage is that there is no need to split the total wear rate into impact and sliding components. This is an important point as it appears that the two wear mechanisms do not obey the same scaling laws. Full scale testing on the other hand, has limitations. In a laboratory, the volumes of fresh slurry may be difficult to handle, and degradation of the slurry can be a problem. In the field, it is very difficult to maintain constant conditions, and even, in some cases, to monitor them accurately. In fact, wear testing is an extremely complex subject; a single approach cannot solve all problems, and each situation must be dealt with in its own way.

Abrasive/erosive wear is a complex process which may be affected significantly by a number of variables including the volume concentration of the solids, the velocity and impact angle of the eroding particles, the abrasivity of the eroding solids and the properties that determine the wear resistance of the target surface. The difficulty level in wearing out a target material through erosion is often described using the term specific energy E_{sp} or specific energy coefficient (both in J/m^3). The specific energy is the erosive energy of particles required to remove a unit volume of material. The lower the value of E_{sp}, the greater the expected wear on the target material for identical slurry flow conditions. A value of specific energy coefficient can be determined empirically for either sliding impact wear mode or a combination. In some cases, the reciprocal of specific energy coefficient may be used; it is known as the wear coefficient (W_c). The value of the wear coefficient may vary significantly with the wear model used.

Under the GIW wear model (Tian *et al.* 2005), the local sliding wear rate can be computed using the multiplication of local friction power by the sliding wear coefficient. The friction power of the slurry layer adjacent to the wear surface is estimated from the solid-liquid flow field. This friction power, the product of wall shear stress and the velocity tangential to wear surface, intrinsically includes the effect of local concentration and particle size. In addition to the variables considered in the sliding wear (including the properties of particles and target material), the specific energy for impact wear (E_i) is also a function of the angle of impact, α. A number of impact wear specimens with different angles of the wear test wedge pieces are tested to characterize E_i as a function of α. The impact power of the particles is proportional to the particle density, concentration and the cube (approximately) of the particle velocity. Similar to sliding wear, the impact wear rate can be computed using the impact power and impact wear coefficient (or the reciprocal of impact specific energy coefficient E_i). For a given slurry/wear surface combination, E_{sp} and E_i (α) are expected to be constant.

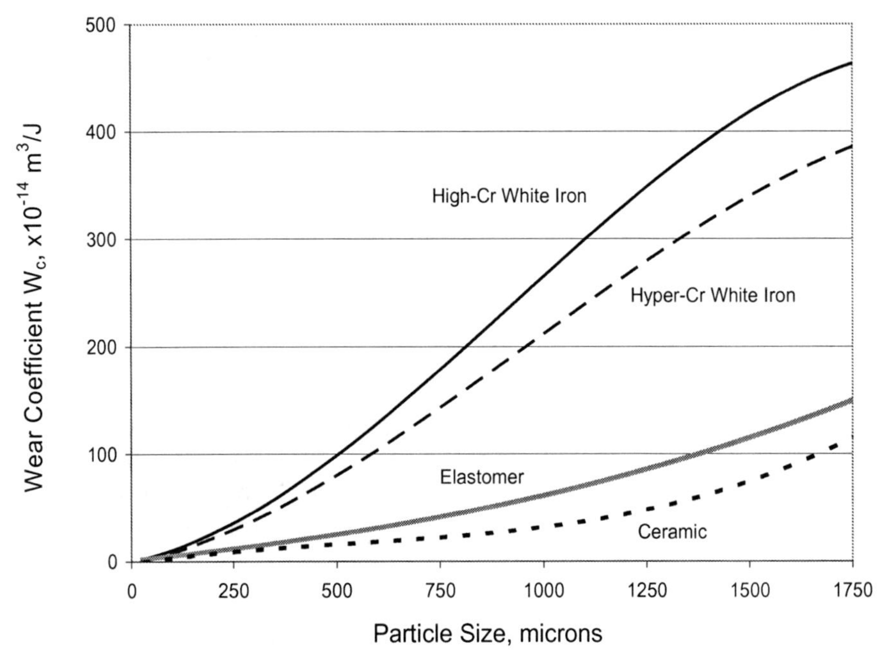

Figure 11.6. Sliding Wear coefficient W_c for different resisting materials in a neutral pH medium and various sizes of abrading particles

Experiments indicate that particle size plays an important role in determining E_{sp} and E_i (α). Fig.11.6 shows sliding erosive wear coefficients at various particle sizes based on the GIW-KP wear model for highly wear-resistant materials including common high-chrome white iron, hyper-eutectic/ultra-high-chrome white irons, elastomers (e.g. natural rubbers) and ceramics (e.g. SiC or SiC-metal composite). It should be noted that the higher the value of the wear coefficient, the lower the wear resistance of the material.

For impact wear conditions, the erosion rate varies for different materials and impact angles. For less ductile materials such as cast iron alloys, normal impact (at right angles to the surface) tends to give the highest wear rate, or the lowest value in E_i. Figure 11.7 shows how ductile (elastomer) and brittle (white iron) materials respond.

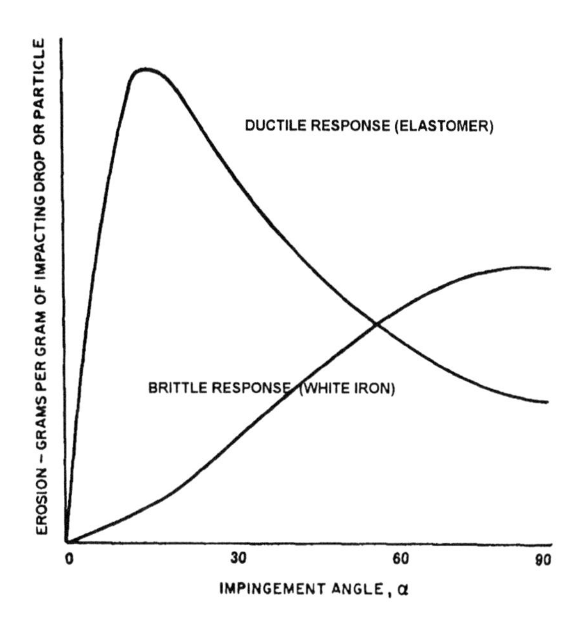

Figure 11.7. Erosion response for different impingement angles and materials

11.5 Numerical Modelling of Flow and Wear

The availability of high-speed computers with ample memory has encouraged many research workers to develop numerical algorithms for analysing flow, and attendant wear, in pumps and other components of slurry pipeline systems. Since wear depends on local values of velocity and concentration, the flow and concentration fields must be computed prior to wear evaluation. Numerical simulation of fluid and particulate flow entails the transformation of the governing partial differential equations of motion into discrete form. The discretised equations are normally non-linear algebraic equations which are solved by an iterative algorithm using the initial and boundary conditions (see Patankar 1980, Poarahmadi & Humphrey 1983). Initial conditions give information about the state of the system before the process begins, but for the steady-state flows of interest here, the initial conditions need not be specified. Boundary conditions give information about what is happening along the boundaries of the flow domain. Together with the equations of motion, the initial and boundary conditions determine the flow within the domain of interest. Thus, in a numerical simulation the boundary conditions must be discretised in the same manner as the governing equations.

One of the major advantages of numerical analysis is that, once a general algorithm has been developed, solutions are easily generated for a wide range of geometries. Conversely, in an experimental situation it can be prohibitively expensive to investigate a large number of variations. In a conceptual sense, the numerical analyst 'turns on' the algorithm and awaits results, just as the physical experimenter starts an experiment and then observes what happens. The numerical analyst has the advantage of being able to alter the variables which represent material properties and boundary conditions, and the ability to test the sensitivity of the studied phenomena to various assumptions or constraints.

However, one should not infer that numerical analysis is a panacea for all flow and wear problems; and major obstacles exist which must be overcome in constructing models which can serve as effective tools for predicting wear. The continuum equations are never exact and thus do not fully represent the physical phenomena, and there is always a truncation error involved in discretised equations, no matter how small the mesh size. In addition to introducing quantitative inaccuracy, the truncation error may lead to computational artifacts which produce qualitatively different results. Another very important limitation of the numerical method is that, with the computing power currently available, it is difficult to include in most models the effects of small-scale turbulence, corner eddies, slip-line effects, and so forth. Although an engineer may not be interested in these small-scale effects as such, they can on occasion affect the overall motion significantly, since dissipation of turbulent energy must necessarily take place at the small-eddy level. Thus, some form of turbulence modelling is required to solve the equations. Turbulence modelling entails a mathematical representation of the Reynolds stresses (which arise from applying averaging techniques to the equations of motion), and the accuracy of any turbulence model must be verified experimentally. The introduction of high concentrations of particles complicates the problem further. This applies both numerically (where generalized models for the interaction of dense solid-liquid flows are still under development) and experimentally (where the presence of solids damages intrusive measurement devices and obscures the field of view). Fortunately, ongoing research in these areas, coupled with technological advances in both computing and experimental hardware, promises to provide progress in this area for some time to come.

Despite these difficulties, the value of numerical modelling can be vast. Although it can never replace theoretical development or physical experimentation, it forms a complementary discipline, supplementing the traditional techniques when investigating flows with complex geometries. Two-dimensional models provide a relatively simple approximation for studying flow and wear in the casings of centrifugal slurry pumps. In these models the plane normal to the pump axis is considered, and it is assumed that the side walls of the pump do not affect the gross flow structure. Numerical analyses of this sort typically employ a potential function for the inviscid fluid velocity distribution, introducing a branch cut in the multiply-connected region as shown on Fig. 11.8. The main advantage of the two-dimensional inviscid code is that it runs relatively rapidly.

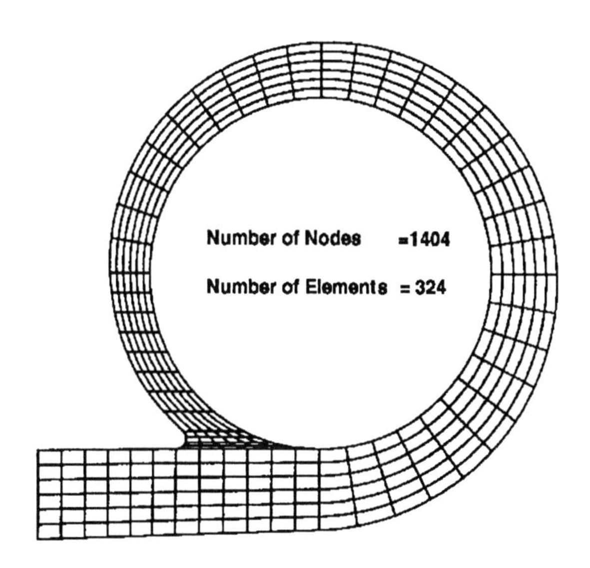

Figure 11.8. Casing geometry used for numerical analysis

Computation of secondary casing flows (quasi-three-dimensional modelling) is much more involved since the flow patterns are more complex and since the viscous forces are of the same order of magnitude as inertial forces and cannot be ignored in constructing the governing equations of the numerical model. Patterns of secondary flow can vary significantly depending on the operating conditions and pump geometry, and the assumptions made about the impeller outlet velocity profile. This profile has a particularly strong influence on performance, and work done by (Hergt *et al.* 1994) has shown that typical profiles in slurry pumps are quite unlike those encountered in conventional water pump designs. The differences are caused by the distortions from the optimal impeller design dictated by the need to pass large solids. Secondary-flow programs are, however, worth the trouble of developing, since they yield detailed results along the casing side walls, enabling prediction of gouging tendencies.

Two and three dimensional numerical methods have also been applied to flow in the impeller passages. As for other pump components, the impeller poses special challenges in the case of slurry pumps. The numerical model must include viscous forces and be able to simulate recirculation and separation within the flow passage. This modelling is a necessity since these effects cannot always be avoided, given the requirements of large solids passage and widely- varying flows that are often imposed on slurry pumps by their users.

Once the velocity field of the flow as a whole has been determined by either two-dimensional or three-dimensional computations, the next step is to calculate the particle trajectories. These calculations are based on the momentum equation, taking into account particle inertia, centrifugal force and fluid resistance (Roco & Reinhard 1980). If the solids concentration is small, the consideration of wear could begin at this point in the analysis, with sliding abrasion occurring where the particles move along the boundary, and particle-impact erosion where the trajectories arrive at an angle. For conditions typical of slurry transport, however, near-wall concentrations are high. The influence of solids concentration on erosion was discussed in Section 11.1, where it was noted that increasing concentration does not necessarily lead to a monotonic increase in wear rate. In these circumstances the question of wear can only be addressed after it is known what fraction of particles is supported by granular contact. Determining the fraction of contact-load solids in a pump casing is basically similar to evaluating the stratification ratio for heterogeneous pipeline flow, a problem which was discussed extensively in Chapters 4, 5 and 6.

As noted in those chapters, even for the simple configuration of pipeline flow the physical mechanisms determining stratification are complex, and achieving better understanding remains a major research aim. For the more complicated configurations encountered in pumps, there is even less certainty as to the action of the various physical mechanisms involved; the field remains rather speculative, and a consensus among research workers is not yet in sight. An approach to this problem which has received considerable attention in the literature is due to Roco and his co-workers. This researcher worked at the University of Paderborn in Germany (Roco & Reinhard 1980) and subsequently was associated with Shook in Saskatchewan (Roco & Shook 1983, and numerous other joint publications). During the mid 1980's, Roco worked in conjunction with the GIW Hydraulic Laboratory (Roco *et al.* 1984a, 1984b, 1987), where the modelling of wear in pumps has an ongoing priority (Pagalthivarthi & Addie 1989; Pagalthivarthi *et al.* 1990, Pagalthivarthi *et al.* 1991).

After the fraction of contact-load solids has been estimated at various locations within the pump, the procedure continues to the remaining step: the consideration of erosion rates. In Roco's algorithm the local erosion rate is taken as proportional to the rate at which energy is lost to friction and other dissipative mechanisms. In this approach, the local energy flux of the particles at the wall is related through empirically-determined coefficients to the predicted wear rate. The total wear rate may also be computed as the sum of an impact component and a sliding component. However, this linear superposition may not be strictly valid. An overall wear coefficient, determined from the wear tests on pump casings which were mentioned in the previous section, helps predict the total wear rate without attempting to break it into its component parts.

Although various remarks have been made throughout this section concerning limitations of numerical wear analysis, this technique remains of significant value to the design engineer. The results of numerical analysis are especially useful when comparing alternative designs, and the capability of predicting local wear rates can be very helpful in avoiding catastrophic erosion failures. For these reasons the numerical and experimental study of liquid-solid flows has been particularly stressed at the GIW Hydraulic Laboratory.

11.6 Pump and Wear Variables

As noted previously, much wear performance is evaluated *post facto*. A good example is the work by Walker (2002). This approach is necessarily limited when it comes to predicting component wear rates, but numerical modelling can help. Although modelling of specific wear calculations cannot be described here, it is worthwhile presenting a general treatment of wear rates, illustrating trends and the effect of different variables. As service conditions vary in practice, pumps do not always operate at the point originally selected, (where the wear and efficiency are near-optimal). Also pump designs themselves diverge from the ideal to permit passage of large spheres and to meet other special conditions. Thus any generalisation should be considered as only a preliminary guide.

Let us consider handling 300 μm solids at a volumetric concentration of 20%, and use this to show the effect of different variables on component wear. Beginning with a number of designs of pumps of different flow capabilities from Addie *et al.* (2005), with each running near its BEP discharge and producing 50 m of head, and using the wear model described in Pagalthivarthi (2001), we obtain Fig 11.9 for the average wear along the vane of an impeller. The various pumps shown in Fig 11.9 represent designs that have been developed and tested by GIW Industries over the last 25 years. They have impellers varying from

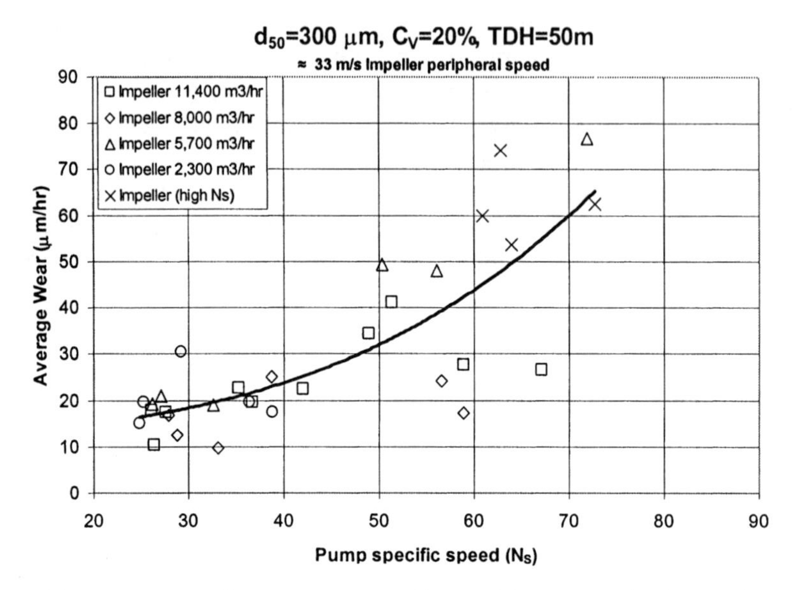

Figure 11.9. Average impeller wear versus specific speed for pumps of various designs and BEP-flow rates

2.3 m to 0.5 m while the suction branches vary from about 1 m to 0.3 m. The pumps of various sizes were selected to operate at the flows noted. The pumps of the group denoted 'high specific speed' were operated at their individual best-efficiency-point flows for the 50 m head. While off-BEPQ operation and design differences introduce significant variability, a clear trend of increasing wear with increasing specific speed is evident, and it is of interest that an order-of-magnitude value for wear rate for the different designs is essentially independent of the capacity (size) of the pump.

A similar graph for suction-liner wear is shown on Fig. 11.10, using the same pumps and the methods outlined in Bross (2001). Despite significant variability a clear trend independent of flow is again evident. The wear rate plotted on this figure is for the "nose face" adjacent the impeller eye. The values shown have been averaged over a standardised thickness described in Sellgren *et al.* (2005). It should be noted that as wear occurs the wear rate (without adjustment of the gap) increases with the increase in leakage flow, and that if the adjustment of the nose gap as shown in Addie (2000) were to be carried out, it would increase wear life significantly. Here the impeller clearing vanes were all set to be small, although as shown in Addie (2002) that deep clearing vanes can halve wear in the nose area of the suction liner, albeit at a cost of some 2 to 5% in pump efficiency.

Figure 11.11 displays the casing wear rate in the same way and to the same scale as the previous wear-rate figures. In this case the wear shown is the two-dimensional maximum (location) wear on the casing centreline as described in Roco (1983) and Pagalthivarthi (1990). As wear of this type occurs, shell areas increase and velocities decrease. This can cause an increase (slight in most cases) of the wear life beyond that obtained from a linear extrapolation based on wear rate and casing thickness.

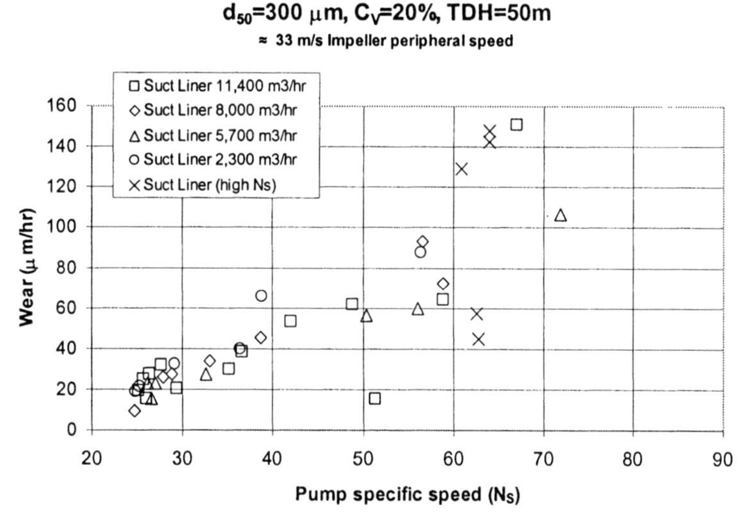

Figure 11.10. Suction liner wear versus specific speed for pumps of various designs and BEP-flow rates

11.7 Variation of Wear with Design Geometry

The wear plots presented in the previous section show considerable variation from one pump to another. This is due in large part to variations in design geometry that are associated with sphere sizes to be passed, number of vanes, and other features. Although a complete investigation of the affect of geometry on wear has yet to be made, it is of interest to consider some key variables that influence wear.

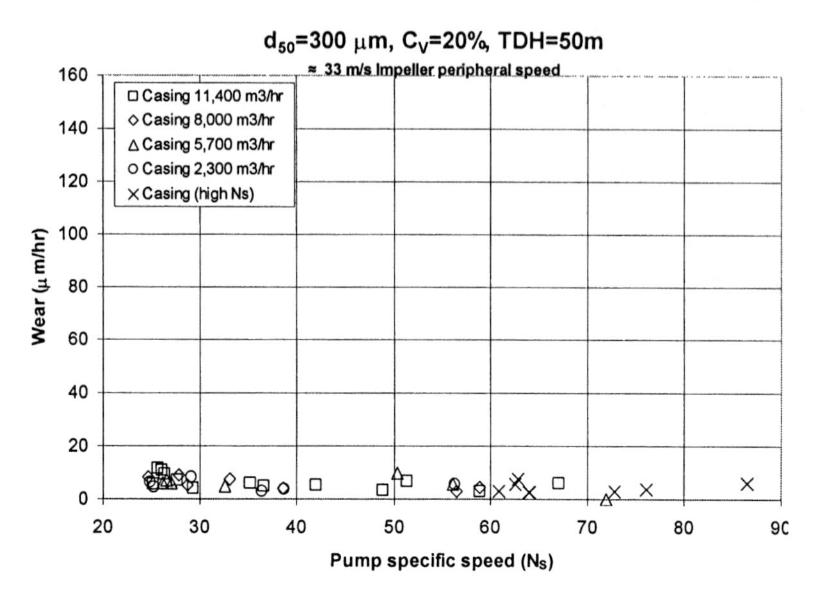

Figure 11.11. Maximum casing wear versus specific speed for pumps of various designs and BEP-flow rates

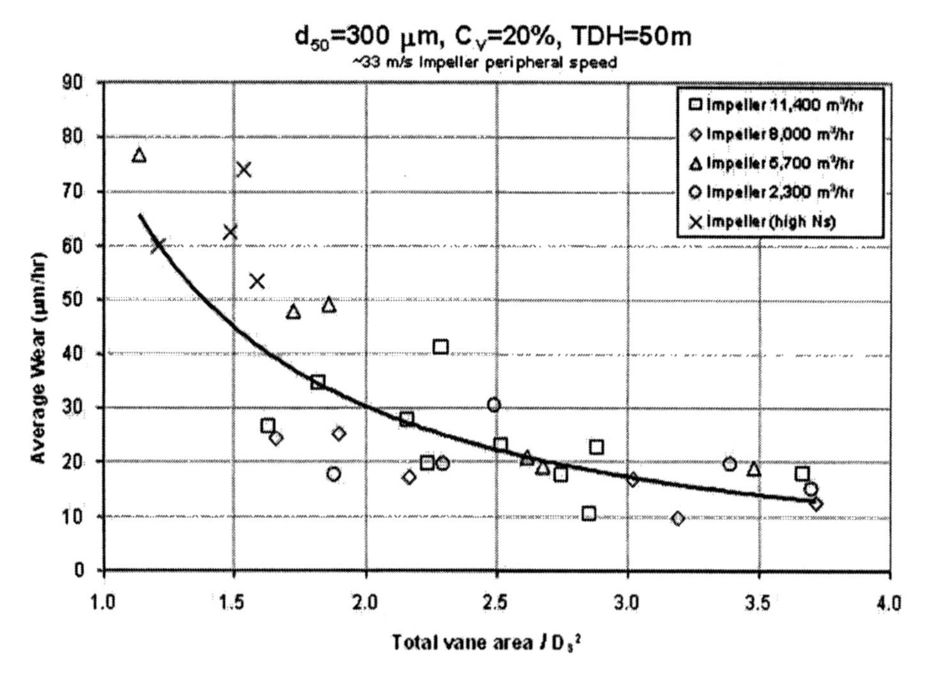

Figure 11.12. Average impeller wear *versus* ratio of total vane surface area to square of suction diameter

For impellers, it is found that there is a weak geometric correlation of the average vane wear rate against the ratio of the vane surface area to the square of the suction diameter D_S. This correlation is plotted on Fig. 11.12, showing that a higher relative vane area leads to considerably lower wear.

For suction liners it is found that the impeller ratio (impeller diameter D/suction diameter D_S) can be used as a wear predictor as shown in Fig. 11.13. This plot shows that it is critically important to keep the suction diameter of the pump as small as possible for a given duty.

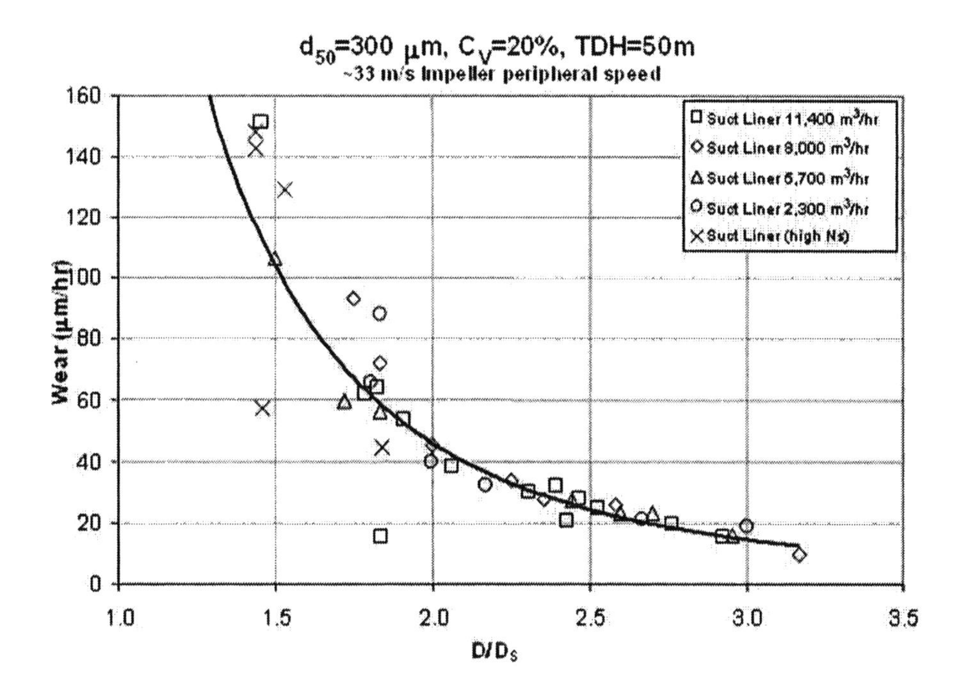

Figure 11.13. Suction liner wear versus ratio of impeller ratio D/D_s

For the casing, the size of the collector (or base circle) relative to the impeller diameter can be varied within reason without affecting hydraulic performance significantly (see definition sketch Fig. 11.14). Shells with ratios larger than 0.9 are said to have large base circles. This ratio can be an important parameter in relation to wear. As indicated on Fig. 11.15, the correlation of this ratio with casing wear rate provides a trend line that may be used to estimate likely shell wear. This figure shows that higher wear tends to occur where the shell ratio R_{T3}/D is less than 0.9.

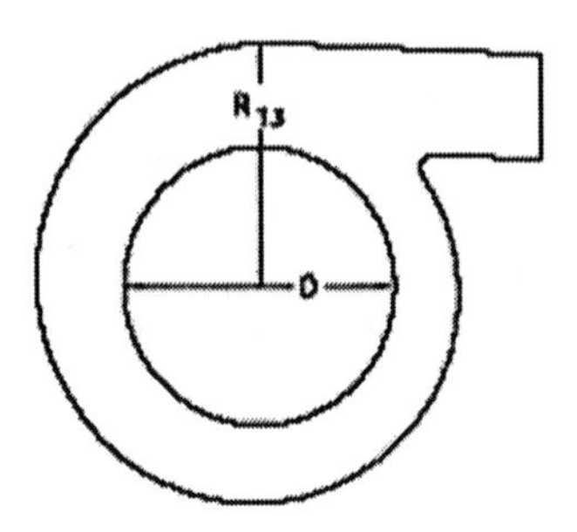

Figure 11.14. Definition of representative shell radius in a pump with impeller diameter D

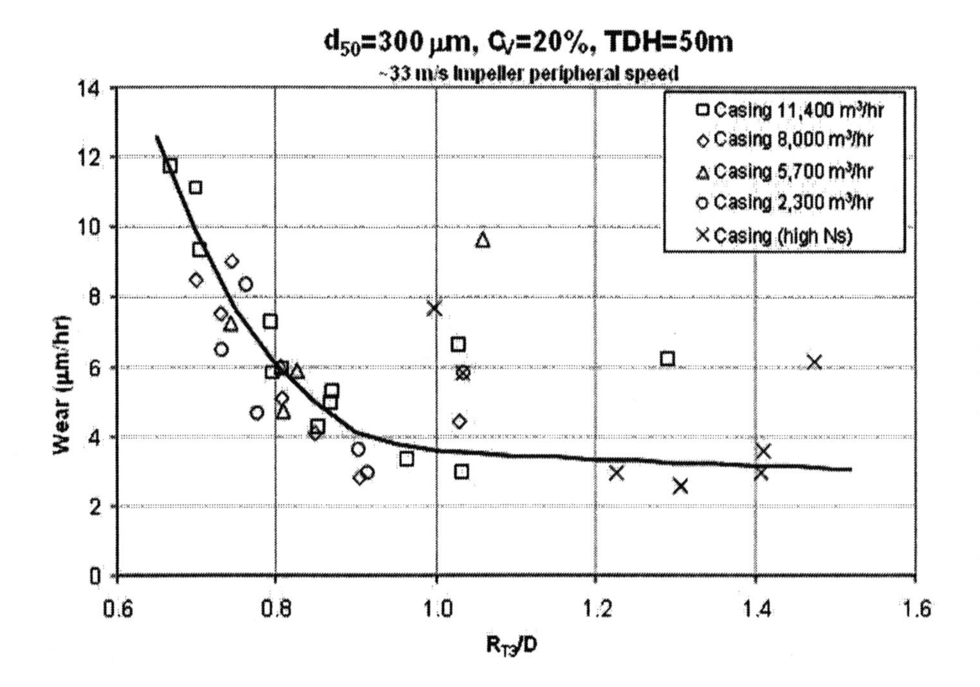

Figure 11.15. Casing wear *versus* the ratio of base circle radius to impeller diameter

11.8 Variation of Component Wear with Different Duties

The data set, assembled to produce the figures of the previous two sections may also be used to show the effect of different pump head and design specific speed, solids size and solids concentration. As outlined in Sellgren *et al.* (2005), the wear for three different heads, solids sizes and concentrations may be modelled by interpolating the wear at three different specific speeds. Figure 11.16 displays calculated wear rates for the nine different conditions for pump casing, liner and impeller components, made of white iron and exposed to a silica sand slurry.

It should be noted that the casing wear shown on Fig. 11.16 is the peak wear along the two-dimensional parting line and is for cases where the pump is operating near its BEP discharge. In the case of the impeller, the wear shown is the average wear over the blade, and in the case of the liner it is the two-dimensional wear in the area adjacent the impeller suction face and assumes that the impeller has small front clearing vanes. The central crossing point on each panel of the figure shows wear rate at a head of 50m, a solids d_{50} size of 300 μm and a concentration by volume of 20%. The top and bottom scales, and those on the right-hand side show the effect of varying head, solid size and concentration. The resulting wear rate is read from the scale on the left-hand side of each panel. Intermediate values of any quantity are to be estimated by linear extrapolation.

11.9 Practical Considerations and Field Experience

In practice, operating duties vary and slurry pumps do not always operate near the optimum points on the curves for wear and efficiency. Pump designs vary a great deal, and, as might be expected, wear varies considerably from pump to pump, and from application to application. Nevertheless, there are several recurrent patterns of wear which are typically observed in slurry pump casings and impellers, and thus merit particular attention. Specifically, slurry-pump casings often experience maximum wear in the outer radius or belly (see Fig. 11.17), a zone where sliding beds of solids form as a result of centrifugal forces. Some specific area on the circumference of the casing generally experiences the maximum wear rate, but the location of this area can shift with the operating conditions and the geometry of the casing. For increasing flowrates, and casings of more annular geometry, it is found that the location of maximum wear tends to shift toward the pump discharge and away from the tongue. These trends are illustrated by Fig. 11.18.

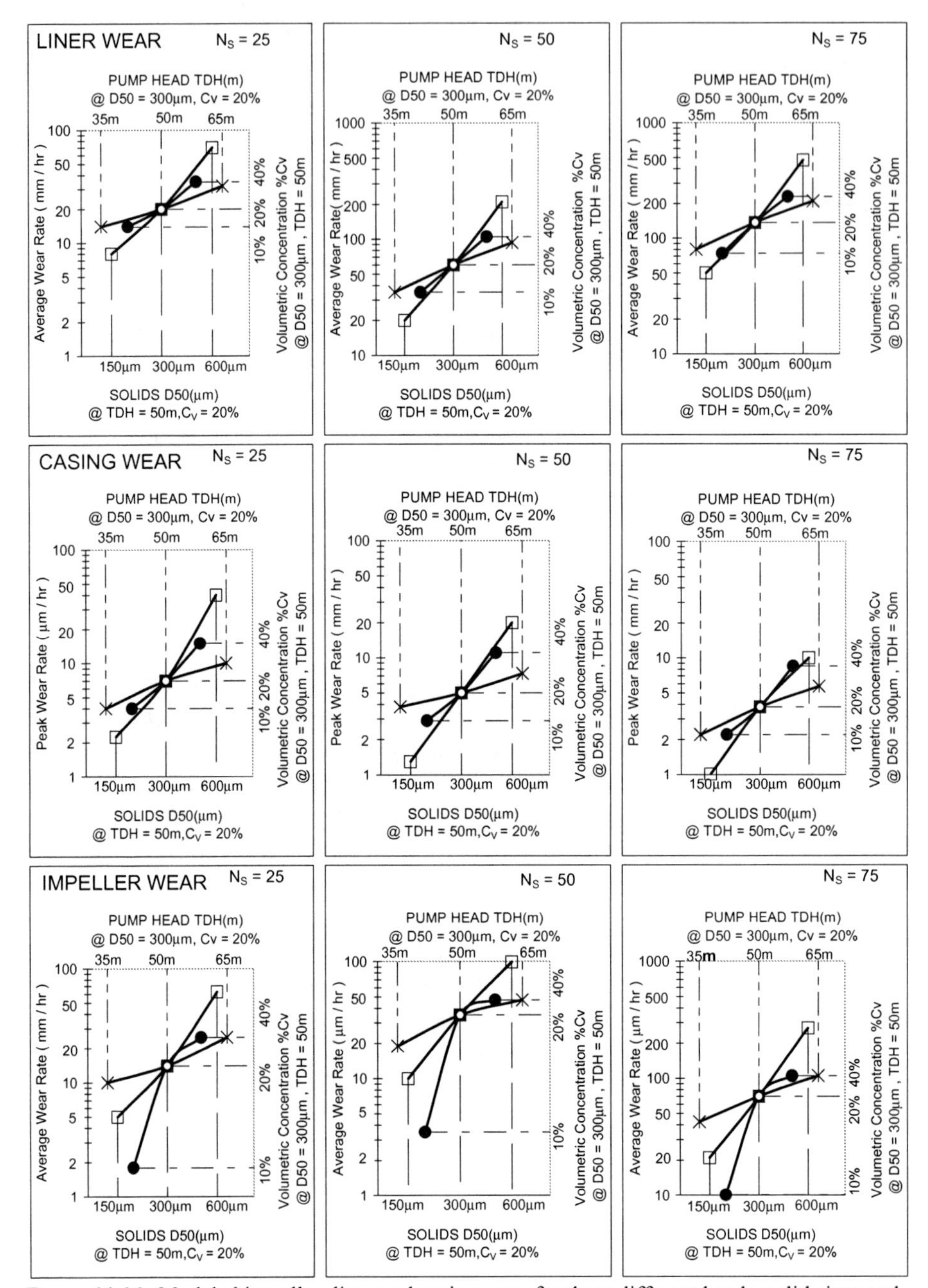

Figure 11.16. Modeled impeller, liner and casing wear for three different heads, solid sizes and concentrations and three specific speeds

Figure 11.17. Typical worn pump casing

Another common wear pattern in slurry-pump casings is gouging or extreme localised wear in the side wall of the casing just downstream of the tongue (see Fig. 11.19). This gouging is initiated by the three-dimensional eddy generated where the tongue parts the fluid. The severity of the gouging depends on the pump design but is typically dominated by the ratio between the operating flowrate and best-efficiency flowrate. Specifically, pumps operating well below the best-efficiency flowrate recirculate large volumes of flow past the tongue, increasing the velocity and size of the gouging eddy.

Gouging eddies may also occur at other locations in the slurry-pump casing, or the impeller. In some instances, such eddies can be identified as having their origin in geometric discontinuities, and can be eliminated by judicial design modifications. In other instances, they may be traced to operational considerations, such as operation well away from BEP flow, as noted above. In all cases, it is vital to have a sound understanding of the large-scale flow patterns within the pump. This understanding depends on careful observation and also on analysis, using both basic fluid mechanics and numerical modelling. A healthy sense of three-dimensional visualisation is extremely useful in suggesting design or operational modifications which can improve overall wear performance.

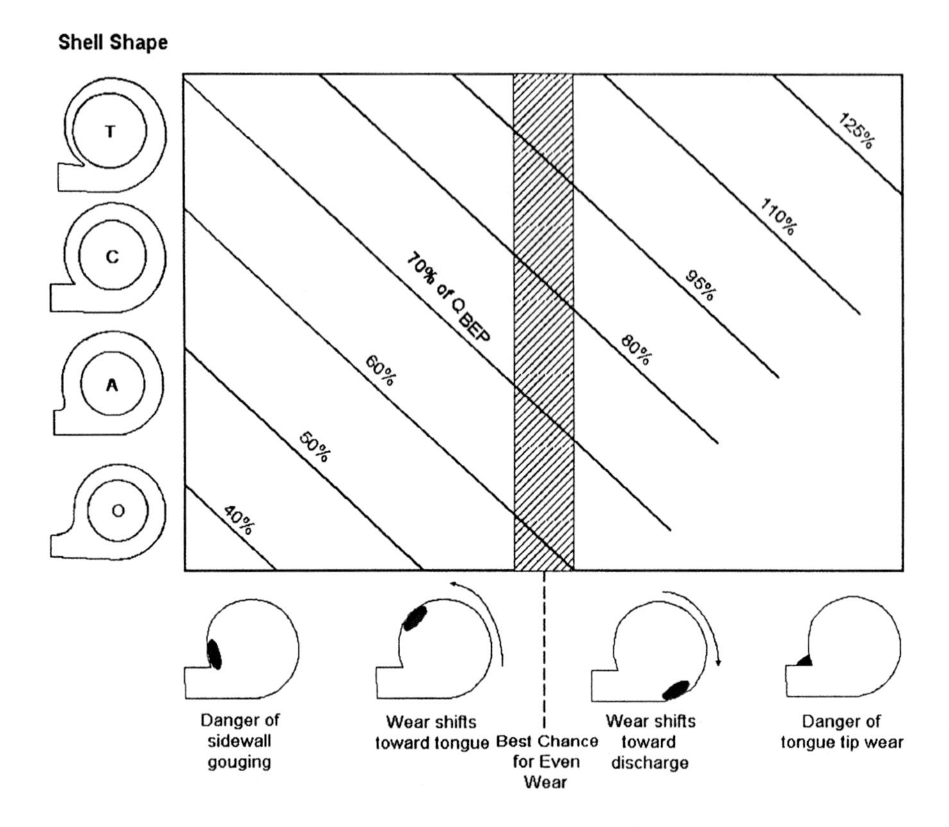

Figure 11.18. Trends in location of casing belly wear

Figure 11.19. Example of gouging wear near tongue

Gouging eddies in the casing may also affect flow patterns in the clearance between the impeller and the suction-side liner, producing secondary gouging here. The localised wear can be severe (see Fig. 11.20), and in this instance a practical method of extending the life of the liner is to rotate it 180 degrees when one-half to three-quarters of its estimated life has elapsed, thus moving the wear zone to a new portion of the surface.

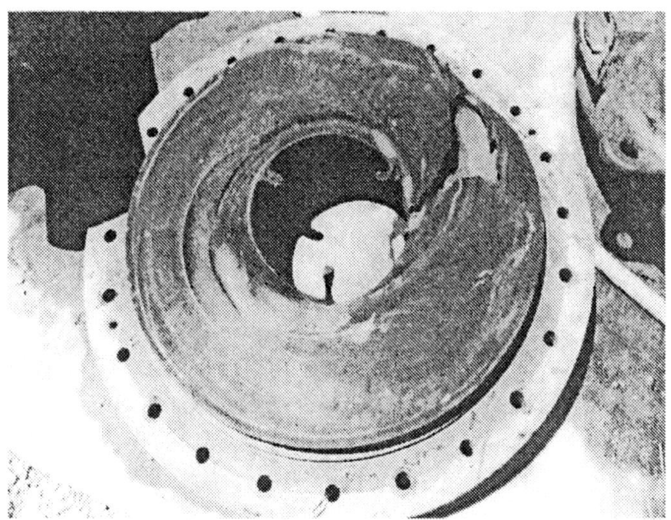

Figure 11.20. Secondary gouging of suction-side liner

As the slurry pump casing is a consumable part, its wear performance can often be improved by noting the wear patterns experienced, and applying the known operational and geometric considerations which affect those patterns. Then, through experimentation, the casing size, geometry and operating conditions can be modified in a direction which indicates increased wear life.

Wear of the impeller of a slurry pump is often closely related to the hydraulic efficiency which it exhibits. This seems reasonable, since improved hydraulic efficiency generally coincides with reduced velocities in the recirculating eddies which cause localised wear. However, a well-designed slurry-pump impeller can often sustain considerable wear before its pumping capacity is reduced to an unacceptable level, and may appear to be worn out long before it needs to be replaced from an operational standpoint (see Fig. 11.21). On an impeller the two areas of worst wear are usually near the inlet and the outlet. At the inlet eddies can be induced by changes in curvature of the flow, or by the intrusion of the recirculating flow from the clearance between the impeller and the suction liner. The high velocities at the impeller outlet are the major cause of problems there.

Figure 11.21. Typical worn impeller

The key to improving the overall wear performance of any slurry-pump component is often found in identifying the causes of localised wear and either eliminating them or spreading them over a wider area. Considerations of design geometry and operating conditions play a role in this process. As discussed earlier, the materials of construction may also be varied in order to improve wear performance. Typically, some balance of strength, toughness, wear resistance, serviceability and cost must be struck, with no one material holding the advantage in all respects.

The wear of other components of a slurry-pipeline transport system is also important. The pipe itself can wear, of course, both by the particle impact associated with fluid turbulence and by sliding abrasion. The latter mechanism is usually dominant in pipe flow, with wear concentrated in the lower portion of the pipe for both fully-stratified and heterogeneous flow regimes. Sometimes the resulting wear can limit pipe life, and in this case it may be worthwhile to rotate the pipe through 120° at the estimated one-third and two-thirds points of its life-span; thus limiting the wear at any one part of the pipe wall. In more serious cases a lined pipe may be considered, but this step is seldom required except when the slurry flow combines with chemically corrosive conditions. A stationary deposit of solids in the pipe produces a characteristic wear pattern, with the wear concentrated in bands at each side of the lower portion of the pipe interior, say at the 4 o'clock and 8 o'clock positions. Sometimes the occurrence of this wear pattern provides the first indication that a system has been operated at too low a mixture velocity, *i.e.* in the inefficient stationary-deposit flow regime (see Section 15.7).

In bends and other fittings wear is usually much more severe than in straight lengths of pipe, with wear rates easily increasing by an order of magnitude or more. The basic reason for this enhanced erosion is the curvature of the stream lines, which produces centrifugal forces that cause the particles to impact on the boundary, a point noted earlier in this chapter. In some cases it is desirable to line the bends in order to avoid frequent replacement, in other instances the replacement cost is tolerated, or else the bends are fabricated of metal that is thicker, and possibly harder, than that used in the straight lengths of pipe.

For systems pumping simple fluids in small-diameter pipes it is quite common to control the flow rate by a throttling valve in the pump discharge pipe. This configuration is highly wasteful of energy, as the pump produces a higher head than is required, and the excess head is dissipated at the valve. The resulting energy loss may sometimes be tolerated in small fluid systems, but designers accustomed to this approach sometimes use the same configuration in slurry applications, generally with unfortunate results. Slurry going through a partly-closed valve can produce extremely rapid wear, and at the same time the pump will be forced to operate well into the under-discharge range, inviting the gouging type of wear described earlier. The appropriate solution is usually to reduce the diameter of the pump impeller (see Section 9.2 and Case Study 9.2), eliminate the throttling valve and use other methods of flow control such as the variable-speed drives discussed in Section 12.3.

11.10 Particle Attrition

Particle attrition can influence both the value of the delivered product and the frictional loss (and operating cost) of the pipeline. For example, the attrition of coal particles during transport gives a product in which water is retained between the fine particles, greatly increasing the cost of subsequent dewatering. The result is a decrease in the value of the coal slurry discharged from the pipeline. On the other hand the fine particles produced by attraction will act as a carrier fluid for the coarser particles (see Chapter 7), causing a reduction in friction loss and in specific energy consumption.

Because coal is highly susceptible to particle degradation, most of the attrition studies found in the literature have dealt with coal. As with the surface wear discussed in earlier sections of this chapter, it is found that a disproportionate amount of particle attrition takes place in pumps, and also in elbows and other fittings. To some extent, the relative effects of pumps and pipeline on attrition may be separated by using various lengths of test loop, by employing a lock-hopper feed (such as that of Fowkes & Wancheck, 1969) or by testing in a wheel-stand (Traynis 1977). A good review of the field was provided by Gillies (1991), drawing on the experimental work carried out over many years at the Saskatchewan Research Council (see, for example, Gilles *et al.* 1982). These publications use a breakage distribution function of the type employed to model grinding in rod and ball mills. The experimental results from Saskatchewan show that the rate of particle attrition is large initially but decreases with time as a test is continued.

An experimental study pertaining to particle attrition was carried out in 1994 at the GIW Hydraulic Laboratory, and subsequently documented by Wilson & Addie (1995). The tests were carried out in a recirculating loop with internal pipe diameter 305 mm. Initially, the particles were coarse gravel, with diameter > 10 mm. There were practically no particles in the size range 0.5 to 10 mm, and few below 0.1 mm. For concentrations in the range of commercial interest, the pressure-drop characteristics for the coarse material are found to be in good accord with the simplified relation for fully-stratified flow presented in Section 5.5, *i.e.* Eq. 5.13.

In the general case, it is necessary to multiply the right hand side of Eq. 5.13 by a coefficient, B′, i.e.

$$\frac{i_m - i_w}{(S_m - 1)} = B' \left[\frac{V_m}{0.55\ V_{sm}} \right]^{-0.25} \tag{11.1}$$

The coefficient B′ is set to 1.00 when the particles are all large, but requires experimental evaluation for particle grading curves showing a significant fraction of sand-size solids.

The experiment continued for 60 minutes, with six series of readings of pressure drops at various throughput velocities taken at intervals. Successive passages through the pump rounded the gravel and diminished its size somewhat, producing a significant sand fraction in the process. The gradient i_m also diminished considerably. The first runs were with a slurry of coarse particles only. In the following series of runs, having up to 60 minutes of exposure time there was no change except for degradation of the solids in the system. With these series the question arose whether the slope M of Eq. 11.1 should be fixed at 0.25 or should be re-evaluated for each set of runs. In practice the data of the initial three series were too jumpy to give a good evaluation of the slope, and only the last two series (with the largest fraction of fines) were suitable for good independent slope determinations. Figure 11.22 is a plot of all runs in the final series which have $V_r > 1.0$, together with the best-fit line. For this line, and also that of the penultimate series, the exponent M was found to be approximately 0.35, i.e. not far from the value of 0.25 for the fully-stratified case.

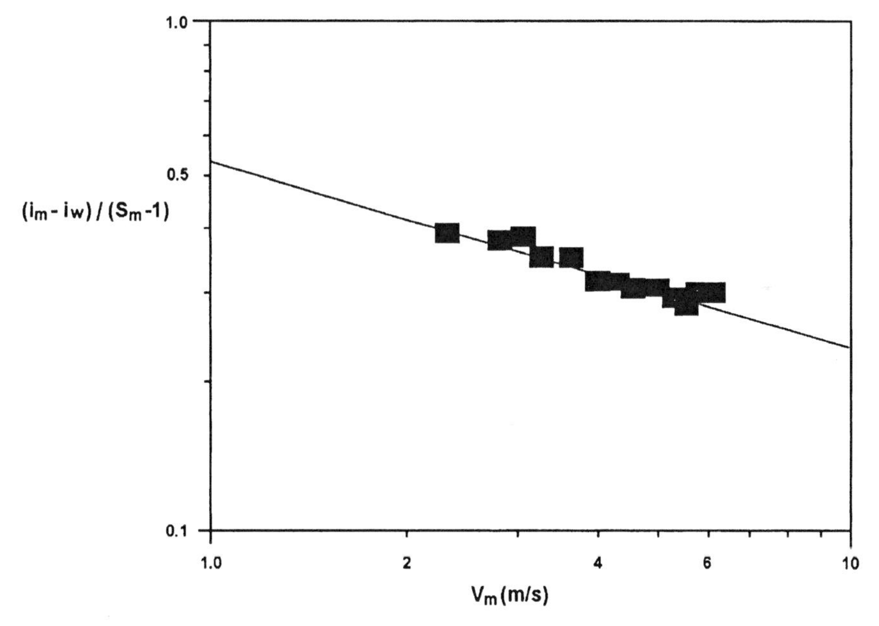

Figure 11.22. Solids-effect plot after 60 minutes of exposure time, from Wilson & Addie (1995)

Thus it was decided to keep M fixed at 0.25 and study only variation of the coefficient B′ in Eq. 11.1. The B′ values range from 1.0 for the first series (freshly-loaded coarse particles only) down to 0.48 for the last two series (after

60 minutes of exposure time, equivalent to about 120 passes through the pump).

A grading curve obtained following the last series of runs indicated that 30% of the original gravel had been degraded to sand during the 60 minute time period. However this 30% reduction in coarse material produced a 52% reduction in B′, showing that the sand and gravel components interact in reducing the solids effect. To see how this interaction might occur, consider a large particle moving by saltation and descending toward a lower layer of heterogeneous sand-water flow (with $C_v < C_{vb}$). In this layer the mixture density increases with depth, and the resulting Richardson-number effect produces a steepening of the velocity profile. The increasing density encountered by the descending large particle slows its approach to the pipe invert, and the steepened velocity profile enhances the rotation of the large particle, and the lift which it experiences. Thus the coarse particle is likely to be deflected upward by the sand layer without fully transferring its axial momentum to the pipe invert (indeed in may well be deflected before contacting the pipe at all). The result of this process is a reduction of $i_m - i_w$ (and B′), and Wilson & Addie (1995) showed that a plot of log (B′) versus time indicates that the lowering of B′ with exposure time can be closely approximated by an exponential decay. The formula for this line, valid only for the specific tests, is

$$B' = 1.0\, e^{-0.012t} \tag{11.2}$$

where the exposure time t is measured in minutes. This time is also a measure of the number of passes through the pump.

In summary, after one hour of testing in the recirculating pipe loop, the coarse solids were extensively degraded and rounded, with 30% of the solids now found to be in the sand size range. At the same time, there was a greater-than-proportional drop in the relative excess gradient, implying that the sand and gravel components are interactive rather than simply additive. The results for various testing times suggest that the excess-gradient parameter decays exponentially with test time or, equivalently, with number of passes through the pump.

REFERENCES

Addie, G.R. & Bross, S. (2000). Modeling of slurry pump nose wear. *15th Annual Regional Phosphate Conference.*

Addie,G.R. & Bross, S. (2001). Prediction of impeller nose wear behavior in centrifugal slurry pumps. *Hydraulic Research Development.*

Addie, G.R., Dunens, E., Pagalthivarthi, K. & Sharpe, J. (2002). Oil sands tailing pump operating cost considerations. *Hydrotransport 15.*

Addie, G.R., Kadambi, J. & Visintainer, R., (2005). Design and Application, Slurry Pump Technology. [*9th International Symposium on Liquid Solid Flows – ASME Fluids Engineering Summer Conference – Houston TX.]*

Bagnold, R.A. (1956). The flow of cohesionless grains in fluids. *Phil. Trans. Roy. Soc.*, London, Ser A, Vol. 249, pp. 235-297.

Finnie, J. (1960)., Erosion of surfaces by solid particles, *Wear*, Vol. 3, No. 3-4.

Fowkes, R.S. & Wancheck, G.A. (1969). Materials handling research: hydraulic transportation of coarse solids, *Report 7283*, U.S. Department of the Interior, Bureau of Mines, USA.

Gillies, R.G. (1991). Particle size degradation in slurries, pp. 227 to 237 of *Slurry Handling Design of Solid Liquid Systems* Ed. Brown, N. and Heywood, N., Elsevier Applied Science.

Gillies, R.G., Haas, D.B., Husband, W.H.W., Small, M. & Shook, C.A. (1982). A system to determine single-pass particle degradation by pumps. *Proc. Hydrotransport 8,* BHRA Fluid Engineering, Cranfield, UK, pp. 415-431.

Hergt, P., Pagalthivarthi, K.V., Brodersen, S. & Visintainer, R. (1994). A study of the outlet velocity characteristics of slurry pump impellers. *Proc. 5th Intern'l Symposium on Liquid-Solid Flows,* ASME, Lake Tahoe, NV.

Hutchings, I.M. (1992). *Tribology: Friction and Wear of Engineering Materials*, Edward Arnold, London, UK.

Hydraulic Institute, (2004). *Rotodynamic (centrifugal) slurry pumps for nomenclature definitions, applications and operations.* ANSI.

Li, J. (2005). Prediction of particle rotation in s centrifugal accelerator erosion tester and the effect on erosion rate, Wear 258, 407-502.

Miller, J.E. (1987)., The SAR number for slurry abrasion resistance. *ASTM STP 946, Slurry Erosion - Uses, Applications and Test Methods.* Ed. Miller and Schmidt, pp. 156-166.

Nishiuchi, S., Yamamoto, S., Tanabe, T., Kitsudo, T., Matsumoto, H. & Kawano, Y. (2003). Development of Stainless Spheroidal Carbide (VC) Cast Iron. *AFS Transactions 2003*, Paper 03-084.

Pagalthivarthi, K.V. & Addie, G.R. (1989). Prediction of dredge pump shell wear. *Proc. WODCON XII, 12th World Dredging Conference,* pp. 481-504.

Pagalthivarthi, K.V., Desai, P.V. & Addie, G.R. (1990). Particulate motion and concentration fields in centrifugal slurry pumps. *Particulate Science and Tech. Vol. 8.* Hemisphere Publishing.

Pagalthivarthi, K.V. & Helmly, F.W. (1990). Applications of materials wear testing to solids transport via centrifugal slurry pumps. *Wear Testing of Advanced Materials, ASTM STP 1167,* Eds. Divakar & Blau, Am. Soc. for Testing and Materials, Philadelphia, PA.

Pagalthivarthi, K.V., Desai, P.V. & Addie, G.R. (1991). Quasi-3D computation of turbulent flow in centrifugal pump casings. *Proc. Multidisciplinary Applications of CFD, ASME Winter Annual Meeting,* Dec. 1991.

Pagalthivarthi, K.V., & Addie, G.R., (2001). Prediction methodology for two-phase flow and erosion wear in slurry impellers. *4th International Conference on Multiphase Flow New Orleans,* May 2001

Patankar, S.V., Pratap, V.S. & Spalding, D.B. (1975). Prediction of turbulent flow in curved pipes. *J. Fluid Mechanics,* Vol. 67, p. 583 ff.

Patankar, S.V. (1980). *Numerical Heat Transfer and Fluid Flow,* Hemisphere.

Poarahmadi, F. & Humphrey, J.A.C. (1983). Modeling solid-fluid turbulent flows with application to predicting erosive wear. *Physio-Chemical Hydrodynamics,* Vol. 4.

Roco, M.C. & Addie, G.R. (1983). Analytical model and experimental studies on slurry flow and erosion in pump casings. *Proc. 8th Intern'l Conf. on Slurry Tech.* Slurry Transport Association, p. 263.

Roco, M.C. & Reinhard, E. (1980). Calculation of solid particle concentration in centrifugal pump impellers using finite-element technique. *Proc. Hydrotransport 7,* BHRA Fluid Engrg., Cranfield, UK, pp. 359-376.

Roco, M.C. & Shook, C.A. (1983). Modeling slurry flow: the effect of particle size. *Canad. J. Chem. Engrg.* Vol. 61, pp. 494-503.

Roco, M.C., Nair, P., Addie, G.R. & Dennis, J. (1984a). Erosion of concentrated slurries in turbulent flow. *Liquid-Solid Flows and Erosion Wear in Industrial Equipment.* ASME FED-Vol. 13, Ed. Roco, pp. 69-77.

Roco, M.C., Addie, G.R., Dennis, J. & Nair, P. (1984b). Modeling erosion wear in centrifugal slurry pumps. *Proc. Hydrotransport 9.* BHRA Fluid Engrg., Cranfield, UK, pp. 291-315.

Roco, M.C., Nair, P. & Addie, G.R. (1987). Test approach for dense slurry erosion. *ASTM STP946, Slurry Erosion - Uses, Applications and Test Methods.* Ed. Miller and Schmidt, pp. 185-210.

Sellgren, A., Addie. G.R. & Pagalthivarthi, K. (2005). Prediction of Slurry Component Wear and Cost, *25-WEDA / 37 TAMU Conference*

Sundararajan, G. (1991). Comprehensive model for the solid particle erosion of ductile materials. *Wear 149,* 111-127.

Tian, H. & Addie, G.R. (1996). Super corrosion-abrasion resistant white iron alloys and their applications. *Spring Conference of American Institute of Chemical Engineers (AlChE),* New

Orleans, Paper 93e.

Tian, H. & Addie, G.R. (1997). Corrosion-wear behaviors of some high alloyed white irons and stainless steels. *AFS Transactions,* 97-126, pp 595-602..

Tian, H. & Addie, G.R. (2003). Experimental study on erosive wear of some metallic materials using Coriolis wear testing approach. *International Conference on Abrasive and Erosive Wear, Cambridge, UK, September, 2003; published Wear 258* (2005), 458-469..

Tian, H., Addie, G.R & Pagalthivarthi, K. (2005). Determination of wear coefficients for erosive wear prediction through Coriolis wear testing. Full length paper accepted for 15th International Conference on Wear of Materials, San Diego, California, April 2005.

Traynis, V.V. (1977). *Parameters and Flow Regimes for Hydraulic Transport of Coal by Pipelines* Translation from the Russian (Ed. Cooley, W.W. & Faddick R.R.) Terraspace Inc., Rockville, MD.

Tuzson, J.J., Lee, J. & Scheibe-Powell, K.A. (1984)., Slurry erosion tests with centrifugal erosion tester. *Liquid-Solid Flows and Erosion Wear in Industrial Equipment,* ASME, FED-Vol. 13, Ed. Roco, pp. 84-87.

Wiedenroth, W. (1984a). An experimental study of wear of centrifugal pumps and pipeline components. *Liquid-Solid Flows and Erosion Wear in Industrial Equipment,* ASME, FED-Vol 13, Ed. Roco, pp. 78-83.

Wiedenroth, W. (1984b). Wear tests executed with a 125 mm I.D. loop and a model dredge pump. *Proc. Hydrotransport 9*, BHRA Fluid Engrg., Cranfield, UK, pp. 317-330.

Wiedenroth, W. (1988). Wear of solids-handling centrifugal pump impellers. *Proc. Hydrotransport 11*, BHRA Fluid Engrg., Cranfield, UK, pp. 507-522.

Wilson, K.C. & Addie, G.R. (1995). Coarse-particle pipeline transport: effect of particle degradation on friction. *Proc. 8th Internat'l Freight Pipeline Soc. Symposium,* Pittsburgh, PA, pp. 151-156.

Wilson, K.C. Streat, M. & Bantin, R.A. (1973). Slip-model correlation for dense two-phase flow. *Proc. Hydrotransport 2*, BHRA Fluid Engineering, Cranfield, UK, pp. E2-23-E2-36.

Chapter 12

COMPONENTS OF SLURRY SYSTEMS

12.1 Introduction

This chapter, together with Chapters 13 to 16, deals with the interrelationships between the elements of slurry systems. The crucial element, the slurry itself, was analyzed in detail in the first half of this book, up to and including Chapter 8. Chapters 9 to 11 considered centrifugal pumps and their interaction with the slurry. The present chapter goes on to consider other system components, including piping, instrumentation and control. Chapter 13 will show how the performance of a slurry handling system can be expressed in terms of pump and pipeline characteristic curves, Chapter 14 is concerned with pump selection and cost considerations, and Chapter 15 gives a number of practical examples of slurry systems whose behaviour went wrong, giving diagnoses and possible remedies. Chapter 16 concludes by considering environmental aspects of slurry systems.

Looking more specifically at the slurry system components to be considered in the present chapter, we see that these exhibit a great range of variation across the industries using slurry transport. As in earlier chapters, it is appropriate to invoke the fable of the elephant, and to construct the holistic view by assembling many individual aspects. The range is not merely in dimensions, though it is true that ocean-going hopper dredges incorporate pumps the size of an elephant's body (impeller diameters up to 2.5 metres), the variations go beyond size alone. The configurations of slurry systems for different applications can differ in almost all respects, including the geometry of pump suction piping, the mechanical arrangement and layout of pumping stations, and the ways in which the system is instrumented and controlled. For example, slurry transport systems that abstract material directly from the environment (such as harbour dredges or open-pit mining operations) differ greatly in the arrangements at the upstream (suction) end from the type of system which is fed with materials processed by an industrial operation. These two types of system also employ greatly different techniques for instrumentation and, especially, for control. A further dimension which complicates system interaction is line length, as long lines require multiple pumps located in a series of stations. As a result, line length increases the problems of mechanical layout and instrumentation, and greatly complicates the control problems.

12.2 Pump Suction Piping Considerations

In the design of suction piping, the prime consideration is to prevent cavitation by ensuring an adequate NPSH at the pump, and a secondary consideration is to locate the pump where it can easily be inspected or maintained. Another consideration in some cases is the elimination or minimization of foam or air in the suction piping and supply sump. NPSH considerations are of lesser interest in the case of booster pumps, for which the suction pressure is provided by the pump or pumps upstream in the series, and the pumping station layout usually provides space for maintenance. Section 12.4 deals with layouts of this type, and booster pumps will considered there.

It is highly undesirable to install valves in the suction piping. If such a valve must be installed, care should be taken to insure that it remains fully open during pump operation, thus causing no restriction to flow. It is especially important to avoid pinch valves in the suction piping, because they may close in a rapid and uncontrolled fashion. Sometimes it is desirable to have a dropout spool-piece attached to the pump suction to provide for repairing the pump without excessive re-piping, and for removal of roots and large particles that may block the impeller eye.

The concern for adequate NPSH at the pump at the upstream end of the system favours a low position for this pump. This consideration may conflict to some extent with the desire to locate the pump in the dry where it can be accessed for inspection and maintenance. As noted above, arrangements differ considerably between in-plant installations where there is usually control over the consistency of the input slurry, and extractive applications such as dredges and open-pit mines where the slurry consistency can vary widely and often uncontrollably. In the in-plant cases, which will be considered first, it is usually possible to locate the pump above ground or in a dry well, and to provide it with positive suction from an open tank or sump. This configuration is typical of plants in the mining industry (except for drainage cleanup service), most tailings pipelines, and some booster stations where sumps open to air are provided (see Section 12.4).

Sump design varies with the type of slurry and the service. Generally, simple designs are preferred because they suffer least wear and require least maintenance. 'Sloughing off' of solids accumulated in the sump is to be avoided, because it imposes a sudden load on the pump and can upset operation of the whole slurry system. Tapered and rounded sides can be used to minimise solids accumulation, but usually there is less likelihood of slough-off if solids are allowed to build up to their natural angle of repose. In this case, a flat-bottomed tank is best. It is desirable to maintain at least two metres of liquid level above the pump centreline, with a minimum volume in the sump equal to at least one minute's discharge. Maintaining an adequate liquid level also helps to prevent air entrainment, which is to be avoided because it causes pump surging, increased wear, and shock loading on the shaft and bearings. To stop air from entering the pump, the flow into the sump should discharge below the surface, as distant as possible from the pump suction pipe. Adequate liquid depth also helps to prevent swirling, which can entrain air into the suction pipe. Baffles can be placed in the sump if entrainment is likely to be a particular problem, but should only be used if really necessary.

Figure 12.1 shows a sump layout which embodies numerous features that may contribute to good operation. The pump suction pipe passes through the side of the sump, and is entered through a downward-pointing bend with a short bell-mouth. A priming jet may be located facing into the bell-mouth entrance to assist in pump priming, and to eliminate blockages on startup. If the slurry has a high solids concentration, or if the solids are very coarse, the priming jet may be allowed to run continuously to prevent sporadic plugging of the inlet with resultant surging. The suction pipe between the tank and the pump should be horizontal and as short as possible, with the same diameter as the pump discharge. In this way, solids deposition in the suction pipe can be avoided, and the pressure loss between the sump and the pump kept low (see Case Study 8.1). Water should be provided to the sump in sufficient quantity to fill the system and act as makeup water. For in-plant systems, tank level control is sometimes used to regulate the water flow.

Phosphate and other matrix-pumping systems normally have a pump which is located above ground level and draws slurry from a pit. Material is moved toward the pump by a dragline bucket and is then drifted into the pit, using high-pressure water guns, through some sort of grizzly screen to remove oversize tramp material, such as the fossil bones and Civil War cannon balls sometimes dragged up in U.S. phosphate deposits. The slurry density is controlled very crudely by an operator raising and lowering an inverted-L section of pipe joined to the pump suction by a flexible rubber connection. The flow rate can be varied by altering the speed of the feed pump. Along with the disadvantage of having the pump above ground, which tends to give difficulties with available NPSH, this type of suction system requires expensive high-pressure water guns to move the slurry to the pit. This arrangement also limits the solids concentration entering the pump, and hence the concentration which can be delivered by the system. As shown in Section 4.7, the result of using low solids concentration is usually high specific energy consumption.

Of the many types of dredge in routine service, the main ones using pumps and suction-piping systems are the cutter dredge and the trailing-hopper dredge. A cutter dredge employs a 'ladder' which pivots off the bow of the vessel and usually carries some sort of rotating cutter or bucket wheel to loosen solids from the channel bottom and move them towards the suction pipe. The entry to the suction pipe is near the end of the ladder, and there may also be an underwater ladder pump. Operation of the cutter or bucket wheel affects the entrance losses, the solids concentration drawn into the suction pipe, and the solids size and shape. Increases in velocity, digging depth and solids concentration can eventually lead to the limiting suction condition at which the NPSH available to the on-board pump becomes insufficient to prevent cavitation. For liquids or, by extension, slurries which exhibit equivalent-fluid behaviour, the onset of cavitation can be staved off by increasing the diameter of the suction piping, and it may be for this reason that pumps are often constructed with a suction branch larger than the discharge branch. The effect of the larger suction is to diminish the mean velocity there, which will reduce the tendency for a liquid to cavitate. However, for a settling slurry this approach may well be counter-productive.

With settling slurries any decrease in the mean velocity can promote deposition, specifically for particles of the Murphian size (see Section 5.3) If an increase in suction piping size results in deposition, the resulting uneven flow pattern and the increased frictional losses in the suction piping will increase the likelihood of cavitation rather than diminishing it. As shown in Case Study 8.1, the situation is even less favourable if the suction piping is inclined, as in the ladder of a suction dredge. In this case a submerged ladder pump is often required.

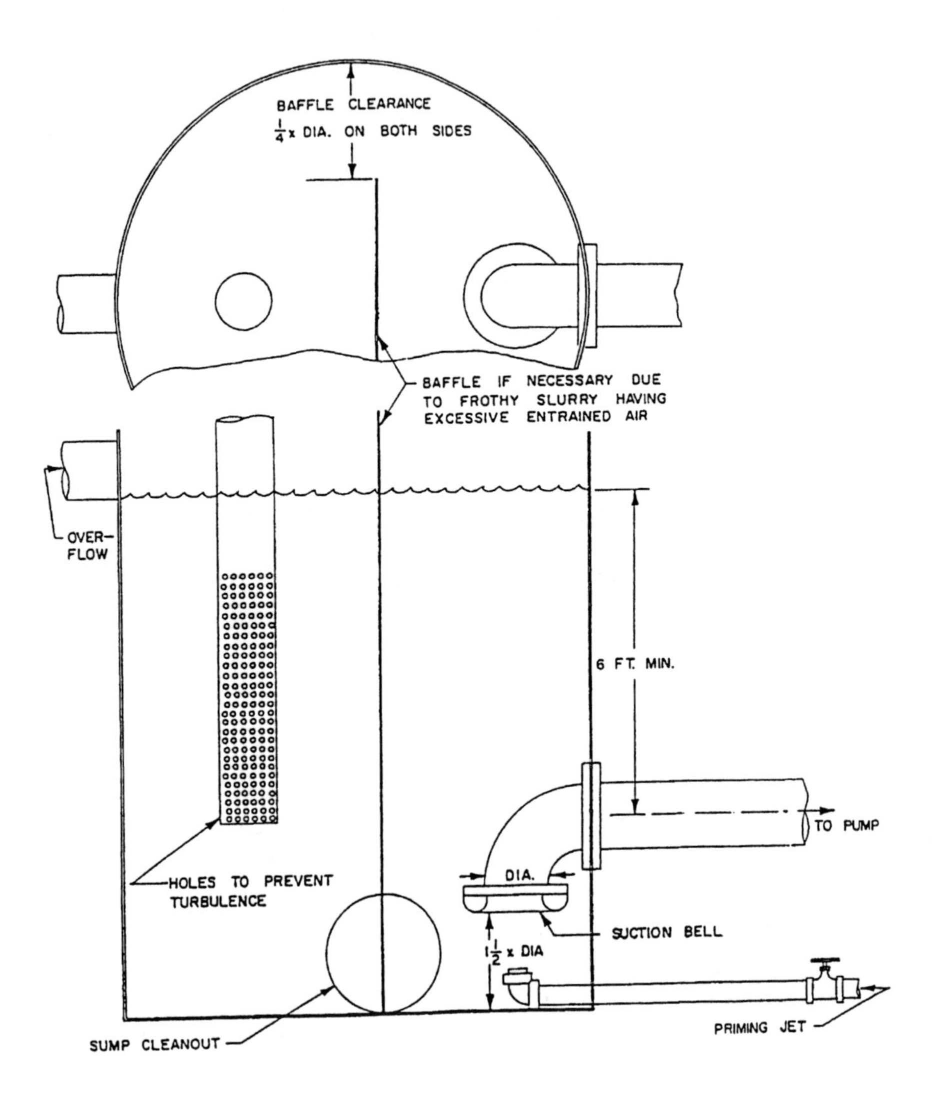

Figure 12.1. Sump layout

The calculation of NPSH was covered in Chapter 9 for the general case. However, some additional explanation is merited for a ladder pump. Figure 12.2 shows the ladder and defines the quantities required for NPSH calculations. The greater the submergence of the ladder pump, the greater the NPSH available to it. Thus, the ladder pump will usually run without cavitation and will prevent cavitation in the next pump in the line, normally on board the dredge. Capital and maintenance cost for the underwater ladder pump is therefore justified by increased production and reduced specific energy consumption associated with operation at higher solids concentration (up to about 35%, see Wilson, 2004). Reliable handling of most solids at volume concentrations up to or beyond 35% can be usually be achieved with the ladder-pump configuration.

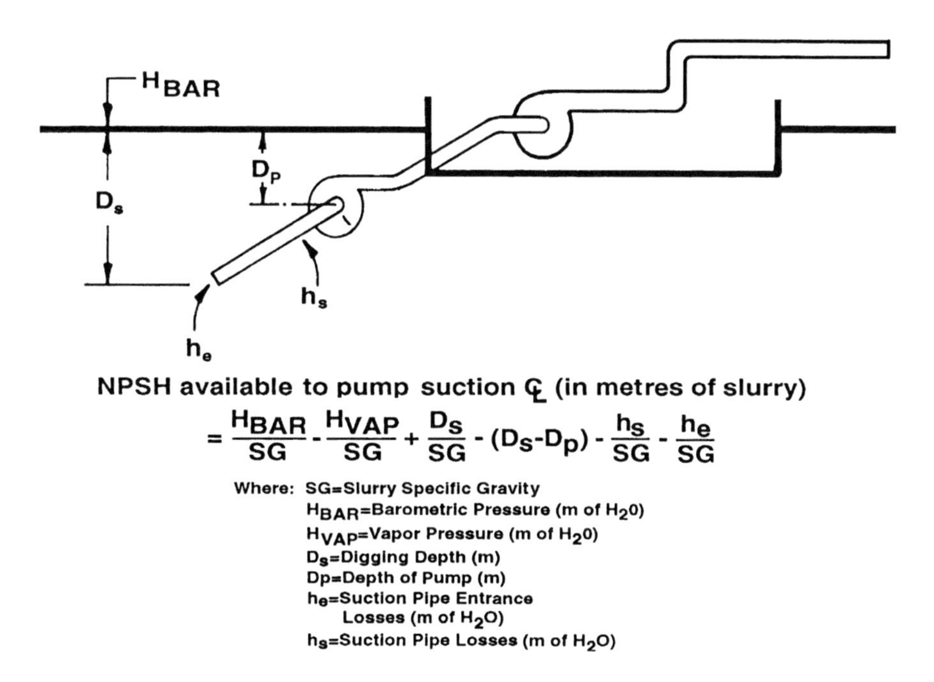

Figure 12.2. Schematic of dredge ladder pump and related variables determining NPSH

Trailing and hopper dredges employ a vacuum-cleaner type of suction draghead in order to maximise the pick-up of fine solids. In these designs, the suction pipe usually passes through the hull at a pivot point on the side of the dredge and is raised and lowered by winches located at the side or stern of the vessel. The pumps are located low in the hull and made large and slow-running to decrease the required NPSH. Slurries with 40% solids by volume can be handled by this type of dredge. The pumps of modern trailing dredges can be very large. A good example is the dredge *Pearl River*. The pumps have suction pipes 1.2 m in diameter and impellers of diameter 2.67 m (Fig. 12.3). They operate at 150 rpm and routinely handle slurries with density 1700 kg/m^3.

Where the suction is from a pit or where a dredge is operated without a ladder pump, it may be necessary to accept moderate cavitation for some operating conditions. Any increase in concentration of solids brought into the pipeline tends to decrease specific energy consumption, as shown in Chapters 4 and 13, and this effect may outweigh the loss in pump efficiency associated with moderate cavitation. In such cases, it is vital to have a pump of the best possible suction performance, not only with small required NPSH at incipient cavitation, but also with the ability to continue pumping when the suction pressure is so low that the head and efficiency are depressed by 10% or more. The skill of the pit operator or dredgeman is very important for this type of operation. If sudden suction blockages do occur, they may well induce water hammer in the pipe, with resulting overpressure failure of pumps or other equipment further down the line (see Chapter 15).

Figure 12.3. The pump room on the *Pearl River* showing one of the two high capacity dredge pumps

Blockages in the pump suction can be caused by rocks, roots and clay balls. The severity of the problem depends on the size of these solids, how frequently they occur, and the size of the pumps. Some sort of screen or 'grizzly' may be useful, as noted above, but it is often necessary to compromise the hydraulic performance of the pump and the location of its design point in order to ensure that it can pass very large solids. Increasing the size of solids which can be passed by a pump usually entails reducing the number of vanes in the impeller. If it is necessary to increase the inside impeller shroud width, the effect will be to shift the best efficiency point of the pump to flows higher than the duty flow. Although it is sometimes possible to counteract this effect with a specially-designed shell, it is not uncommon to find large pumps operating extremely inefficiently on the underdischarge, and wearing excessively, because they have been oversized to eliminate the chance of blockage.

Some operators use a so-called 'root cutter', intended to keep impellers from clogging. This device consists of a bar bolted inside the suction pipe, running parallel to the pipe axis, and extending into the eye of the impeller. Figure 12.4 shows an example of a root cutter. Tests show that the associated reductions in head and efficiency are generally negligible, but the effect on suction performance can be important. Tests at the GIW Hydraulic Laboratory showed that the required NPSH at a given flow rate increases approximately 20 percent when a root cutter is installed.

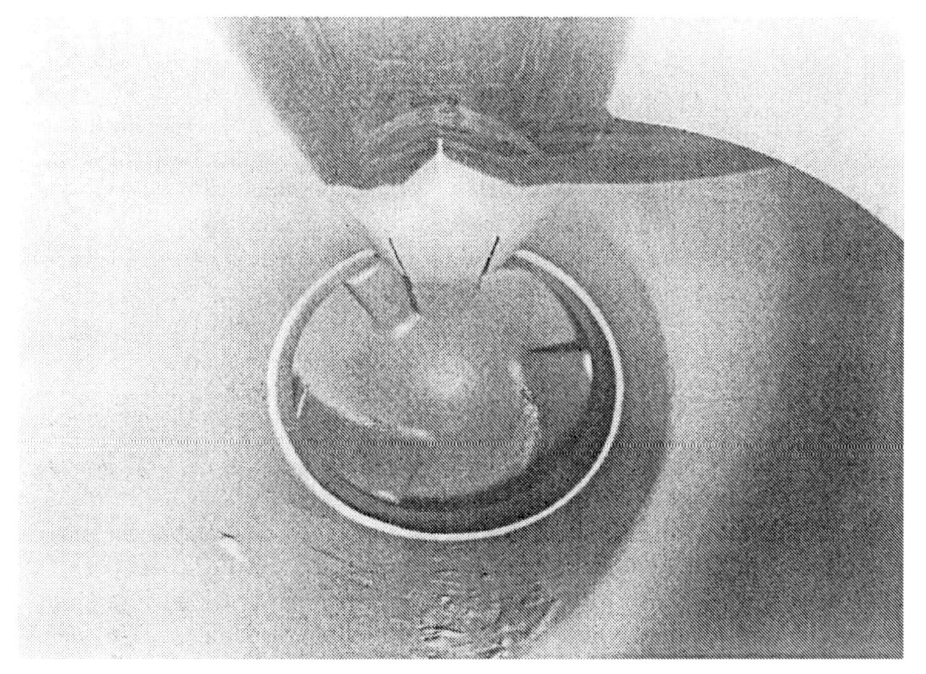

Figure 12.4. Root cutter.

Jet pumps may be used at the entrance to a suction piping system, or even at the suction of the pump, to improve solids feeding. A jet pump adds to the TDH, but such devices are usually less than 50% efficient when the effect of the supply water pump and its drivers are included, and the water they add reduces the solids concentration and thus has an adverse effect on specific energy consumption. Nevertheless, jet pumps do eliminate some wearing parts, and have a role to play in handling difficult slurries. They are most likely to be useful where dilution can be tolerated, operating efficiency is unimportant, and the system head is low.

12.3 Pump Drive Trains

Selection of drive components must be based on matching the drive characteristics to the power demand of the pump. From Eq. 9.2, the power drawn by the pump is

$$Power = \rho_m \, g \, Q \, H \, / \, \eta \tag{12.1}$$

where ρ_m is the slurry density, Q the discharge, H the total developed head in height of slurry, and η the pump efficiency expressed as a fraction. Equation 12.1 applies in S.I. or any other consistent set of units. In the mixed units in customary use in the U.S.A., a numerical coefficient is required. For many water pumps, H falls off as Q increases so that the pump power goes through a maximum. However, slurry pumps typically have flatter head-discharge characteristics, so that the power demand increases monotonically with Q. The drive must therefore be sized for the power at the maximum discharge which can occur. In a mining or dredging operation where the length of the discharge piping is varied to meet placing requirements, the greatest discharge and therefore greatest power demand corresponds to the least system resistance, i.e. to the shortest line length. As the power demand also increases with slurry density, by Eq. 12.1, the power rating for the drive should be based on the maximum solids concentration. The value of η must include the effect of solids on pump efficiency, discussed in Chapter 10. Furthermore, the motor rating must allow for losses in the motor and drive.

For a direct drive to be used, the motor speed must be suitable, the pump and motor must be closely matched, and operation must be steady. These conditions are rarely met in slurry systems. Therefore drive trains for slurry pumps typically comprise a motor operating at relatively high speed, a speed reducer, and a coupling. The pump and drive must be able to respond to changes in operating conditions. Some implications for system design and pump selection are considered in more detail in Chapters 13 and 14. It must be noted here that the head produced and power drawn by the pump will vary if the system is not operated under steady conditions. The variability dictates the type of drive to be used. Where the solids throughput or size varies widely, the power requirements will also vary widely and may change suddenly. This is typical of many mining and dredging operations, where substantial power changes can occur on a minute-by-minute basis. When there is a high degree of variability or uncertainty in the input conditions, the pump speed must also be variable. Characteristic curves for a variable-speed pumping system are shown on Fig. 12.5.

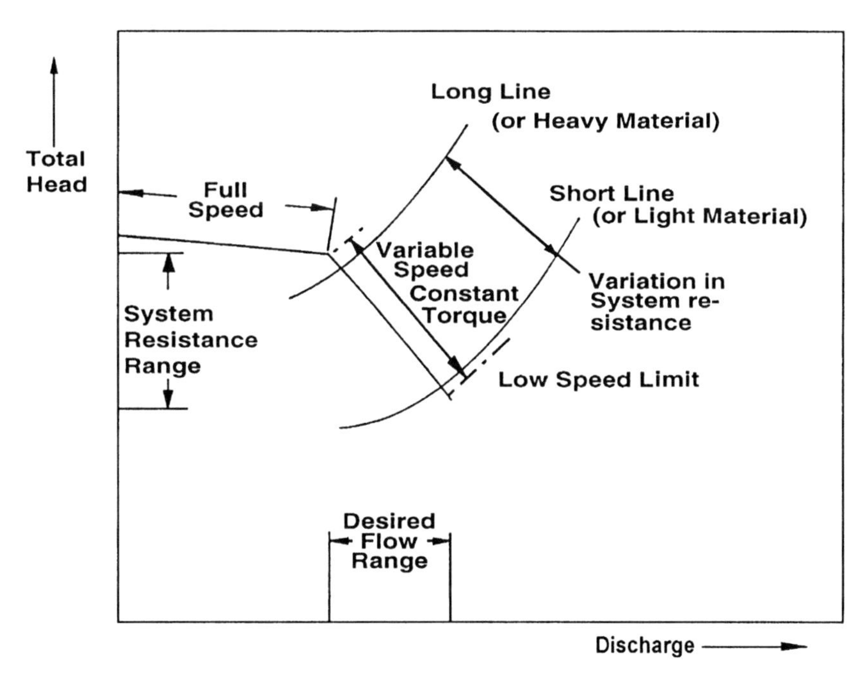

Figure 12.5. Characteristic curves for a variable-speed pumping system

Four main types of drive are in common use for slurry pumps: fixed-speed electric motors, hydraulic motors, variable-speed electric motors, and piston engines. With fixed-speed motors, pump shaft speed can be varied using gears or V-belts. V-belts are relatively inexpensive and simple to maintain, and are able to alter speed easily and rapidly. Fixed-speed electric motors of three-phase squirrel-cage design are common. Up to about 250 kW, the commonest drive is a 4-pole synchronous motor with a V-belt drive train.

The hydraulic motor offers the advantages of variable speed. It is usually relatively inexpensive and easy to maintain, but limited in power. Its major disadvantages are its limited speed range and low drive efficiency. This drive system is normally confined to small dredges.

Insulated Gate BiPolar Transistor (IBGT) type variable speed drives using a high speed motor, usually implemented with a gearbox or V-belt drive, are becoming more common, replacing older inefficient designs involving slip-ring or DC-shunt motors. IBGT drives may be either AC or DC. The AC type is becoming more common because it uses a near-standard motor, but the DC type has traditionally had the advantage in that its operation can provide constant torque to lower speeds. The latest AC drives, however, offer this feature when used with special inverter duty AC motors, which are standard AC motors modified to provide superior cooling at lower speeds. High efficiencies are claimed for both types, but this claim may not be justified for smaller power ratings. The cost of such units is trending down, however, which continues to justify their use in more applications. When larger power is required, say in excess of 400 kW, some slurry-pump applications, such as phosphate mines in Florida, use direct-drive fixed-speed squirrel-cage electric motors for most pumps in the system, reserving wound-rotor motors specifically for the variable-speed units. The larger slurry pumps in mining operations use high-speed squirrel-cage motors with 4 or 6 poles, employing a gearbox for the fixed-speed units. Large IBGT type variable-speed units are expensive, and thus are less widely used than fluid-drive units, despite the inefficiency of the latter type.

In dredges, the main pump is usually driven by a diesel engine. Some manufacturers provide these engines to cover a wide range of power rating, and the adaptation to a dredge pump is usually through a gear reducer. In choosing an engine of this type, some attention must be given to its speed-torque characteristics. If properly configured, the diesel engine can function as a variable-speed constant torque drive over a very wide range of conditions. When a wide operating range is required, turbo-charged four-stroke diesels appear to have an advantage over the two-stroke types. Some engines can operate at full torque between 80% and 100% of full rated speed. This 20% speed range is usually equivalent to a variation in pump head of about 35%, allowing operations over a wide range of conditions without the need to change impeller size. This arrangement is able to provide virtually the full available power over a broad range of conditions. Correct controls on the engine are also important, and the engine manufacturer should be fully briefed as to the requirements, so that correct governor settings can be established.

Drives for ladder pumps on dredges are sometimes arranged with the motor mounted above the water surface and a long drive shaft passing down the ladder. Modern arrangements tend to use submerged electric motors, usually variable-speed DC, with a characteristic of the type shown on Fig. 12.6.

Full consideration must always be given to the environment in which the pump operates. For the rough and variable operating conditions often encountered in mining and dredging, speed reducers may be used. In this case, an extra service factor on the installed drive power is normally needed, and values of 1.5 to 2.0 times the delivered power may be appropriate. Single-reduction gearboxes are most commonly used on larger units, and the designer must then remember that the pump will rotate in the opposite direction to the drive motor. Maintenance of very large units is helped by providing a small reversible auxiliary drive for mounting and dismounting the impeller.

As shaft misalignment and rough operation inevitably occur, couplings should be chosen to be able to cope with these eventualities. The most popular couplings are the double engagement gear type, but Falk and other types are also in common use. Particularly for dredge pumps, couplings should be chosen to permit the considerable movement required to adjust the impeller clearance as the pump wears. On large electric-drive installations, especially for underwater dredge pumps, it may be wise to have a torque-limiting coupling to minimise the damage to the drive if large solid objects enter the pump and jam it.

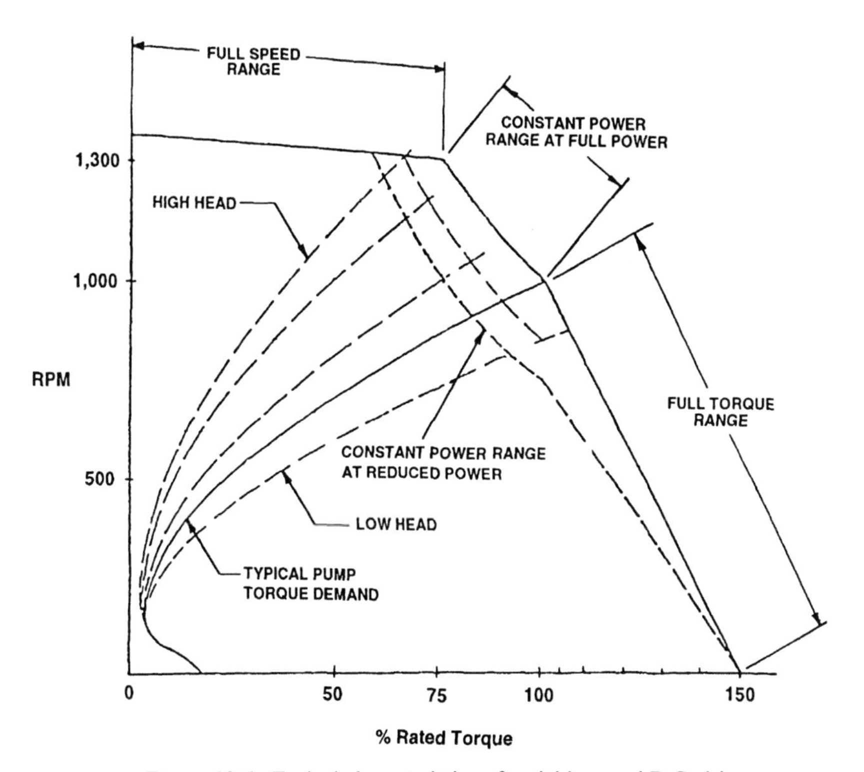

Figure 12.6. Typical characteristics of variable-speed D.C. drive

Pump bases perform an important role in maintaining the alignment of the pump and the components of the drive-train. They provide the stiffness to resist torques from the drive train and loads imposed by the piping. Bases are usually constructed of heavy-duty steel beams welded together, and are held in position by grout and bolts. In many matrix pipelines in Florida the bases, or 'sleds' as they are called, simply sit on the ground, and thus can easily be spaced out along the line and moved (with the help of a bulldozer) when required. In these circumstances, it is desirable to provide bases somewhat more robust than usual. An example is shown on Fig. 12.7.

Figure 12.7. Phosphate-matrix pump and drive mounted on 'sled'

12.4 Pump Layout and Spacing

Considerations of wear and ease of maintenance usually dictate that a centrifugal pump must be a single-stage machine. Furthermore, because of the limitations on pressure rating, a single centrifugal pump cannot be used to transport a slurry over a distance of more than a kilometre or so. Several pumps in series may therefore be needed for a medium-length pipeline, although centrifugal pumps as prime movers are generally limited to pipelines not much longer than about 10 kilometres or, say, 6 miles. For this configuration, the total head generated by pumps in series is simply the sum of the heads developed by each pump at the common flow rate. Several alternative selections are often considered, based on different numbers of pumps in series. The most cost-effective selection usually involves fewest pumps, and this should be the initial assumption. When a selection has been made on this basis, the operating and capital costs of alternative multiple-pump systems can be compared before a final decision is made. Ideally, the pumps should be the same size. Problems which can arise in series pumping with pumps of different sizes are considered in Section 16.2.

To limit the pressure each pump must withstand, it is desirable to space the pumps at roughly equal distances along the pipeline. However, this configuration increases the difficulties associated with start-up and control, and adds to the cost of power supply and the cost of providing pump houses, if these are required. Conversely, if each pump house contains a group of pumps, all but the first pump will experience elevated operating pressures, which can affect the design of casing, shaft and seals. With pumps spaced along the line, simple flexible pipes or expansion joints are sufficient to prevent piping loads from damaging the pumps.

In locating the pumps along the line it is desirable to have about one atmosphere of positive pressure at the suction of each pump. In lines with three or more pumps, it is not uncommon to break the line with an open sump to prevent hydraulic transients being transmitted down the line.

Because of the higher pressures involved, multiple-unit pumping stations must use bases bolted to concrete foundations. The pressure loads from internal and external piping must be taken into account, as well as the loads on the pumps, the method by which the pipes are fixed and the magnitude of any thermal-expansion loads which they carry. In such cases a full stiffness analysis of the pumps and associated piping may be required, potentially involving three forces and three torques at each of the two flanges of every pump.

The layout of pumps within a series station depends on the number of units and requirements of maintenance, access and safety. Figure 12.8 shows a photograph of a 6-pump station with two lines of three pumps in series. In this case, access to all pumps is excellent, and reasonable work areas are provided adjacent to each pump. Note that the use of different levels makes it possible to employ simple single-bend pipe sections between pumps. Though not readily visible on the photograph, large launders are provided to each pump, connecting down the centre to a single wet sump. More compact arrangements can be achieved by rotating the pump branches and using straight pipes between adjacent discharge and suction. An example of this arrangement is shown on Fig. 12.9.

Operation of pumps in parallel is not common for pipelines transporting settling slurries, but is sometimes used to deliver a large flow against a relatively low system resistance or where the slurry to be pumped is non-settling. Parallel pumps may also be installed where back-up spares are needed, where a very wide range of flows is to be handled and each individual pump is to be kept down to a manageable size, or where wear limitations and suction requirements make it preferable for each pump to handle less than the total flow. In parallel pumping, the 'combined pump characteristic' is obtained by adding the individual pump flows at each value of head. The overall system operation is defined by the intersection of this combined characteristic with the system characteristic. Each individual pump operates at the flow, efficiency and NPSH corresponding to this head. Even more than for series pumping, it is generally important to ensure that all the pumps are identical. Some of the problems arising from imbalance in parallel pumping systems are reviewed in Section 16.3.

Figure 12.8. Interior of pumping station showing three pumps in series

The profile of the pipeline can cause difficulties if there are any large changes in level. When the line is shut down, any local high spots are susceptible to sub-atmospheric pressures, which are sometimes severe enough to cause vaporisation. On subsequent start-up, the collapse of the vapour pocket could initiate severe hydraulic transients. To avoid these, vacuum relief valves should be fitted at all high points, or else the line must be completely drained after each shutdown. Vacuum relief valves can also be used on the suction of a pump that is prone to cavitation, to let air into the suction, thereby reducing the effects of cavitation. This same idea is applied to dredge suction pipes, where a hoffer valve is used, letting in water rather than air.

Figure 12.9. Interior of pumping station showing compact arrangement of pumps in series

12.5 Flow Control and Valving

It was noted previously that in dredging and some mining operations, the solids throughput and size may vary widely, sometimes in a matter of minutes or less. In dredging operations, the system resistance may halve or double over a period of days, as the dredge is moved about. Control of pump and system operation is required to maintain reliability and efficiency. For this reason variable-speed drive is usually provided, as outlined in Section 12.3 above. Longer-term adjustments can be made by changing the drive to a different fixed speed or range of variable speed. As a more extreme alteration, the impeller can be changed for one with a smaller diameter.

When a system is being primed and filled there is little or no system head. Some of the problems that occur are examined in Chapter 16. Particularly for a long system, operation at fixed speed can lead to excessive overdischarge, so that a variable-speed drive should again be used. In systems with several pumps in series, it is common practice to provide variable-speed drives on the first and last units. The speed of the first pump is varied during filling, with each subsequent fixed-speed pump brought into service as its suction pressure shows sufficient available NPSH for it to be operated without cavitation. The variable speed on the last pump is used to control flow during subsequent operation. This method is preferable to varying the speed of any of the upstream pumps, because reducing speed could lower the discharge pressure to the point where the NPSH available to the next pump falls below the NPSH required, causing it to cavitate.

Dredges with diesel or electric motors normally have variable speed drive, for control purposes. However, full control extends beyond simple adjustments to speed and level, and requires more comprehensive instrumentation to measure flows and concentrations. Some available instruments are reviewed in Section 12.6 below. Many dredges now have full instrumentation with on-board computer systems for sensing, monitoring and controlling operation, and, in more sophisticated application, for producing a historical production record and for on-line maintenance planning. This kind of system is also becoming common in big mining operations and is starting to penetrate into the remainder of the mining industry.

It has already been noted that flows in in-plant systems and tailings lines are typically less variable than other applications. Variable-speed drives can sometimes be avoided in these cases by allowing the liquid level in the sump to vary, provided the range of variation in sump level is significant in relation to the total head change through the system. The action of this type of 'passive control' is illustrated in Fig. 12.10. If the flow into the sump reduces, the level falls. The static lift required increases correspondingly, so that the system curve moves up until operation stabilises at point B1 with reduced throughput. Similarly, if flow into the sump increases, the level rises and the system requires less head, so that operation stabilises with the increased flow at point B2. For a pump with a flat characteristic, as in Fig. 12.10, relatively large variations in flow can be accommodated in this way, provided that the pump is able to operate over a wide enough range.

Flow control by one of the above approaches is preferable to using throttling valves, which cause high energy dissipation and are subject to rapid wear. Nevertheless, throttling valves are sometimes used as a last resort, and a number suitable for slurry use are available. One example is shown in Fig.

12.11. To avoid cavitation problems, a throttling valve should never be installed on the suction side of a pump. In principle, slurry systems should be designed with as few valves as possible, but in practice, the need to isolate for maintenance usually requires isolation valves on both sides of a pump. Such valves may also be needed to take lines in and out of service and to isolate backup equipment. For slurry systems, parallel-slide gate valves with hardened metal, rubber or urethane seats are preferred for this purpose.

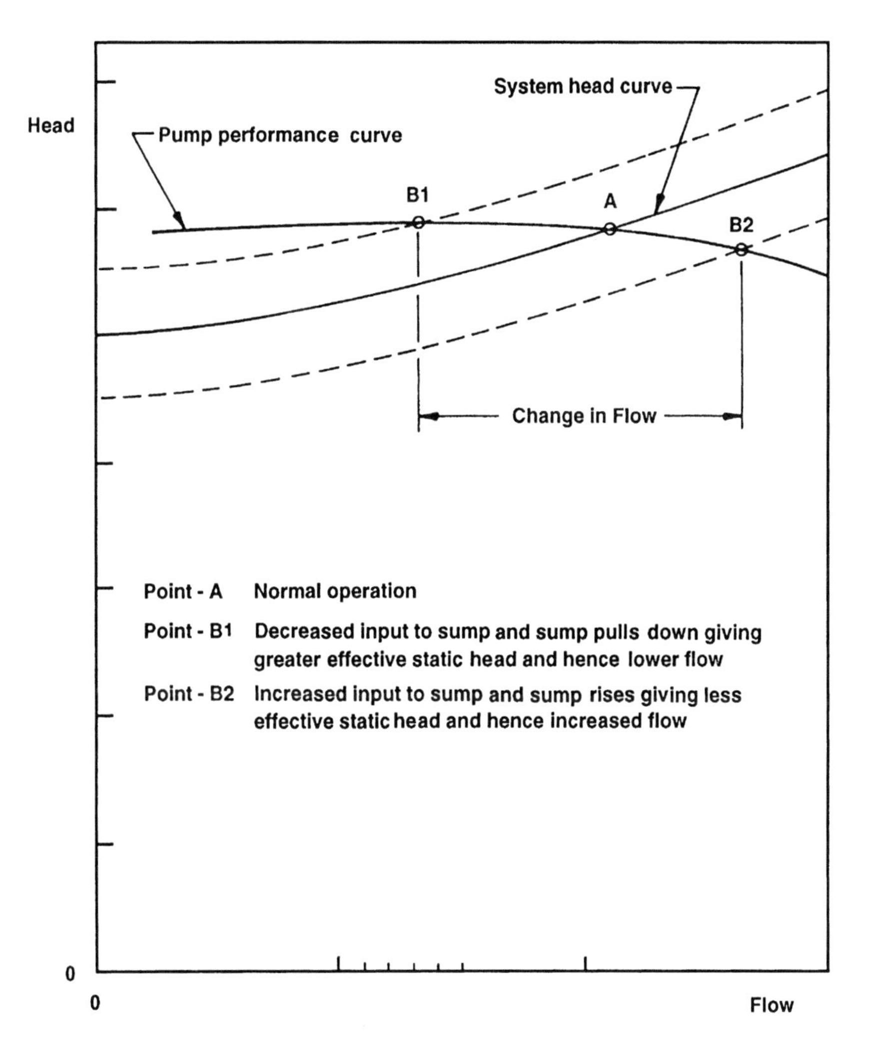

Figure 12.10. Schematic illustration of stabilising influence of sump on intercept of pump and system characteristics

Pipelines in mountainous terrain often have steeply-descending reaches that require downstream chokes to eliminate slack flow and excessive velocity. Such chokes, often combined in series at choke stations, are usually orifices of highly-resistive ceramic material (Derammelaere & Shou, 2002; Betinol & Jaime, 2004). Similar chokes may be required when placing backfill in worked-out areas of deep mines (Wilkins et al., 2004).

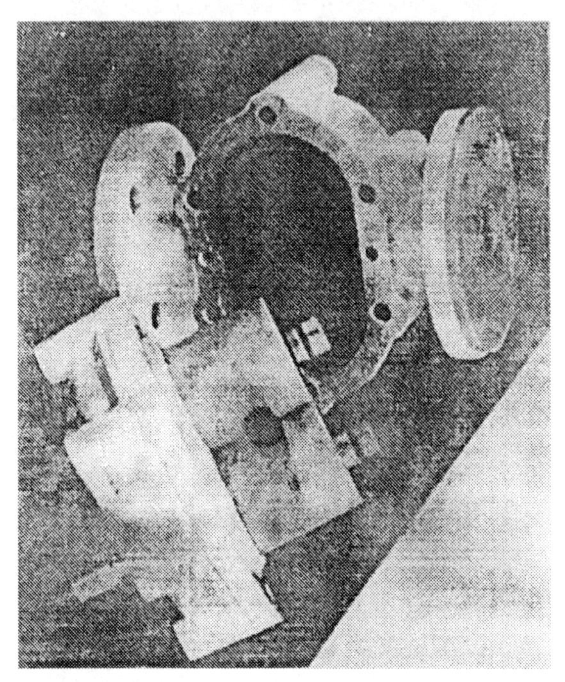

Figure 12.11. Control valve

12.6 Instrumentation

The abrasive and sometimes corrosive nature of most slurries is such that it is difficult to make instrumentation accurate and reliable. Great care must be taken with selection, installation, calibration and maintenance if instruments are to perform as required. Some available instruments were reviewed in Chapter 9, in the context of pump testing. The discussion in this chapter relates to operation and field testing. Safe operation and trouble-shooting dictate the bare minimum of instrumentation needed. More extensive instrumentation is necessary to maximise output and achieve cost-effective operation. As noted above, instrumentation can also play a vital role in the maintenance of system components and can be used to provide a record of operation.

To ensure stable pump operation, it is essential to monitor suction conditions, indicated by sump level or suction pressure. Pressure at the pump discharge can be used to give an immediate indication of changes in operating conditions, but otherwise has limited value as a diagnostic or control measurement because it depends on several variables. The pressure increase across a pump depends on delivered slurry density (see Chapter 10), but also depends weakly on discharge rate because most slurry pumps have gently drooping head-discharge characteristics (see Chapter 9). Taken together, the suction and discharge pressures and the power drawn by the pump can provide useful diagnostic information. For example, a rise in pump power and discharge pressure together usually indicates a rise in solids concentration, one of the transient conditions considered in Chapter 13. However, systematic control requires measurement of other parameters. The most important are flow rate and solids concentration.

The abrasive properties of slurries exclude orifice plates for flow measurement. Venturi and nozzle meters can be used, but they are still subject to wear even when constructed from a wear-resistant material (see Chapter 11) so that the calibration must be checked regularly. A simpler device, which lends itself readily to slurry use, is the bend meter shown schematically in Fig 12.12. The pressure difference is measured between the inside and outside of a bend. A particular advantage of this technique is that it can be applied to an existing bend, and requires no additional disturbance to the slurry. The precautions explained in Chapter 9 must be used to ensure that the pressure readings are reliable. This type of device becomes less reliable if the flow is stratified, and it is therefore preferable to apply the measurement at a bend where the entering flow is vertical, i.e. at the top of a 'riser' or the bottom of a 'downcomer'. Experience at the GIW Hydraulic Laboratory has shown that, provided these precautions are taken, the bend meter is reliable and gives reproducible measurements (Addie, 1981). As for related devices (such as an orifice plate or venturi) which obtain flow rates by measuring the pressure difference due to a change in the momentum of a flowing fluid, the bend meter has flow rate proportional to the square root of the pressure difference (corrected for any difference in elevation between the two measurement points). To a first approximation (Bean, 1971) the slurry flow rate for a 90° bend, with pressures measured at the 45° positions as shown in Figure 12.12, is

$$Q_m \approx \frac{\pi D^2}{4} \sqrt{\frac{R\,\Delta p}{D\,\rho_m}} \tag{12.2}$$

Here D is the pipe diameter, R is the bend radius, and Δp is the pressure difference from the outside to the inside of the bend. While Eq. 12.2 may be used to calculate the range of a bend meter and the range of Δp to be determined, it is essential to calibrate any specific bend meter against some 'absolute' measurement such as a magnetic flow meter.

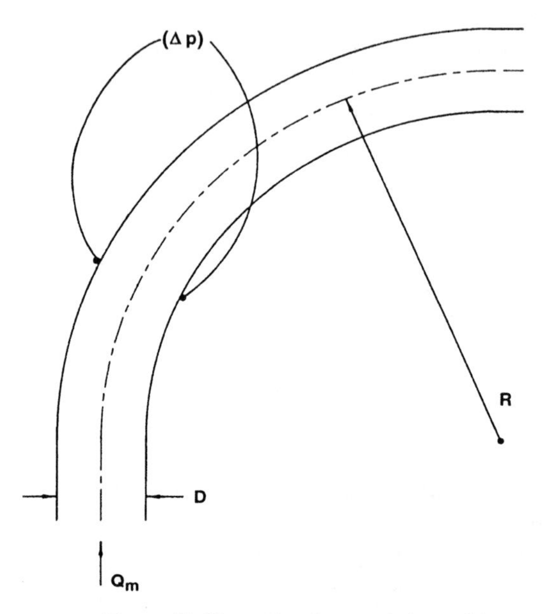

Figure 12.12. Bend meter (schematic)

For precise measurement, the most widely used flow instruments are magnetic flow meters and Doppler meters. A magnetic flow meter measures total volumetric flow and is non-intrusive: i.e. it does not disturb the flow. Readings from a magnetic flow meter can be accurate to within 0.5 to 1%, of full-scale reading, although if the solids are significantly ferromagnetic the device will indicate an incorrectly high flow. Furthermore, magnetic flow meters of the AC type respond anomalously to large particles, and are therefore affected by stratification. Entrained air also causes a magnetic flow meter to give an incorrect reading of the slurry flow rate. For these reasons, it is desirable to locate a meter of this sort in a straight vertical section of pipe.

Doppler meters are also popular, partly because of their relatively low cost. A Doppler meter uses a sensor in the pipe wall, transmitting ultrasonic waves into the flow and measuring the frequency shift of waves reflected by the transported particles. Thus a Doppler meter is also non-intrusive. Because the device responds to the particle velocity, the reading can be distorted by stratification, to the point where the velocity indicated depends on the sensor location in a stratified flow (Addie & Maffett, 1980). Again, it is usually preferable to locate the instrument in a straight vertical pipe.

To measure solids concentration, or the equivalent parameter of mean mixture density, a particularly useful device is the U-loop, shown schematically in Figure 12.13. The pressure gradient is measured in two vertical pipes, a 'riser' and a 'downcomer'. The theory behind this type of device was covered in section 8.2, leading to Eq. 8.5 for the average *in situ* solids concentration. It was also shown that, in most cases of practical interest, this relationship can be used to calculate the relative density of the delivered slurry and the delivered solids concentration. Thus:

$$S_{md} = \frac{(p_1 - p_2) + (p_4 - p_3)}{2gz\,\rho_w} \tag{12.3}$$

$$C_{vd} = \frac{1}{(S_s - 1)}\left[\frac{(p_1 - p_2) + (p_4 - p_3) - 1}{2\,gz\,\rho_w}\right] \tag{12.4}$$

where p_1 to p_4 are the static pressures in the slurry at the four measurement locations shown in Fig 12.13. If the pressure gradients are actually measured by connecting the tappings to a pair of transducers with the connecting lines filled with water then, as shown in Section 2.2, the recorded pressure differences will be

$$\Delta p'_A = (p_1 - p_2) - \rho_w gz \tag{12.5}$$

and

$$\Delta p_B' = (p_4 - p_3) - \rho_w gz \tag{12.6}$$

In terms of these measured values,

$$S_{md} = \frac{\Delta p'_A + \Delta p'_B}{2} + 1 \tag{12.7}$$

and

$$C_{vd} = \frac{\Delta p'_A + \Delta p'_B}{2(S_s - 1)\,gz\,\rho_w} \tag{12.8}$$

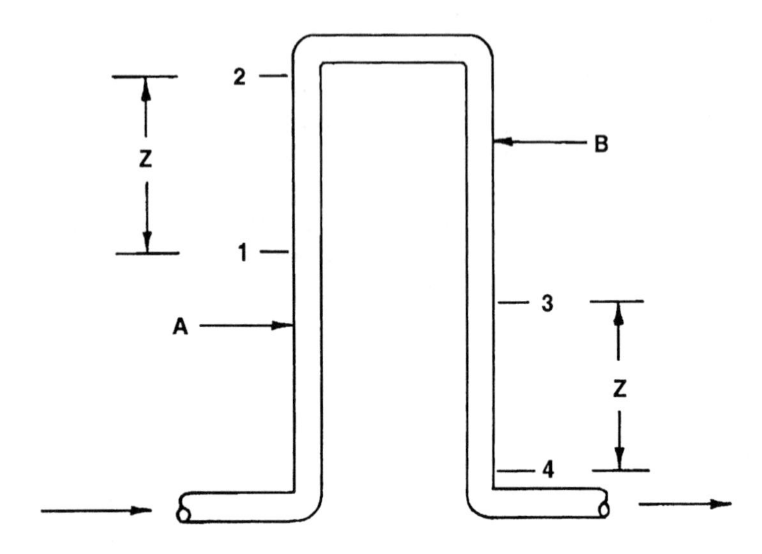

Figure 12.13. Schematic of inverted U loop

The limits to the accuracy of these equations, examined in detail by Clift and Clift (1981), are reviewed in Section 8.2. The U-loop concept lends itself well to field measurements, and the device is often used on dredges. Figure 12.14 shows an example of such a device used in a field test on a phosphate matrix pipeline.

As an alternative, slurry density may be measured using a meter which relies on the attenuation of some form of radiation passed through the slurry. Gamma-rays are most commonly used, and this type of instrument is sometimes referred to as a nuclear densitometer. The parameter measured is the mean density along the radiation path, which indicates the *in situ* density rather than the delivered value: this inconvenient feature is sometimes overlooked by operators who use the results from this type of instrument to calculate throughput. Furthermore, in a stratified flow the reading will depend on the location of the gamma-ray beam within the slurry. Therefore, as for most of the other instruments reviewed here, a radiation densitometer is best installed on a straight vertical section of pipe.

An interesting future development rests on the possibility of mapping the local density within the slurry by making gamma-ray attenuation measurements along a number of paths at different orientations through the pipe (MacQuaig *et al.,* 1985). A time-averaged image of the density profile in the pipe is then reconstructed by computer-aided tomography (CAT), in exactly the same way as medical CAT-scanners are used to obtain non-intrusive sections through a living body.

Figure 12.14. Inverted U loop used in a field application

Although tomographic imaging is not likely to become widespread, it promises to become a valuable research tool for slurries and other multi-phase flows. Imaging technologies based on γ rays and magnetic resonance imaging (MRI) are also gaining in popularity (see Graham et. al. 2002, and the references therein). For example, the velocity profiles shown on Fig.7.1 were obtained by MRI.

As noted above, pump power can be a useful performance diagnostic. For direct measurement of pump input power, it is preferable to use a separate torque bar installed in the drive, coupled with magnetic speed pickup. The torque bar may conveniently be of the strain gauge type, set up with a conditioner unit and multiplier unit to show power directly. Separate torque bars require extra couplings and a space in the drive to accommodate the bar, and are therefore usually used only in laboratory tests or specially-set-up field tests. Some strain transducers are available with wireless transmitters. If a torque bar cannot be installed, it is possible to mount strain gauges on the drive shaft and use the wireless type of signal transmitter. However, sensitivity is then bound to be poor, because a well-designed drive shaft will experience little strain.

Where pump power cannot be measured directly, it may be inferred from the motor power. If an electric motor is used, the current (I) indicated by an ammeter enables the motor power to be estimated as:

$$P = \sqrt{3}\,\eta\, E\, I \cos\theta \qquad (12.9)$$

where E is the voltage of the supply (assumed in Eq. 12.9 to be three-phase), θ is the phase angle between voltage and current, and η is the efficiency of the motor. For a fully loaded motor, $\cos\theta$ can be taken as 0.8 and η as 0.9. Two-element type wattmeters, using current and potential transformers, give P more accurately, because they automatically correct for changes in θ and E; however, the reading will still have to be corrected for the motor efficiency. Furthermore, drive train losses must be subtracted in estimating the power input to the pump; these are usually roughly constant, and typically 3 to 10% of the motor power. Dredges with diesel engines are often equipped with built-in sensors and indicators to evaluate power.

Measurements of pipe friction loss are routinely carried out in hydraulic laboratory work, and may also be made in the course of field trials. The pipe in which the losses are measured must be straight and should be horizontal. Ideally, the nearest upstream bend or fitting should be at least 100 pipe diameters from the first pressure tappings. The measurement section should be as short as is compatible with the sensitivity of pressure measurement, so that the pressure loss can be read simultaneously with flow rate and slurry density. Two tappings should be provided at each of the upstream and downstream measurement locations, ideally at 45° on either side of the upward vertical. As noted in connection with pump performance measurement in Chapter 9, care must be taken to ensure that the tappings are flushed before each measurement. An arrangement which has proved reliable at the GIW Hydraulic Laboratory is to provide a small liquid-filled chamber or 'pot' at the pipe connection, to collect any solid particles which enter the pressure measurement system.

Differential pressure transducers of the capacitance type are recommended for measurement of pressure loss along the test section, although diaphragm gauges or manometers can also be used. Ideally, the differential pressure sensor should be connected to the pots at the pipe by clear plastic lines, which must be kept full of water and which should therefore be purged occasionally.

Multi-channel analogue-to-digital converters and computer data logging systems are now available at reasonable cost, allowing real-time visualisation of the full performance of a system. As noted previously in dealing with pump testing, this technique has great advantages. One is that the stability of the system can be monitored continuously, to ensure that measurements are only made when operation is steady.

Selection of the system components discussed in the present chapter is important, but careful overall design of the system, together with correct selection of the pumps, is absolutely necessary to ensure satisfactory operation. The matching of pumps to the operating characteristics of the system is considered in Chapter 13, and pump selection is covered in Chapter 14. Some of the problems which can arise if pumps or other components are not properly selected and operated are reviewed in Chapter 15.

REFERENCES

Addie, G.R. & Maffett, J.R. (1980). Doppler Flow Meter Tests at Georgia Iron Works, *8th Annual Mining and Metallurgy Division Symposium and Exhibit,* ISA, pp 125-133.

Addie, G.R. (1981). Slurry Pump and Pipeline Performance Testing at Georgia Iron Works Hydraulic Laboratory, *Proc. Sixth International Technical Conference on Slurry Transportation,* Las Vegas NV.

Addie, G.R. & Hagler, T.W. (1990). Positive Feed System Project, *Report to Florida Institute for Phosphate Research,* GIW Industries, Grovetown, GA.

Bean, H.S., Ed.(1971), *Fluid Meters - Their Theory and Application,* 6th ed., pp.75 -77 and 255.

Betinol, R.G. & Jaime, H.E. (2004). Startup of dual concentrate pipeline for Minera Escondida Limitada, *Proc. Hydrotransport 16*, BHR Group, Cranfield, UK, pp 363-377.

Clift R. & Clift D.H.M. (1981). Continuous Measurement of the Density of Flowing Slurries, *Int. J. Multiphase Flow,* Vol 7, No 5, pp 555-561.

Derammelaere, R.H. & Shou, G. (2002). Altamina's copper and zinc concentrate pipeline incorporates advanced technologies, *Proc. Hydrotransport 15*, BHR Group, Cranfield, UK, pp 5-18.

Graham, L., Hamilton, R., Rudman, M. Strode, P. & Pullum, L. (2002). Coarse solids concentration profiles in laminar pipe flows, *Proc. Hydrotransport 15*, BHR Group, Cranfield UK, pp 149-158.

MacQuaig N., Seville, J.P.K., Gilboy, W.B. & Clift, R. (1985) Application of Gamma-Ray Tomography to Gas Fluidised Beds, *Applied Optics,* Vol.24, pp 4083-4085.

Wilkens, M., Gilchrist, C., Fehrsen, M. & Cooke, R. (2004). Boulby Mine backfill system: design, commissioning and operation, *Proc. Hydrotransport 16*, BHR Group, Cranfield, UK, pp 349-362.

Chapter 13

SYSTEM DESIGN AND OPERABILITY

13.1 Introduction

Reference has been made in earlier chapters of this book to the need to match the characteristics of the pipeline system and pumps to ensure stable operation of a slurry system. The present chapter illustrates how understanding of slurry behaviour can be applied to analyse system operability, by examining both steady-state operation and transient responses to changes in slurry properties.

For the operator of a slurry system, it is important to understand how the system will respond to variations in the properties of the slurry fed to it. It is also important for the designer to appreciate fully the implications of variation in particle size, concentration and throughput. As for all handling and processing operations, design must reflect the range of variability which must be anticipated. For slurry systems, this implies finding the economic balance between controlling slurry consistency on the one hand, and on the other operating the pipeline under conditions which are not optimal but which can accommodate variations. For long-distance freight pipelines, the dominant capital and operating costs are usually associated with conveying. Therefore much effort is devoted to preparing and controlling slurry properties. At the other extreme, short in-plant conveyors and dredging systems may have to be designed for wide and sometimes uncontrollable variations in solids size and throughput. Nevertheless, in our experience some industries using hydraulic transport can make significant savings by recognizing that the stability and efficiency of conveying operations - particularly the energy efficiency - can be improved by limiting operating variability.

In this chapter, we consider the two commonest kinds of variation: changes in solids concentration with constant particle size distribution, and changes in particle size at fixed solids concentration, primarily with fixed-speed pumps. However, these are by way of example only, and the same kind of reasoning can be extended to more complex effects. Homogeneous slurries are considered in Section 13.2. The more complicated effects arising with settling slurries are discussed in Sections 13.3 to 13.5. A case study, to show how variability affects system design, is given in Section 13.6. The design is posed in the way which arises most commonly in practice: for a given solids throughput (or range of throughputs), what concentration and pipe size should be used and what does this imply for the pump specifications?

13.2 Homogeneous Slurries

Figure 3.1 showed, in simple generalised terms, the effect of changes in solids concentration on the resistance of a piping system to flow of a homogeneous, non-settling slurry. This kind of behaviour is again shown schematically in Figure 13.1. For simplicity, only the frictional contribution is shown; i.e. the curves refer to horizontal transport. At velocities sufficiently high for the slurry to be in turbulent flow, the head lost in flowing through the system (measured as head of a fluid with the slurry density) can be estimated by treating the slurry as a fluid with an effectively constant friction factor (see Section 3.7). This leads to the 'turbulent' system curve in Figure 13.1, for which the influence of the solids concentration is usually negligible. However, in laminar flow, to the left of the turbulent curve in Figure 13.1, the system head varies strongly with concentration, as shown schematically in Figure 13.1 where $C_1 < C_2 < C_3$. The transition between laminar and turbulent flow is indicated roughly by the intersection of the laminar system characteristic with the turbulent curve. Therefore, as illustrated by Figure 13.1, the transition velocity increases somewhat with solids concentration.

To examine the operability of such a system with centrifugal pumps, we now superimpose the pump head-discharge characteristic on to the system characteristic. As shown in Section 10.5, for fine slurries the head delivered, evaluated in terms of the slurry density, is virtually unaffected by the solids. Therefore the characteristic of a fixed speed pump appears as a single curve in the co-ordinates of Figure 13.1, independent of slurry concentration.

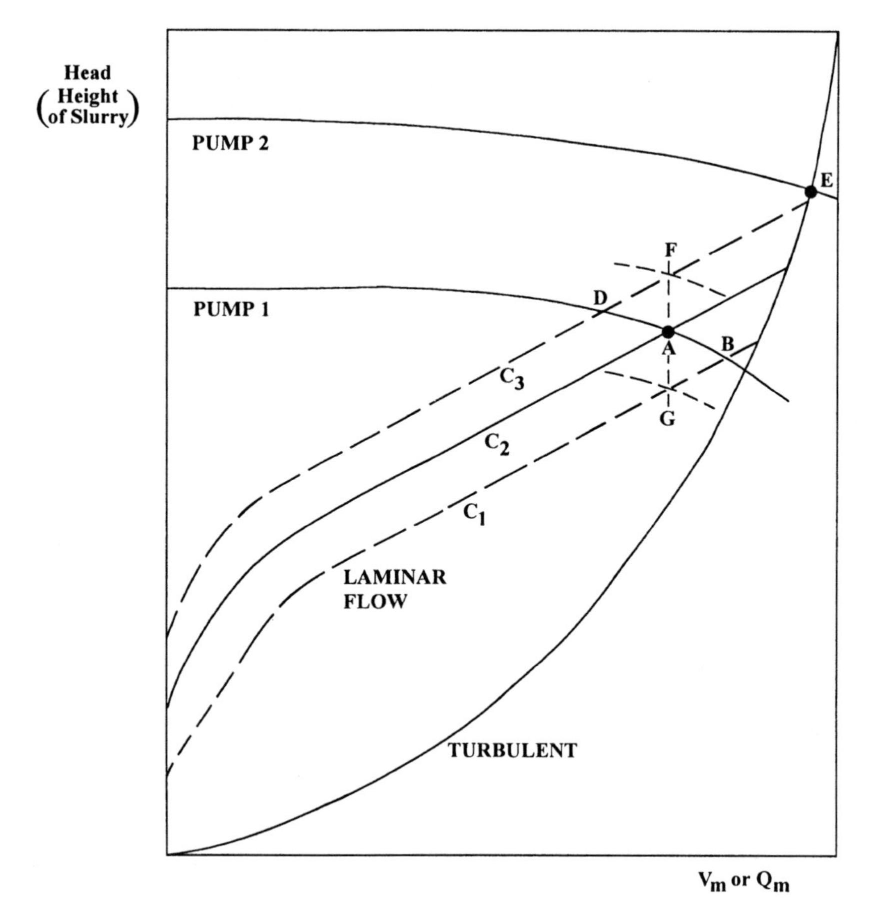

Figure 13.1. System and pump characteristics for flow of a homogeneous slurry at three concentrations (schematic)

It was indicated in Chapter 3 that the most economic system operation for a homogeneous slurry may lie in the laminar range. Consider first a system which has been designed for such a case, corresponding to point A in Figure 13.1 with slurry concentration C_2. The curve labelled 'Pump 1' represents the characteristic of a pump selected for this operating point.

Figure 13.1 shows the effect of varying solids concentration. If it decreases to C_1 then the system operation shifts to point B, the new intersection of the system and pump characteristics. Similarly, a concentration increase to C_2 shifts operation to point D. It was shown in Chapter 3 that relatively small variations in concentration can have strong effects on slurry rheology and hence on the laminar flow curves. Thus the shifts from point A may result from quite small changes in slurry consistency. Nevertheless, as shown by Figure 13.1, the associated changes in mean velocity (and hence solids throughput) are amplified by the way the system and pump characteristics intersect.

Thus steady operation in laminar flow with a fixed speed pump and with no flow control valve is only possible if the slurry consistency is tightly controlled.

For variable solids concentration, two alternatives are available. One is to operate in turbulent flow, at a point like E in Figure 13.1. 'Pump 2' shows the head-discharge characteristics of a centrifugal pump selected for this duty. By operating on the turbulent system characteristic, variations in slurry consistency can be accommodated without significant changes in mixture velocity. This is probably the main reason why many designers prefer to use turbulent flow even for homogeneous slurries. The design point is then selected to be in the turbulent range for the highest solids concentration expected. However, it should be noted that this will not correspond to the lowest specific energy consumption. This is an example of the general point that design for variable operation is usually incompatible with design for minimum energy consumption.

The other option is to use a variable-speed pump. This is illustrated by points F and G in Figure 13.1. If the solids concentration increases from C_2 to C_3, then the pump speed is increased to keep a constant flow rate. The corresponding increase in pump speed raises the pump head-discharge characteristic to the broken curve shown passing through F. Similarly, if the concentration falls from C_2 to C_1, the pump is slowed down to give the characteristic passing through point G. Again, it must be noted that there is an economic penalty for designing for variable concentration, this time represented by the cost of the variable-speed drive and the energy losses in the drive train.

The effects of particle size distribution can be investigated in a similar way. Generally, decreasing the particle size of a homogeneous slurry tends to increase the effective viscosity. Thus it is broadly equivalent, in terms of the coordinates of Figure 13.1, to increasing concentration. A similar conclusion then follows: if the system will have to handle variations in slurry rheology, then it must be designed either for turbulent flow or with variable-speed pumps. If the former option is used, the system must be designed for turbulent flow with the most 'viscous' slurry anticipated.

The preceding discussion concerns long-term system stability, in response to slow changes in slurry properties. To examine the short-term effects of rapid transients, it is necessary to return to coordinates in which the system and pump heads are measured in terms of the density of the carrier liquid. Figure 13.2 shows an enlarged plot of the region around point E of Figure 13.1 in these terms.

Because the carrier-liquid head is S_m times the slurry head (Eq. 2.46), both the system curve and the pump characteristic shift when the slurry density changes. Point E_2 represents the design point, at concentration C_2. The long-term operating points for concentrations C_1 and C_3 now appear as points E_1 and E_3 respectively. Following the preceding discussion, all these points correspond to the same mixture velocity.

Now consider the effect of a sudden change in concentration from C_2 to C_3. This shifts the pump to the upper characteristic in Figure 13.2. However, the pipeline still contains solids at concentration C_2, and is therefore still represented by the middle system curve. Therefore, the immediate effect is to shift the intercept from point E_2 to point H, representing a slight increase in slurry throughput and velocity. As the new concentration propagates through the line, the system curve shifts from C_2 to C_3. Thus the operating point moves along the pump characteristic from H to E_3, representing a gradual return to the operating velocity. Similarly, as can readily be shown using Figure 13.2, sudden reductions in solids concentration will result in short-term reductions in mixture velocity. In each case the time required to return to the design velocity is essentially the slurry residence time in the pipe. The same arguments apply to pseudo-homogeneous flow of fine sand slurries.

Now consider laminar flow. Figure 13.3 shows the operating region around points A and F of Figure 13.1, plotted in terms of the head of the carrier liquid as in Figure 13.2. To avoid complexity, only variations in concentration from C_2 to C_3 are shown. We start with operation at the design point, A, with concentration C_2.

The feed concentration now rises suddenly to C_3. If the pump were to operate at fixed speed, this would raise the pump characteristic from 1 to 2 in Figure 13.3. With the pipeline full of slurry at concentration C_2, operation would immediately shift to point J, representing an increase in slurry throughput. As the new solids concentration propagates along the line, the system characteristic gradually moves up to C_3. As a result, the operating point shifts back along pump characteristic 2, finally reaching point K when the line contains only slurry at the new concentration C_3. This new steady operating point represents a substantial reduction in slurry throughput.

Now consider the case where a variable-speed pump is used, controlled to maintain constant slurry throughput. The first effect of the increase in solids concentration will be to *reduce* pump speed, to prevent 'overshoot' towards point J and maintain operation at point A. As the concentration front moves through the pipe, the pump will speed up to move operation from point A to point F. When the system reaches point L the pump will have returned to its original speed. As noted above, the final operating point, F, requires higher pump speed.

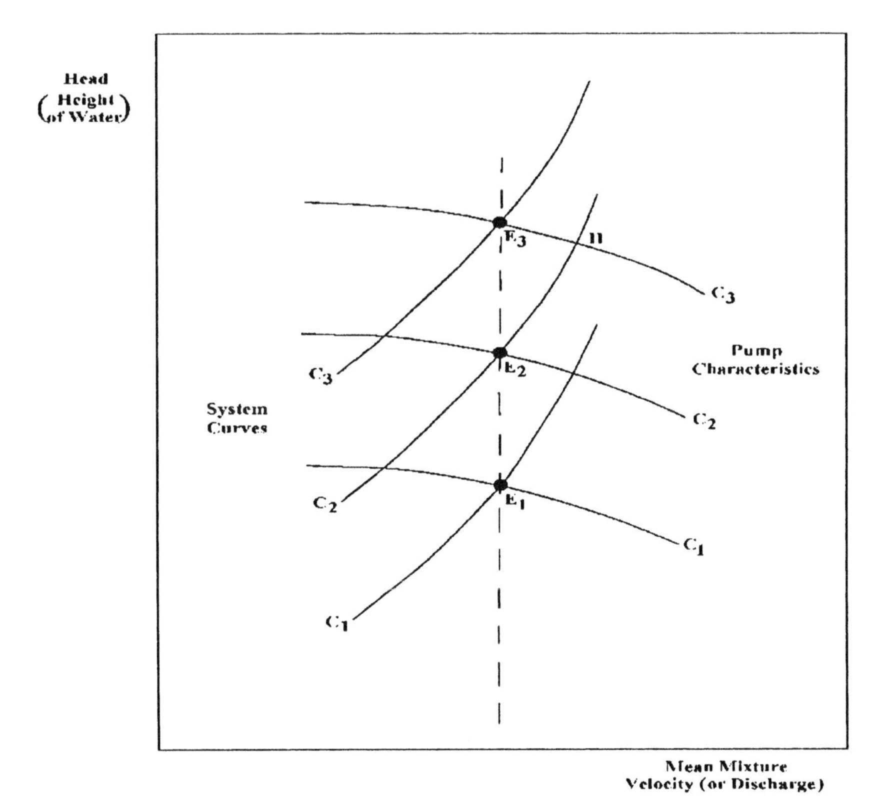

Figure 13.2. Effect of rapid changes in slurry concentration, for a homogeneous slurry in turbulent flow with a fixed-speed pump

Figures 13.2 and 13.3 refer only to the effects of changes in solids concentration. Changes in slurry rheology due to changes in particle size do not cause the effects of 'overshoot' and recovery.

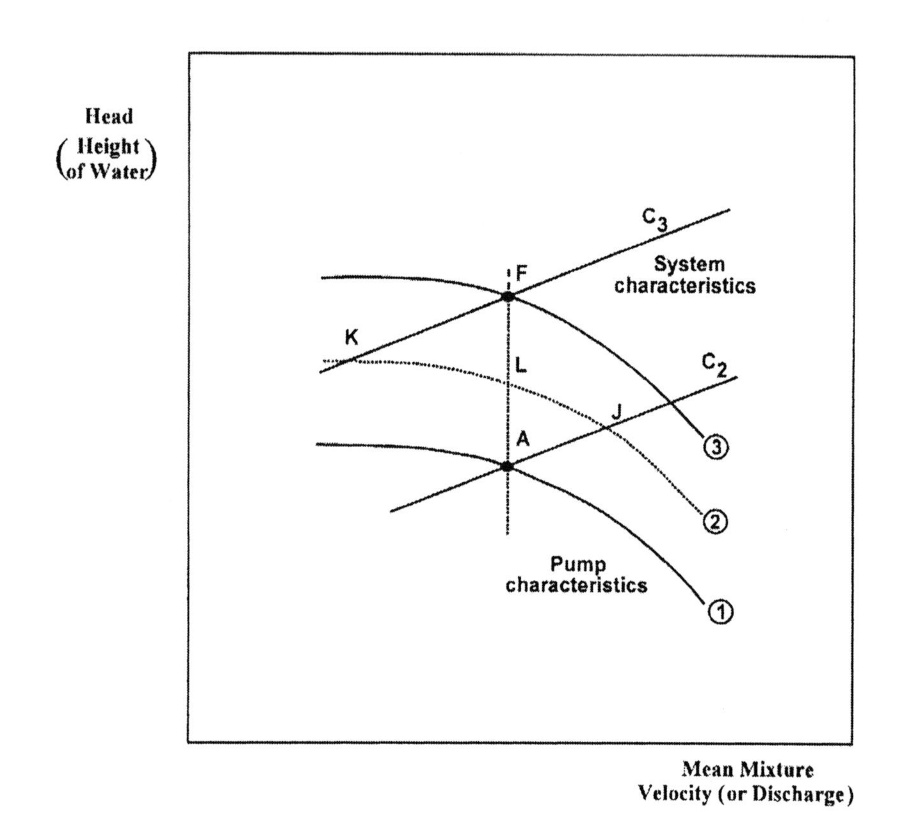

Figure 13.3. Effects of rapid change in slurry concentration, for a homogeneous slurry in laminar flow with fixed- and variable-speed pumps

13.3 Effect of Solids Concentration in Settling Slurries

We now turn to problems of system operability for settling slurries. Figure 13.4 shows typical 'system characteristics' for a settling slurry at three delivered concentrations, in the two forms introduced in earlier chapters: in Fig. 13.4a, the friction gradient is expressed as head of carrier liquid, i_m, while Fig. 13.4b gives the same information in terms of head of slurry, j_m. As in the previous section, only the frictional contribution for horizontal transport is considered here. The form of the system characteristics is

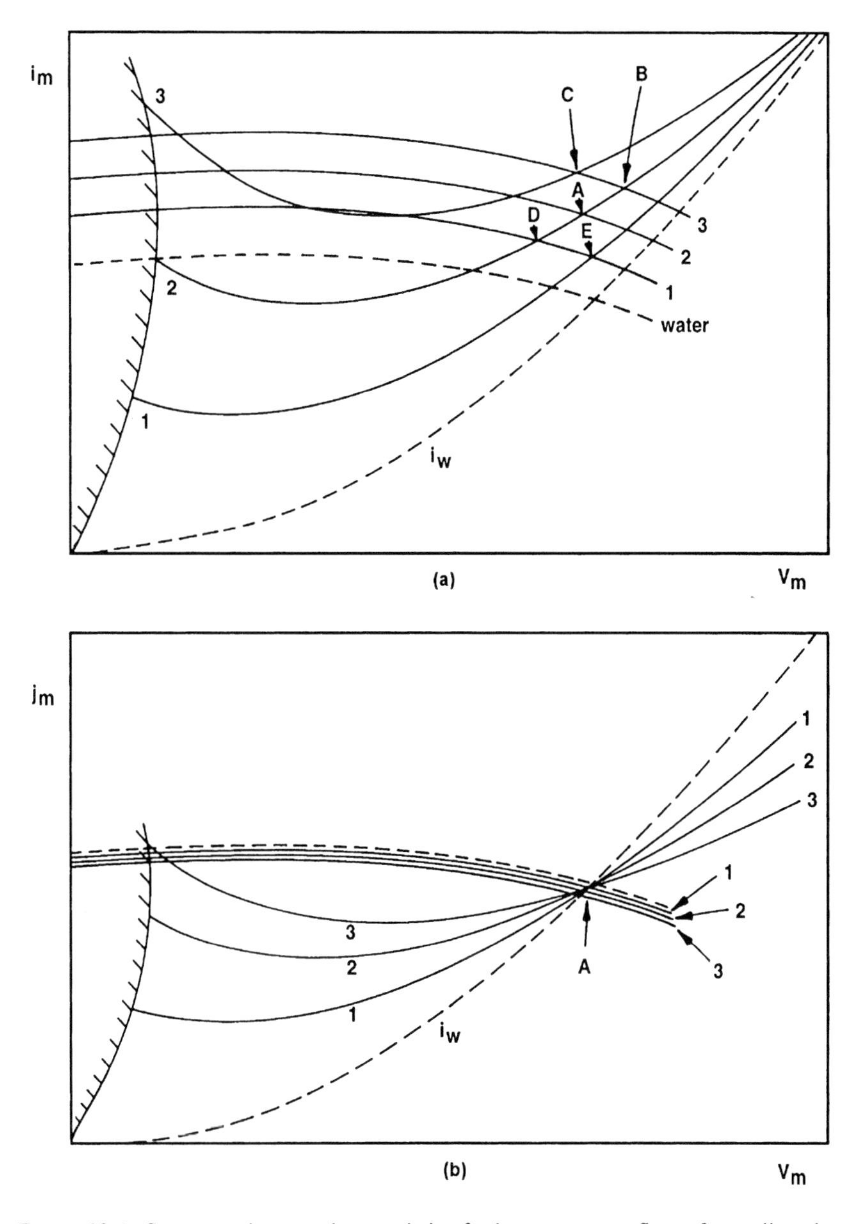

Figure 13.4. System and pump characteristics for heterogeneous flow of a settling slurry at various delivered concentrations (schematic)
(a) in terms of head of carrier liquid
(b) in terms of head of slurry

typical of heterogeneous flow of a settling slurry, as discussed in Chapter 6. Although the magnitude of the solids effect does not correspond to any specific slurry, the three slurry characteristics are scaled for three notional delivered relative densities, i.e. three values of S_{md}. The characteristics in the two forms are related, as usual, by $j_m = i_m/S_{md}$. Figure 13.4 is shown with linear coordinate axes, but absolute values are deliberately not given. Thus the system curves may be interpreted as total head loss for a fixed pipe length, while the curves in Fig. 13.4a can equally well be interpreted as total pressure drop along the pipe. The slurry curves all show the minima typical of a heterogeneous slurry flow. As implied by the relations given in Chapter 6, the position of the minimum moves to higher velocities as the solids concentration increases. Also, as is typical for heterogeneous slurries, the minimum friction gradient occurs at velocities above the deposition point. As noted in Chapter 6, expressions for the minimum friction gradient j_o and the associated velocity V_o were derived by Wilson et al. (2002). These will be repeated below.

Another important velocity shown on Fig. 13.4b is that at the point where j_m equals i_w. At this point, and this point alone, the heterogeneous slurry behaves like an 'equivalent fluid'. It was seen in Chapter 3 that equivalent-fluid behaviour corresponds to a value of 1.0 for the coefficient A' in the homogeneous-flow equation (Eq. 3.3). This use of 'equivalent fluid' is the same as the 'equivalent true fluid' in the terminology of Spells (1955), who also defined the 'standard' velocity as that at which, in present notation, $j_m = i_w$. The term 'standard' velocity will also be used here with reference to the point at which $j_m = i_w$. As shown on Fig. 13.4b, this standard velocity is not sensitive to changes in C_{vd}. Equations for the standard velocity V_x and the corresponding gradient j_x, derived by Wilson et al. (2002) are:

$$V_x = [[0.22V_{50}{}^{M}]2gD/f_w]^{1/(M+2)} \qquad (13.1)$$

$$j_x = [f_w/2gD]^{M/(M+2)}[0.22V_{50}{}^{M}]^{2/(M+2)} \qquad (13.2)$$

The corresponding quantities at the minimum point, mentioned above, are best defined in terms of V_x and j_x. They are:

$$V_o = V_x\,[M(S_m - 1)/2]^{1/(M+2)} \qquad (13.3)$$

$$j_o = j_x\,[(M+2)/(MS_m)][M(S_m - 1)/2]^{2/(M+2)} \qquad (13.4)$$

It is now appropriate to consider the behaviour of this system with one or more fixed-speed centrifugal pumps, whose head-discharge characteristics are shown in Figure 13.4. The total developed head measured in terms of delivered slurry density (Fig. 13.4b) decreases slightly with increasing concentration, due to the effect of solids on pump performance discussed in Chapter 10. Therefore the pump discharge head, measured as the water column equivalent to the discharge pressure of the pump, increases with slurry concentration. This increase is in slightly less than direct proportion to S_{md}. For the case illustrated by Fig. 13.4, where the pump has been selected for operation close to the standard velocity at point A, the system can accommodate variations in solids concentration from zero up to the maximum shown: there will be some reduction in mean velocity as C_{vd} increases, because of the effect of the solids on the pump characteristic (Fig. 13.4b), but the variation in steady-state operating conditions is slight.

However, the transient behaviour is again more interesting. Consider the case where the system has been operating steadily at concentration 2, and the slurry presented to the pump suddenly changes to the higher concentration 3. Referring to Fig. 13.4a, the system characteristic is now as 2, but the pump is handling a higher-density material so that its discharge pressure increases to characteristic 3, as in the case considered in Figs. 13.2 and 13.3. The immediate effect is to shift the system operating conditions to point B, increasing both the mean slurry velocity and the power drawn by the pump. As the higher solids concentration propagates along the line, the system resistance moves up to characteristic 3, so that the velocity decreases and system operation moves back to point C. Conversely, if the system has been operating steadily at point A and the slurry entering the pump is suddenly reduced to concentration 1, the mixture velocity is reduced as the system moves to point D. As before, the system resistance now moves gradually back to characteristic 1, and operation moves back to point E. Much as for the turbulent non-settling slurry in Figure 13.2, the system at first ‘overshoots' in response to the change in slurry concentration, but then returns to the design slurry throughput. For this settling case, this automatic compensation is a result of operating close to the standard velocity, i.e. point A in Figure 13.4 b.

The above discussion refers to a system with one or more fixed-speed pumps at a single location. Where long pipelines with several pumps are used, the common practice (discussed in Chapter 12) is to space the pumps along the line and to use a variable-speed drive for the first pump, and sometimes for the last, so as to reduce the surges arising from variations in solids concentration.

Figure 13.5 illustrates operation of the same system but with pumps selected to operate further back on the system characteristics, giving a velocity below the standard value at concentration 2. As discussed in Chapters 3 and 6, and illustrated by Case Study 6.2 in Section 6.9, this can be more economic: operating with higher concentration (and hence lower velocity) can reduce the specific energy consumption. This also emerges from Case Study 13.1 in Section 13.6 below. However, we must now investigate operation at velocities significantly below the standard velocity, showing the response to variations in solids concentration and throughput.

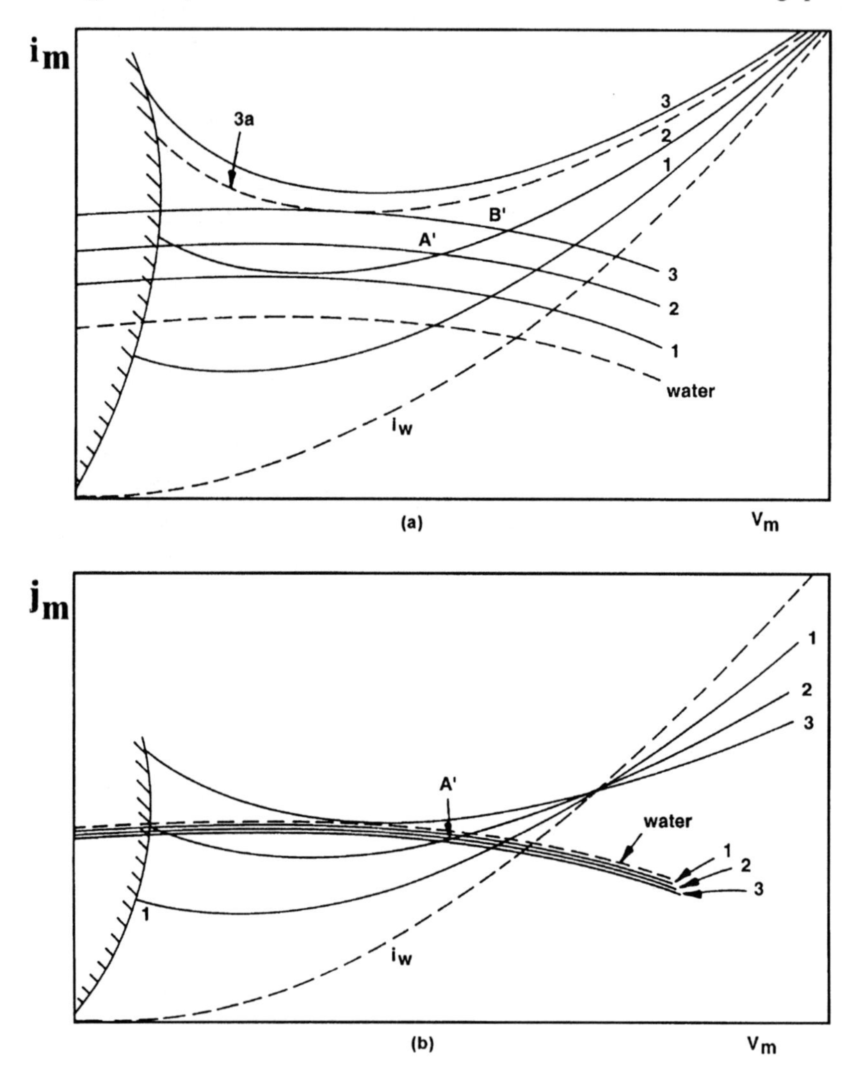

Figure 13.5. System characteristics of Fig. 13.4 with different pump characteristics
(a) in terms of head of carrier liquid
(b) in terms of head of slurry

Consider first the effect of increasing concentration to characteristic 3. As before, the effect on the pump occurs before the new concentration has propagated along the pipeline, so that the immediate effect is to shift operation from A′ to B′. The system again responds more slowly, and the pipe velocity therefore decreases from the maximum at B′. However, in this case, steady operation at concentration 3 is not possible with fixed-speed pumps, because they cannot generate sufficient head. Thus, when the system reaches acharacteristic corresponding to 3a, the velocity abruptly reduces back into the deposit region. In other words, the line becomes 'plugged'. Figure 13.5a shows that reducing the solids concentration, even to the point of pumping water alone, cannot clear the plug: higher pump speeds are needed, or alternatively a slurry of fine particles may shift the deposit (see below). If a variable-speed pump or a clay slurry is not available, the only recourse will be to open up the line at some intermediate point and pump the solids out.

Two general conclusions can be drawn from the foregoing discussion. Comparing the system and pump characteristics is essential, because it enables qualitative but very informative assessment of operating stability. For most systems driven by centrifugal pumps, operation at velocities below the standard velocity is feasible only for relatively fine slurries (see below) or for systems where the solids concentration is not liable to wide variations. This represents another case where design for variations in solids throughput implies a penalty in operating and capital costs.

Figure 13.5 also illustrates why the velocity at the limit of deposition is often unimportant for settling slurries: although operation led to a 'plugged line', the cause was poor matching of the pump and system characteristics, rather than operation too close to deposition. This also illustrates why field data (and some laboratory data) indicate deposit velocities much above the values estimated from the analysis in Chapter 5: they actually correspond to the limit of stable operation with centrifugal pumps, rather than the limit of operation without a stationary deposit. In practice, centrifugal pumps permit operation near the deposition point only for relatively fine particles; this point is discussed further below.

13.4 Effect of Particle Size

Figure 13.6 shows schematic system and pump characteristics for the case where the slurry concentration remains constant. Three different particle sizes are considered, designated 'fine', 'medium' and 'coarse'. The same simplifications have been made as in Figs. 13.4 and 13.5; the curves for the medium slurry M are identical to those for concentration 2 in the earlier figures, and conversion between i_m and j_m has been carried out using the same value of S_{md}.

Several features of these characteristics are of interest. For the 'fine' slurry, the velocity at the threshold of turbulent suspension is typically much smaller than the velocity at the limit of deposition (see Chapter 6). Therefore, above the deposit point, there is very little stratified load, and the system characteristic has a positive gradient throughout. If the slurry is so fine as to contain particles sufficiently small to affect momentum transfer in the laminar sub-layer (see Chapter 6), then for homogeneous flow A' is greater than zero in Eq. 3.3, and the system characteristic at high velocity approaches an asymptote above the clear water curve. The standard velocity for such a slurry depends on the value of A', and is not necessarily as shown in Fig. 13.6, i.e. slightly below the value for a coarser material.

Because there is more stratification in the 'coarse' slurry, the corresponding system characteristic shows a pronounced minimum, at a velocity higher than at the minimum for the medium slurry. Similarly, the standard velocity is high for the coarse slurry. This combination of coarse particles and high velocities may lead to excessive wear (see Chapter 11).

For the reasons outlined in Chapter 10, the coarsest particles have the greatest effect on pump performance. However, unless the pump is small, the solids effect is minor. As a result, if the slurry density is constant, the head developed by the pump is almost constant whether measured as water head, delivery pressure, or slurry head. Thus the pump characteristics will have the form shown schematically in Figure 13.6.

The transient effects of variations of head-end particle size will now be considered. Figure 13.6 illustrates the case where the system is designed for operation with coarse solids at their standard velocity. Reducing the particle size has negligible effect on pump performance, although the system characteristic changes gradually towards curve M. With fixed-speed pumps, the mixture velocity will increase. With a variable-speed drive, the pump can be slowed down slightly to maintain constant mixture velocity.

Figure 13.7 shows the case where the pumps are sized for operation with medium to fine solids near their standard velocity. If the particle size now becomes 'coarse', the system characteristics gradually move up to curve C. If fixed-speed pumps are used, operation is now unstable. Much as for the effect of concentration illustrated by Fig. 13.5, a point will be reached at which the pumps are no longer able to sustain the flow, so that the velocity drops sharply into the deposit region. If the pumps cannot be speeded up, possible action to 'remove the plug' is then to pump fine solids through the line (to aid transport of the coarse material - Chapter 6) at higher concentration (to increase pump discharge pressure - Chapters 9 and 10).

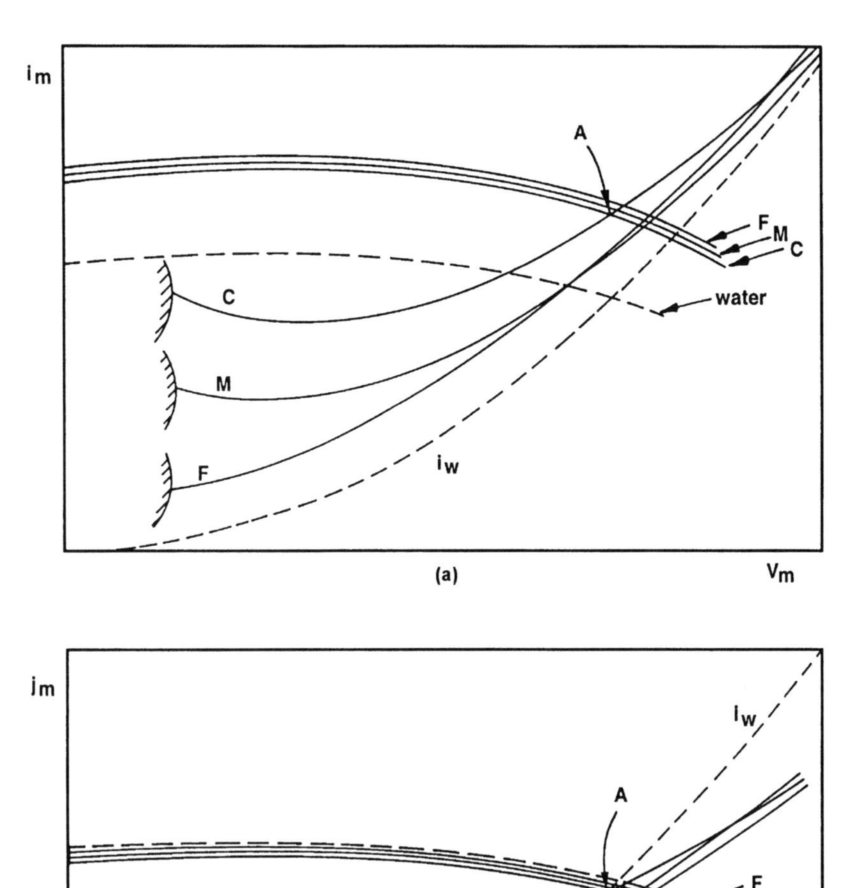

Figure 13.6. System and pump characteristics for heterogeneous flow of slurries of various particle sizes F - Fine; M - Medium; C - Coarse
(a) in terms of head of carrier liquid
(b) in terms of head of slurry

The cases presented here to illustrate analysis of system operability have all been based on the relatively flat head-discharge characteristics typical of centrifugal slurry pumps with relatively low rotational speed. Higher rotational speed gives a more steeply falling pump characteristic. Therefore, in principle, the system should be operable at lower mixture velocities, as illustrated in Fig. 13.8. In practice, unless the system is to operate with steady particle size and concentration, the scope for reducing mixture velocity in this way is limited, even if it is possible to increase pump speed without incurring excessive erosion.

Even if stable operation can be achieved at velocities below the standard value for medium slurries, the economic incentive to do so is small, because the minima in the system characteristic curves ensure that the economic transport velocity is clear of the deposit range. However, for slurries like that designated 'fine' in Figs. 13.6 to 13.8, operation is most economic close to deposition; only for these materials, then, is the deposit velocity of practical concern in horizontal lines. These characteristics are typical of long-distance slurry freight systems where, as noted in Section 13.1, it is economic to 'tailor' the particle size distribution to reduce transport costs by reducing pipe friction and to ensure that the slurry has good 'restart' characteristics. To achieve stable operation close to the deposit range requires a very steep pump characteristic. Long-distance systems usually use positive displacement pumps for which the discharge rate is fixed - in terms of Fig. 13.8, the pump characteristic is a vertical line - so that operation is stable at any velocity above deposition. However, this technique involves departing from the use of centrifugal pumps; and further discussion would not fall within the scope of this book.

Because of the complex nature of slurry flows, problems in operating slurry pipeline systems can arise from a variety of causes. The sources of unstable or inoperable conditions in this chapter have deliberately been simplified, with only one change considered at a time to illustrate the principles involved. However, the reasons for problems in operating slurry systems can be more complex and sometimes far from obvious. Chapter 15 covers a selection of operating problems examining how they arose and how they could have been avoided.

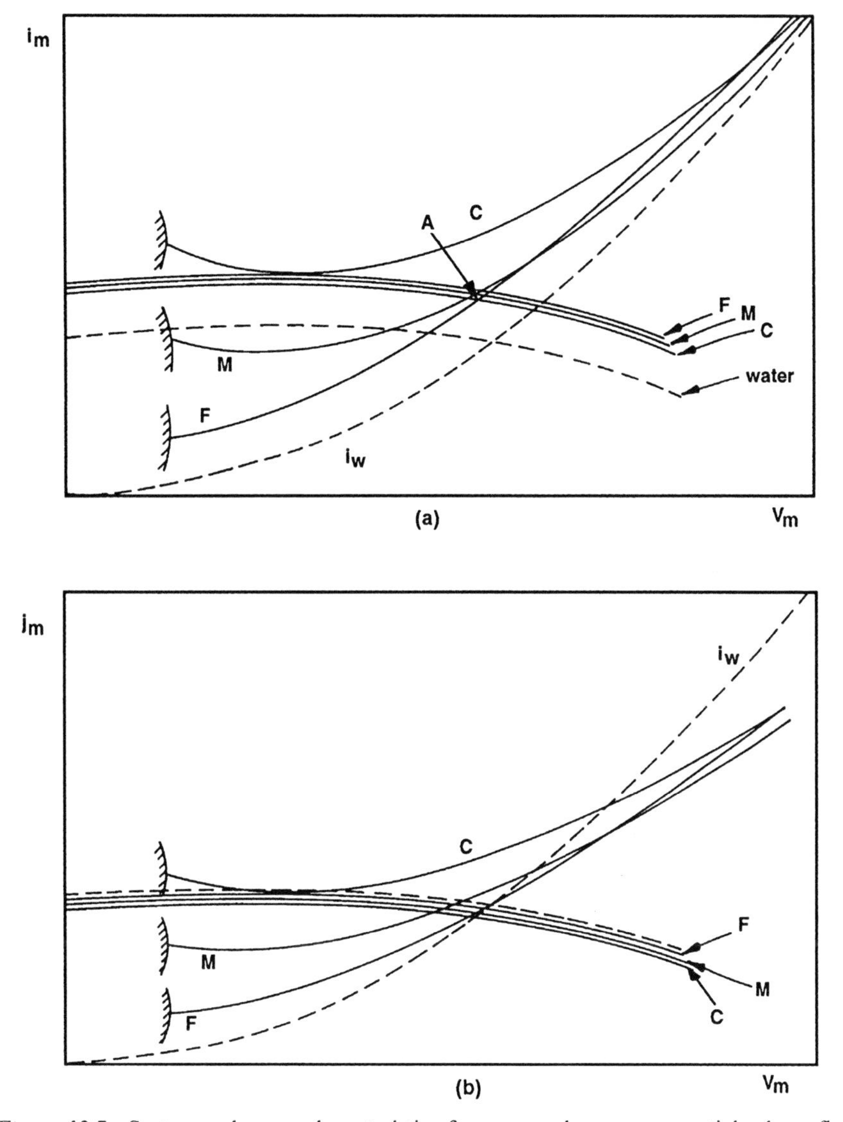

Figure 13.7. System and pump characteristics for a case where coarse-particle slurry flow causes difficulties. F - Fine; M - Medium; C - Coarse
(a) in terms of head of carrier liquid
(b) in terms of head of slurry

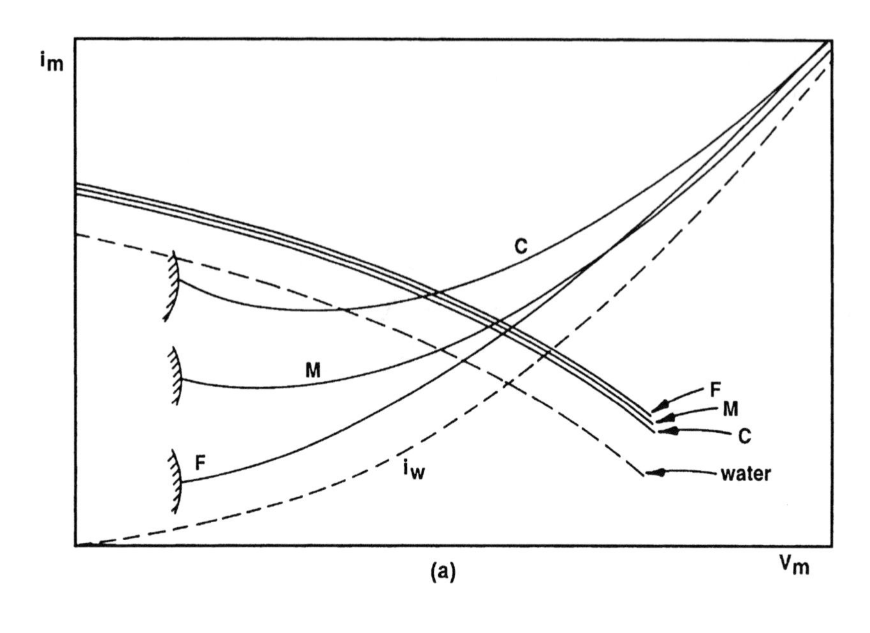

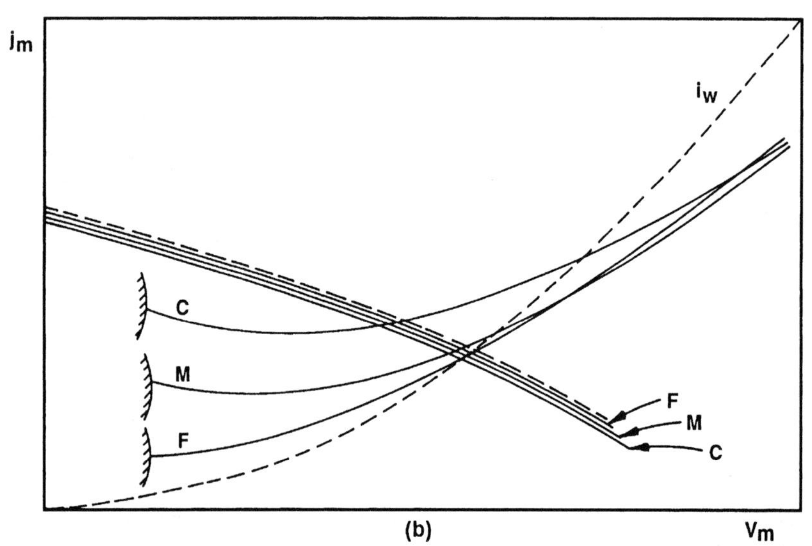

Figure 13.8. System and pump characteristics for heterogeneous flow of settling slurries showing effect of pump type and speed

F - Fine; M - Medium; C - Coarse

(a) in terms of head of carrier liquid

(b) in terms of head of slurry

13.5 Case Study

Case Study 13.1

Case Studies 5.1 and 6.1 were concerned with the transport of 4000 tonnes per hour of coarse sand ($S_s = 2.65$) as a slurry in water, at delivered concentrations up to $C_{vd} = 0.24$. In this case study the design will be completed by examining system operability. Because the earlier studies all represent stages in the design, they will be summarised here.

Case Study 5.1 was the preliminary selection of pipe size, not needing information on the particle size beyond noting that the slurry has settling characteristics. Three pipes sizes were identified as suitable for more detailed study, keeping $V_m > V_{sm}$ to avoid deposition. The results are shown on Table 13.1.

Table 13.1

Table 13.1 Condition	D (m)	V_{sm} (m/s)	C_{vd}	V_m (m/s)
A	0.55	4.45	0.24	7.35
B	0.60	4.65	0.24	6.17
C	0.65	4.84	0.20	6.30

In Case Study 6.1 the particle size distribution was introduced, and used to calculate the friction gradient and specific energy consumption. It was shown that, for the 0.65 m pipe with 20% solids of the type specified (Condition C above) the friction gradient is 0.0612 m. water/m. pipe with specific energy consumption (SEC) of 0.315 kWh/tonne-km. It was also found that increasing C_{vd} and decreasing V_m reduced the specific energy consumption. A revised operating point (Condition D) was therefore chosen for further analysis.

D = 0.65 m (V_{sm} = 4.84 m/s)
C_{vd} = 0.221
V_m = 5.70 m/s
i_m = 0.0619 m.water/m.pipe
SEC = 0.289 kWh/tonne-km

The feasibility of operating at this point was not investigated in Case Study 6.1. Note that for these conditions the 'standard velocity' V_x is 7.52 m/s (from Eq. 13.1) and the velocity at minimum point V_o is 5.48 m/s (from Eq. 13.3). Thus the proposed operating velocity V_m lies between V_o and V_x. It is presumed that the pumps have been selected to operate at V_m and that they have the usual falling characteristics: i.e. head developed decreases with discharge.

We now pursue this study further by examining system operability. Figures 13.9 to 13.14 show the system characteristics and SEC curves for three pipe sizes: 0.65, 0.75 and 0.55 m.

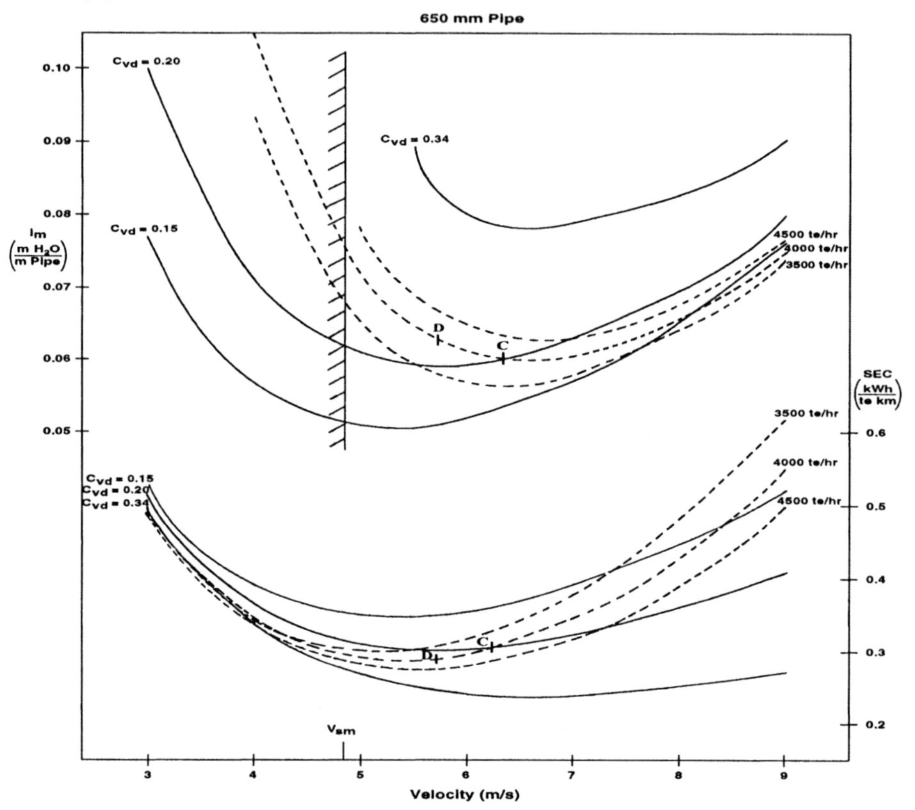

Figure 13.9. Friction gradient (i_m) and specific energy consumption (SEC) for transport of coarse sand in 0.65 m pipe

The curves are all calculated using the approach illustrated in Case Study 6.1, with values for the deposit velocity (V_{sm}) obtained from Eq. 5.10 as in Case Study 5.1. We consider first the pipe size which appeared to be attractive in the earlier case studies : D = 0.65 m. Figure 13.9 shows the system characteristics (expressed in terms of i_m) and SEC curves. The full curves are drawn for constant delivered concentration. The broken curves refer to constant throughput: throughput can be kept constant by increasing mixture velocity while decreasing delivered concentration. Conditions C and D considered in earlier case studies correspond to the points indicated on this figure.

Only now that the system characteristics are plotted is it evident that Condition D is too far back on the system curves to represent a feasible operating point for this material - the system characteristic shows i_m decreasing as V_m increases, so that operation will be unstable. Even Condition C is marginal. Given that the booster pump will have variable - speed diesel drive, operation at C might be feasible if solids consistency and throughput do not fluctuate too widely. However, if the variations are significant - for example, throughput variations from 3500 tonne/hr to 4500 tonne/hr (see Figure 13.9) - then operation will become unstable.

To achieve truly flexible operation, it is preferable to operate around the 'standard' velocity, where $i_w = j_m$. Figure 13.10 shows the same system characteristics, but now plotted in terms of j_m (i.e. i_m/S_m in m. slurry/m. pipe). The recommended operating velocity is therefore around point E, i.e. at V_m of 7.54 m/s, with C_{vd} = 0.168 and with SEC increased to 0.421 kWh/tonne-km. Thus, once operability is considered, it is seen that lower concentrations should be used, with correspondingly higher energy consumption.

We consider now whether going to a larger pipe would improve matters. Figures 13.11 and 13.12 show the curves of i_m, SEC and j_m for a pipe of 0.75 m diameter. To achieve operation at the standard velocity, it is now necessary to reduce the concentration further still: V_m = 7.83 m/s, with C_{vd} = 0.121. Again because of the cost of pumping the conveying water, the specific energy consumption is high: SEC is 0.511 kwh/tonne-km at a throughput of 4000 tonnes per hour. Furthermore, increasing the pipe size will increase the capital cost. Thus the 0.75 m pipe is not recommended.

A smaller pipe size should also be considered. Figures 13.13 and 13.14 show the curves of i_m, SEC and j_m calculated in the usual way for a pipe 0.55 m in diameter. Condition A from Case Study 5.1 is now seen to be quite close to the standard velocity. It therefore represents a flexible but stable operating point. Furthermore, because the delivered concentration is relatively high, at about 24% solids, the specific energy consumption is quite attractive, at 0.353 kWh/tonne-km.

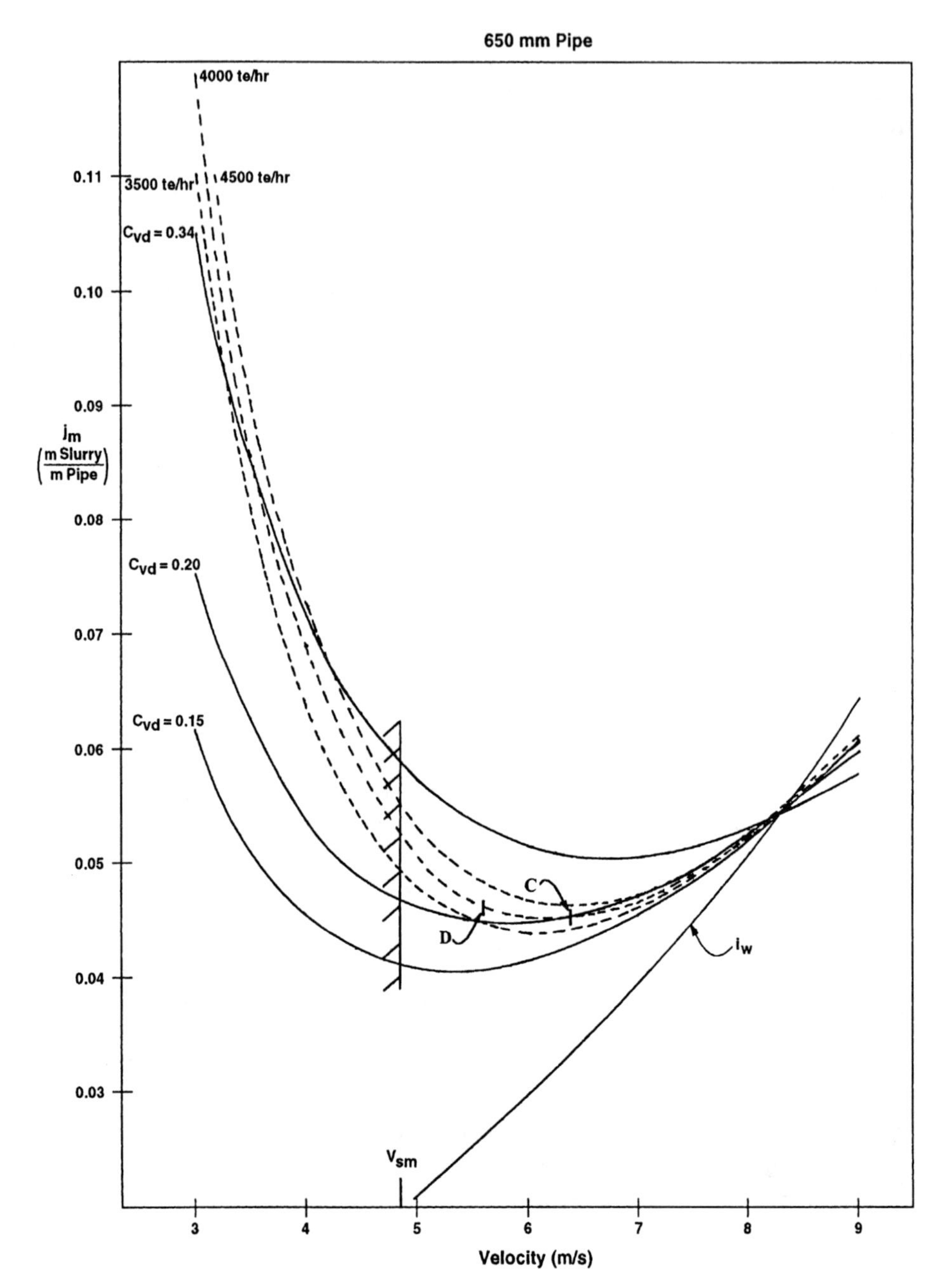

Figure 13.10. Friction gradient in terms of head of slurry (j_m) for coarse sand in 0.65 m pipe

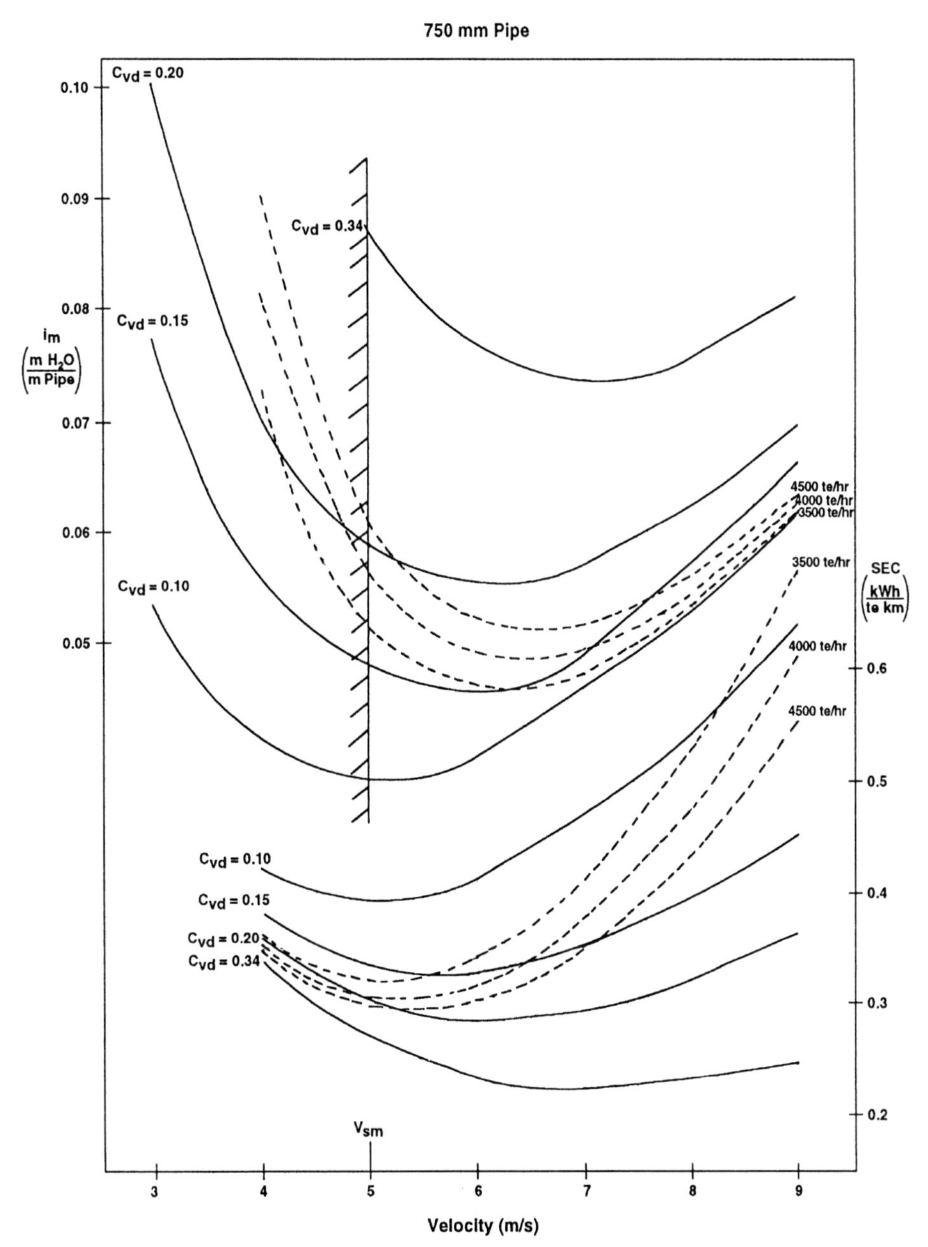

Figure 13.11. Friction gradient (i_m) and specific energy consumption (SEC) for transport of coarse sand in 0.75 m pipe

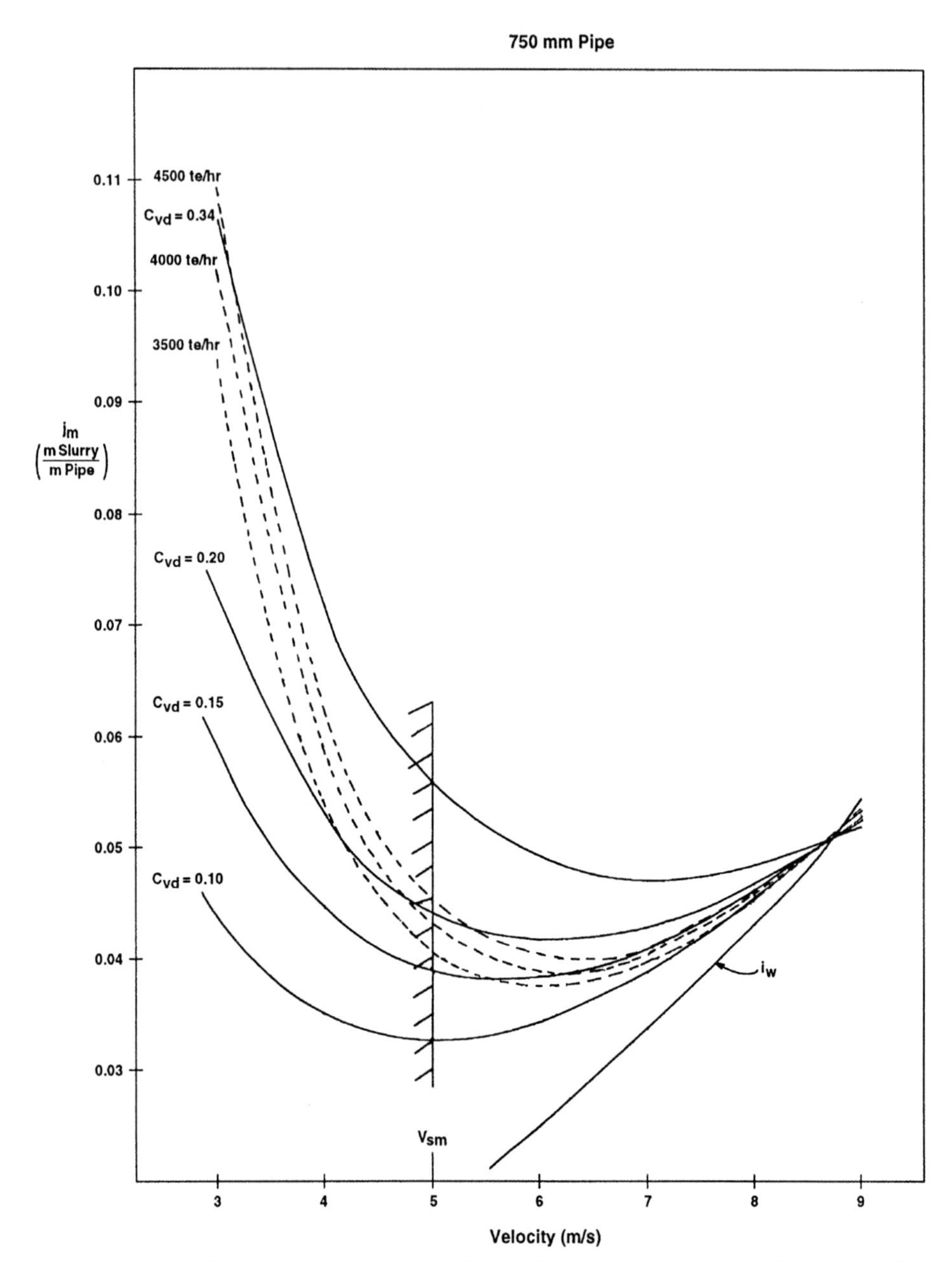

Figure 13.12. Friction gradient in terms of head of slurry (j_m) for coarse sand in 0.75 m pipe

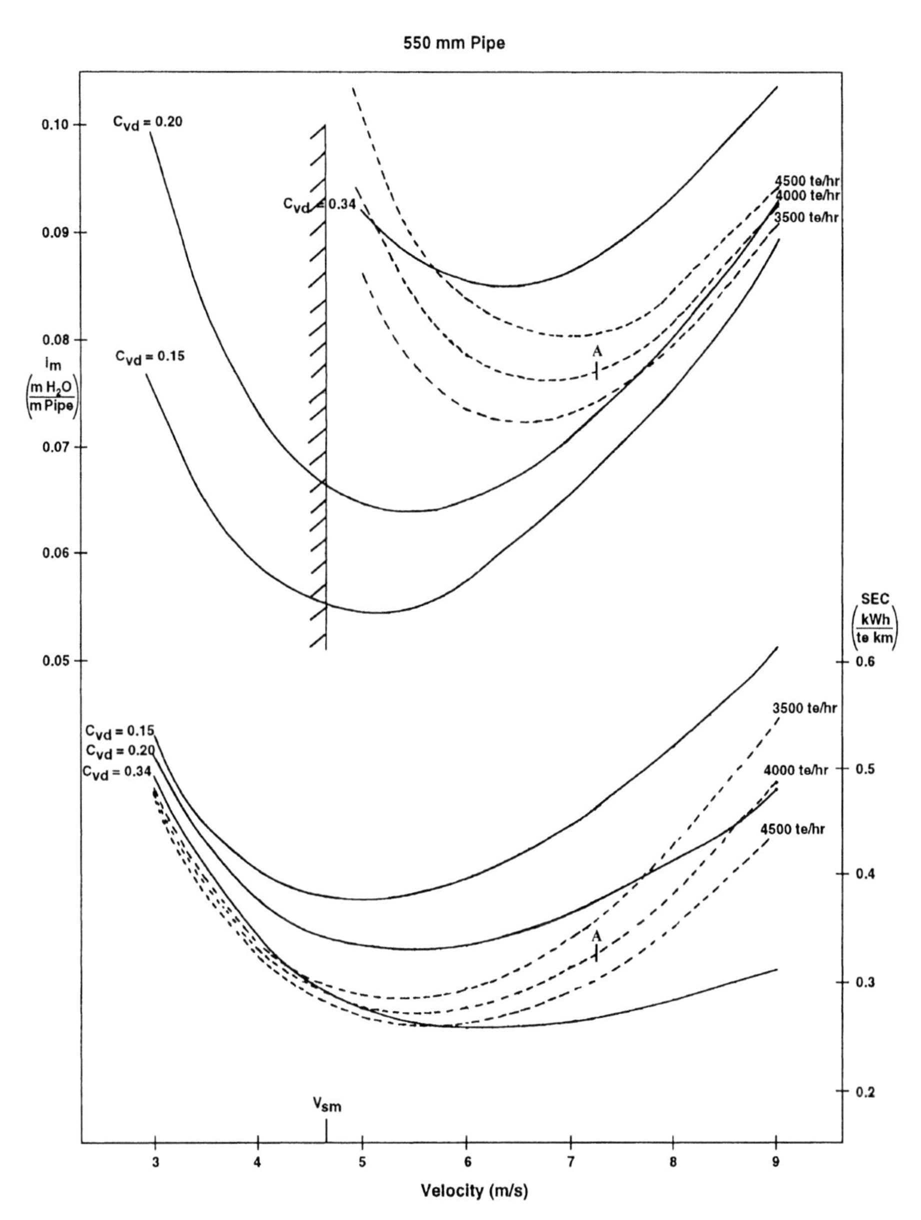

Figure 13.13. Friction gradient (i_m) and specific energy consumption (SEC) for transport of coarse sand in 0.55 m pipe

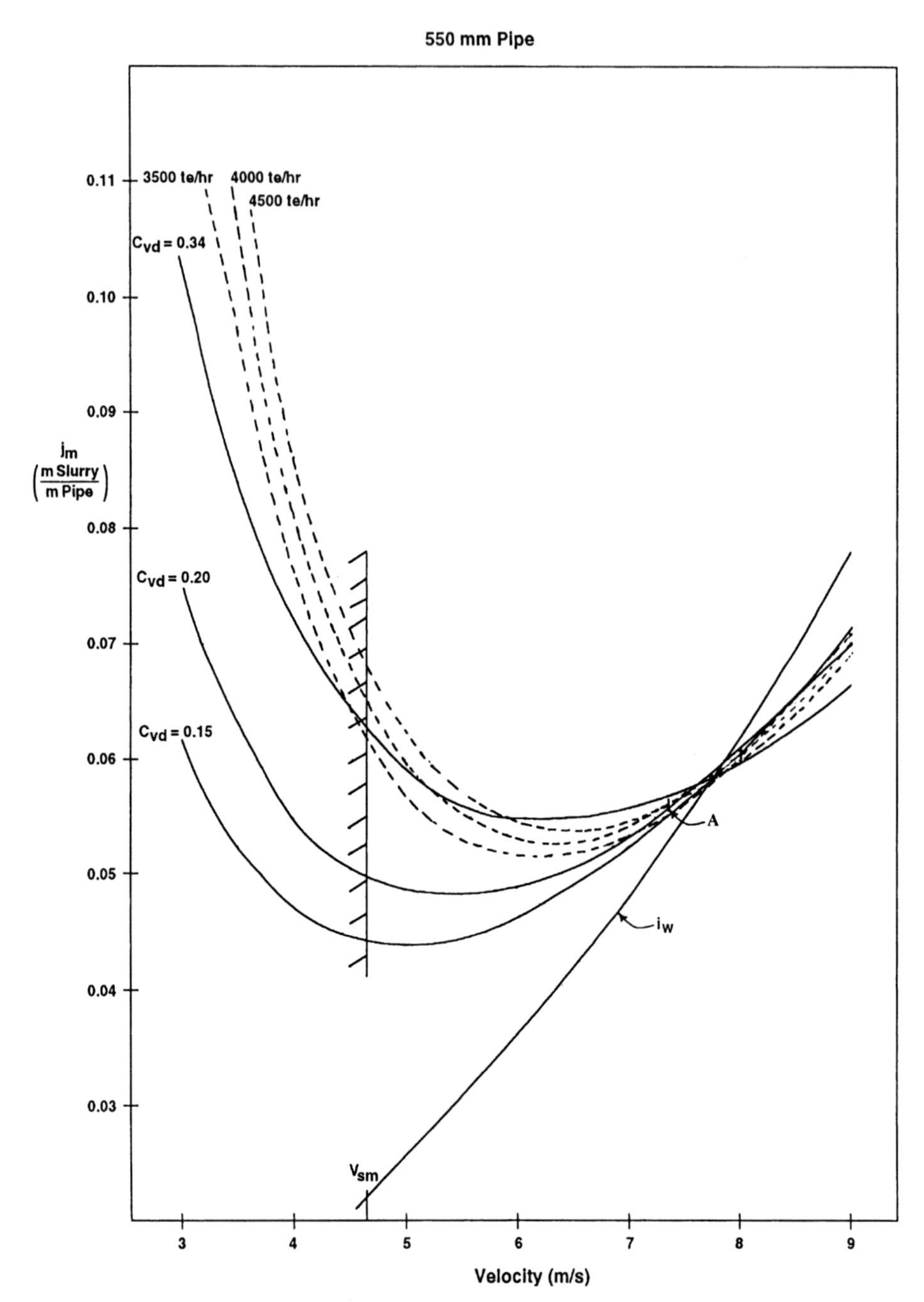

Figure 13.14. Friction gradient in terms of head of slurry (j_m) for coarse sand in 0.55 m pipe

Thus, of the three pipe sizes considered in this case study, the smallest would be preferred. In addition to its favorable specific energy consumption, it will have the least capital cost. It is left to the reader to carry out the comparison for the intermediate size, D = 0.60 m. A full design would now proceed to consider pump selection again, essentially repeating Case Study 12.1 for the revised pipe size.

The conclusions illustrated by this case study are general: for slurries with settling characteristics, considerations of energy consumption together with operational stability favour use of high solids concentration and small pipe diameter. Also, as indicated throughout this chapter, stable operation rather than deposition controls the operating velocity for this type of slurry.

On the negative side, the mixture velocity can be too high. For example, Condition A, as selected above, has a design mixture velocity of 7.35 m/s, and the pipe will experience substantial wear. As noted in Chapter 11, for some applications, e.g. conveying of phosphate matrix, wear is accommodated by rotating the pipe at intervals, usually through 120°, so as to expose new parts of the pipe wall to the stratified solids which cause the abrasion.

REFERENCES

Spells, K.E. (1955). Correlations for use in transport of aqueous suspensions of fine solids through pipes. *Trans. Instn. Chem. Engrs.*, Vol. 33, pp. 79-84.

Wilson, K.C., Clift, R. & Sellgren, A. (2002).Operating points for pipelines carrying concentrated heterogeneous slurries. *Powder Technology*, Vol. 123, pp. 19-24.

Chapter 14

PUMP SELECTION AND COST CONSIDERATIONS

14.1 General Concepts of Pump Selection

As indicated in earlier chapters (for example, Section 2.3) pumps must be selected by matching their head-discharge performance to the requirements of the piping system. Figure 2.12 shows that the intercept of the pump characteristic with the system characteristic defines the operating head and discharge. For settling slurries in particular, selection of appropriate operating conditions raises special considerations, discussed in Chapter 13. In the case of long lines, the total system head will be more than can be handled by a single pump. It is then necessary to use several pumps in series. This case, and the less common case of several pumps in parallel, was considered in Chapter 12 and will be dealt with further in Chapter 15.

Once the operating conditions have been established, pump selection in its simplest sense amounts to determining the performance of each available pump for the head and flow required, and selecting the one best suited to the duty. In general, the smaller the pump, the less costly it will be. However, reducing pump size implies higher speed for a given discharge. Thus, in a water pump application, NPSH and mechanical considerations tend to limit speed and set the overall pump size. In slurry pump applications, the NPSH determines limits only in special cases (to be covered later), and the shaft speed (and velocities within the pump) is limited by the wear life, which decreases as speed is increased. Therefore, slurry pumps typically have larger size and lower rotational speed than those of water pumps for equivalent heads and discharges. Actual wear data will not readily be available for a specific pump type, shaft speed, and slurry duty. However, the modelling techniques discussed in Chapter 11 can be used to transfer information on wear from one configuration to another, and in certain cases to calculate expected wear lives. Where exact details of the service are known, and can be relied on, calculations using the techniques discussed in Chapter 11 are recommended for estimating the lives of shells, impellers and suction liners.

Slurry concentration and particle size cause reductions in the theoretical head, efficiency and NPSH characteristics. This topic was discussed in detail in Chapter 10, and will not be repeated here.

In general practice, pump selection begins with semi-empirical selection guide, such as the charts on Fig. 14.1 and 14.2 and the associated tables (Table 14.1 and 14.2). The initial classification as to Service Class (1, 2, 3 or 4) is based on Fig. 14.1 with if necessary, an adjustment for the abrasivity of the slurry by altering the size of solids in Table 14.2 for the same wear using Fig. 11.6 in Chapter 11.

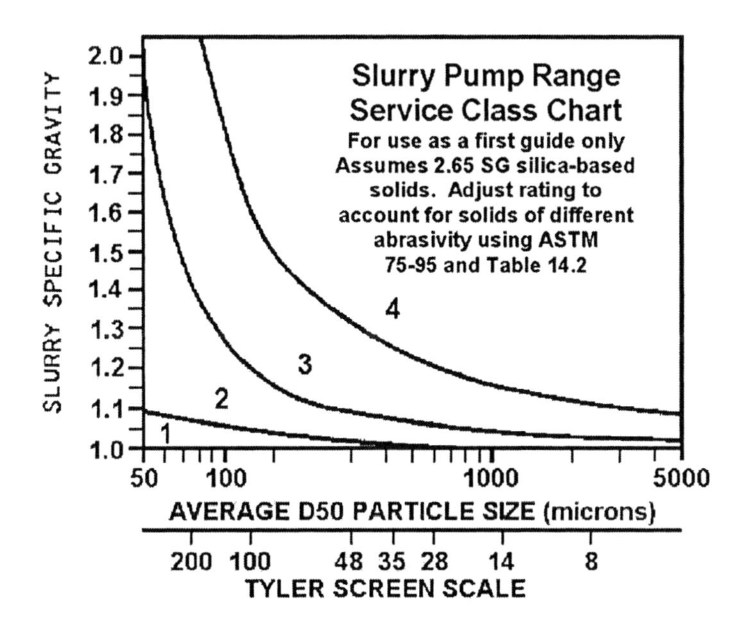

Figure 14.1. Service class chart for slurry pumps referencing Table 14.2

From all of the available pumps, a set of acceptable selections is established using the discharge branch velocity, suction branch diameter (Fig. 14.2) and other limits set out in the tables. Particular attention should be paid to the impeller peripheral speed (and the head it implies), and the percentage of BEP operation for the various shell types. As a final step, the efficiencies, power requirements and available driver speed limitations are considered and a selection made. If several hundred possibilities are to be considered, this process is tedious. Initial branch checks may be used to limit the available choices and the amount of evaluation. A more common approach now, however, is the use of a computer program to scan the possible choices and come up with acceptable ones in order of efficiency.

Table 14.1 Recommended operating limits for slurry pumps

		Service Class			
Maximum discharge velocity	Shell type	1	2	3	4
m/s		12	10	8	6
ft/s		40	32	27	20
Maximum throat velocity					
m/s		15	12	9	6
ft/s		50	40	30	20
Recommended range:	Annular (A)	20-120	30-130	40-100	50-90
percentage of BEP	Semi-volute (C)	30-130	40-120	50-110	60-100
flow rate	Near volute (T)	50-140	60-130	70-120	80-110
	Annular/oblique Neck (OB)	10-110	20-100	30-90	40-80
Maximum impeller peripheral speed:					
(a) All-metal pump:					
(m/s)		43	38	33	28
(ft/min)		8500	7500	6500	5500
(b) Rubber-lined pump					
(m/s)		28	25	23	20
(ft/min)		5500	5000	4500	4000

Table 14.2 ASTM G75-95 Miller Number slurry abrasivity adjustment

Mineral	**Miller Number**	**Abrasivity**
Limestone	34	0.60
Raw Coal	56	0.75
Phosphate	90	0.90
Sand	112	1.00
Tailings	219	1.33
Copper Ore	287	1.50

In addition, the pressure rating of each pump should be checked against the duty required and the pump spheres passage checked against the largest size of solids. The size and capacity of the shaft and bearings must be checked, while the wear resistance of the wetted-end materials must be assessed as outlined in Chapter 11. The NPSH available from the system must be checked, to ensure that it exceeds the NPSH required by the pump. The basic suction considerations were explained in Section 9.3. Some applications, notably dredging, involve special suction-side considerations, which were examined in Chapters 8 and 12. For pit pumps and some dredge pumps, avoiding cavitation can limit pump speed to values below the wear limits in Table 14.1. It is sometimes necessary to change selection in favour of a slower-running pump, or one with better suction characteristics.

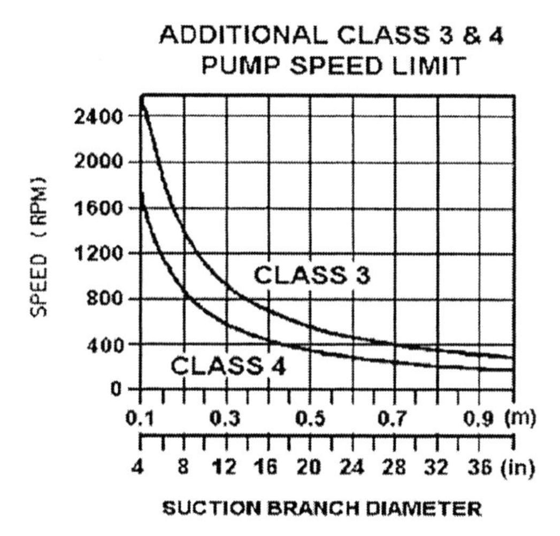

Figure 14.2. Pump Speed Limit Chart for Classes 3 and 4

Checks on the rotating assembly pressure and the NPSH are made using information provided by the manufacturers for any specific configuration being considered.

14.2 Wear Considerations

The previous section proposed general guidelines or, more particularly, limits that are in general use. These provide selections that will give acceptable wear lives. Nevertheless, the appropriate wear life has not been defined, nor has the cost of operation of different options been considered. Likewise, the selection procedure of the previous section provides no guide to the actual wear life of the different wet-end components (shell, liner and impeller), nor does it deal with the time between parts replacement or the downtime cost.

While not specifically stated, the service class chart for slurry pumps (Figure 14.1) shows lines of approximately constant service condition (maximum wear potential) for the solids size and concentration of the different service classes detailed on Table 14.1. The service-condition wear potential may be defined as $W_c \cdot C_v$, where W_c is the wear co-efficient defined in Chapter 11 and C_v is the volume concentration of solids.

Figure 14.1 shows increasing impeller peripheral speeds (and head) corresponding to the maximum class 1, 2, and 3 conditions at a constant wear intensity. These are such that the smaller solids size and lower concentration (wear intensity) for the class 1, 2 and 3 conditions result in decreasing wear, so that the lower number classes of service have higher overall lives that are roughly doubled for each class change.

The increase in life of the lower service classes reflects the practical reality that pumps which move large solids at high concentration must be run slowly at a low head per stage in order to get acceptable wear. When operating with smaller solids or lower concentrations, higher speeds and heads per stage are possible, giving acceptable wear lives for the pump components. In the end, economic considerations decide what an acceptable wear life is, and what selection parameters should govern. The next section investigates these concepts.

14.3 Economic Considerations

Figure 11.16 in Chapter 11 shows that if high concentrations (40% by volume), of large size (1,000 μm or greater) solids at high (per stage) heads (60 metres), are to be pumped then the pump wear will be high and the time between replacement of parts will be short, regardless of the pump design. It is obvious that the result will be high wear parts cost and, if applicable, a downtime cost penalty.

Depending on the circumstances, the size of the solids to be pumped can be reduced by crushing or other means. This will undoubtedly require some cost which might be offset by a reduction in wear parts and downtime costs. Ideally, an evaluation of this type should be carried out as part of the selection process to determine which alternative provides the lowest total operating cost. An example of a slurry of large solids is shown in Sellgren & Addie (1993). A reduction in the concentration of solids pumped will also reduce wet-end component wear. As noted by Sellgren et al. (2005) the increase in the cost of water pumped for the same dry tonnes per hour of solids transported will usually be greater than the reduction in wear parts cost, implying that this option usually is undesirable.

From an application view-point, using more pumps (in series) and reducing the head per stage will reduce the wear on the individual pumps. On the other hand, the number of wearing pumps (and parts), will increase and hence more capital, interest and rotating-assembly cost will be incurred; and a total operating cost comparison should be carried out to determine the best alternative.

From a design point of view, slower-running pumps with larger diameter impellers and lower specific speed will have longer impeller and suction-liner lives. Slower-running pumps with larger diameters are more expensive, however, and will increase capital cost.

Figure 11.16 in chapter 11 provides some guidance for predicting wear rate in a given application. What makes best economic sense, when looking at pump designs and how many pumps to use in a given system, is another question. It would be nice if one pump design gave the lowest cost of ownership to the customer for all services, but unfortunately, different wear services require different pump designs, Addie & Sellgren (2001), Addie *et.al.*(2002), Sellgren *et.al* .(2005).

Thus it will be necessary to consider a number of representative applications. We look first at an application with a class 3 duty service condition, pumping solids with d_{50} of 150 μm at C_v of 20% and a head of 50 metres. From Fig. 14.3, which uses data from Sellgren *et al.* (2005) at four different transport rates, we see that, for a range of designs, operating costs vary with the size of the pumps. Interestingly, yearly operational costs are roughly constant for different pump specific speeds, and hence for a range of operating speeds.

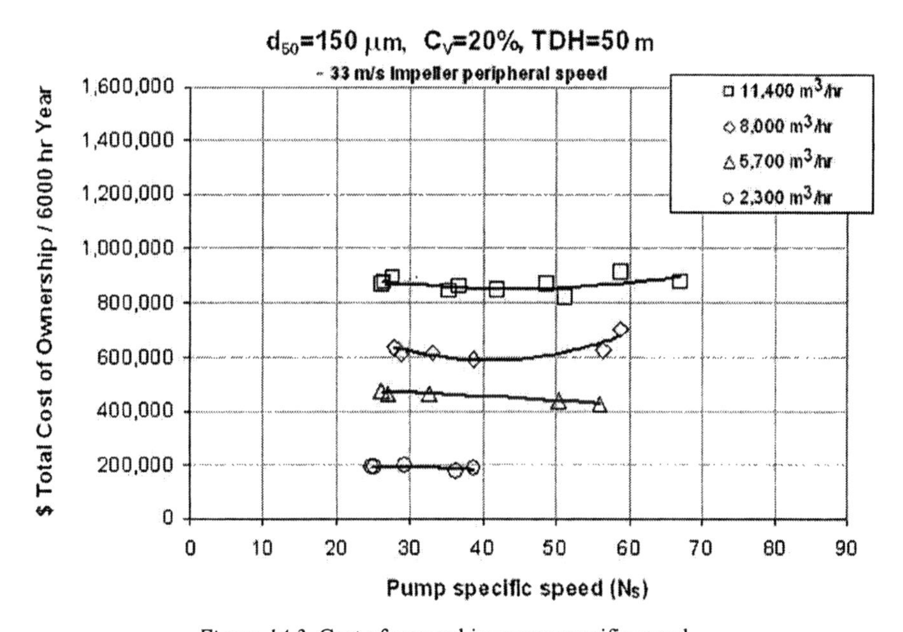

Figure 14.3. Cost of ownership versus specific speed

The cost calculation includes the total of capital, rotating assembly, energy and wear parts costs for the 50 m of head produced, as noted in Sellgren *et al.* (2005).

If we now consider a heavier-duty case where the solids size is 600 μm (but head and concentration remain at 50 m, and 20% by volume) then it can be seen from Fig. 14.4 that operating costs have increased by about one-third. This increase is due mostly to the increase in wear parts cost and, to a lesser degree, to the extra energy absorbed by the higher solids effect on the pumps. The breakdown into wear, capital, rotating assembly and energy costs is shown in Fig. 14.5 for the 600 μm case.

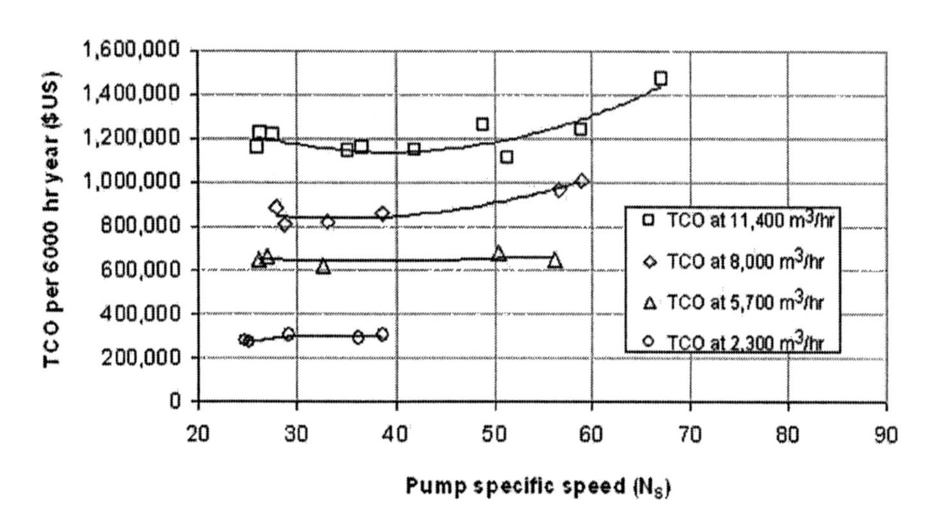

Figure 14.4. Cost of ownership versus specific speed

Figure 14.5 shows energy costs averaging 65%, wear costs being about 33% and capital and rotating operating costs being 2-3%. While not shown, the case for 150 μm solids had energy cost at around 90% with wear-parts cost at about 7%.

If we consider the parts cost (impeller, suction liner, casing and wet-end wear parts) for the case of d_{50} = 600 μm, C_v = 20% and 50 metre head for the large pumps, we get the result in Fig. 14.6. This shows parts cost varying with pump specific speed with a minimum parts cost around a specific speed of 40.

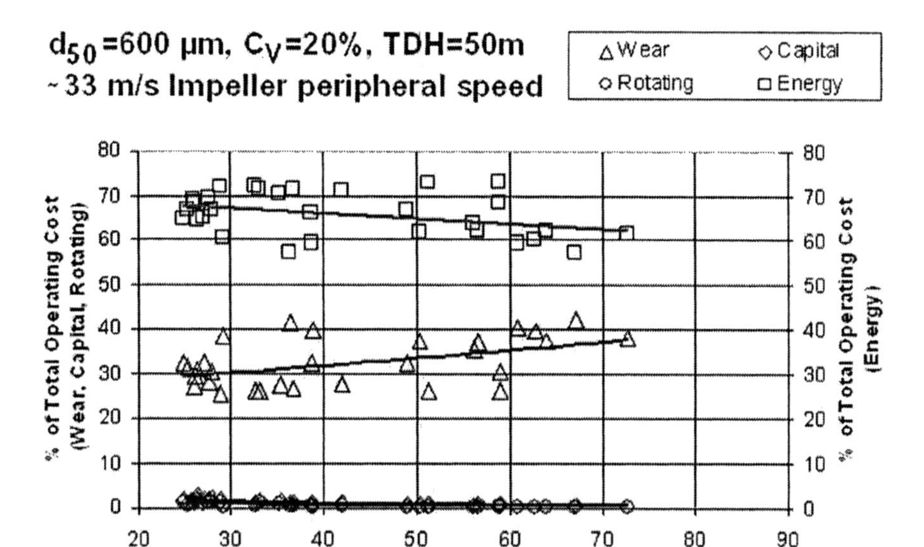

Figure 14.5. Operating cost breakdown versus specific speed

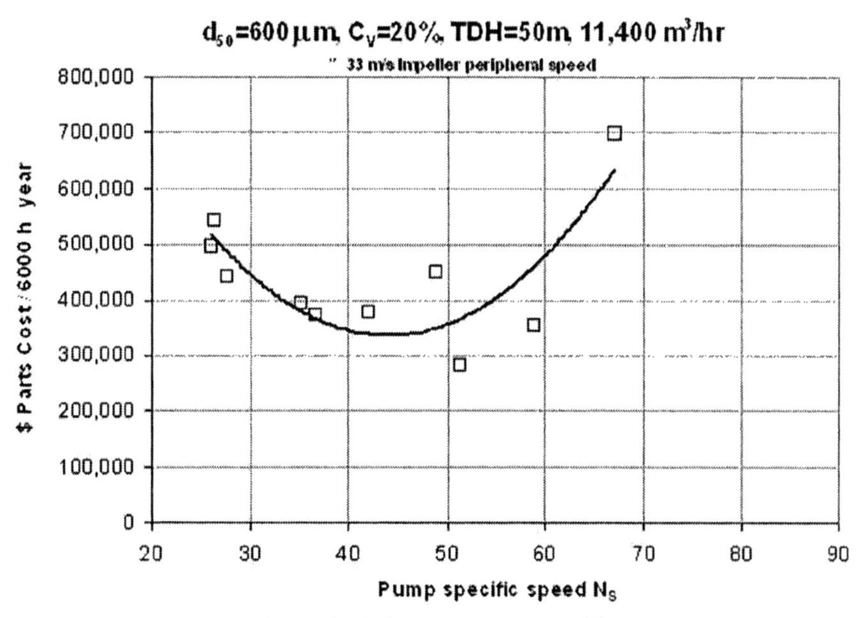

Figure 14.6. Parts cost versus specific speed

Figure 14.7 shows a similar curve for the case of d_{50} = 300 μm, C_v = 10% and TDH = 50 m. Here, parts cost is about one-fifth that for the former case.

While the previous plots showed significant differences in cost with different size solids and other different operating conditions, the difference in the operating cost of different designs was small, except in the case of the larger pumps where pump specific speeds around 40 showed minima in wear-parts cost represented in Figs. 14.6 and 14.7.

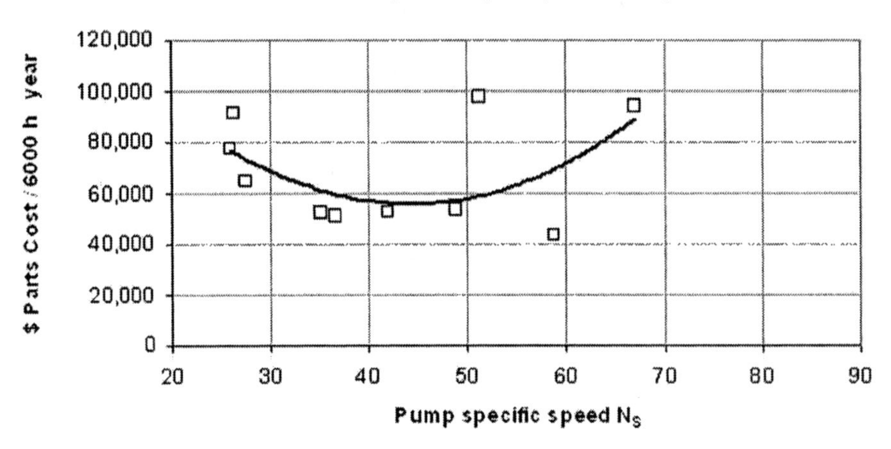

Figure 14.7. Parts cost versus specific speed

On a per tonne basis, the size effect seen in Figs. 14.3 and 14.4 is less pronounced. Figure 14.8 from Sellgren et al. (2005) shows the costs per tonne for the case with $d_{50} = 300$ μm, $C_v = 10\%$ and TDH = 50 m. As might be expected the larger pumps tend to give a lower cost per tonne than the smaller pumps do, although the difference is not so great as might be expected.

Taking the average cost per tonne for pumps of various sizes and designs results in Table 14.3, which presents cost per tonne for various values of d_{50}, C_v and the TDH of the pump. The values in parentheses are unit costs corrected back to the 50 metre head of the base case. For this base case the unit cost is about 2.8 cents (US) per tonne. Note that at low concentration ($C_v = 10\%$) the unit cost is nearly double in spite of the lower wear. On the other hand at $C_v = 40\%$, the unit cost is 44% less than the base case. Operating with a d_{50} of 600 μm increases the cost by 32%, while at 150 μm, it is 8% lower than the base case. When corrected to produce 50 metres of head, the unit costs for the 65 and 35 metre cases (shown in parentheses) are about the same as that of the base cost.

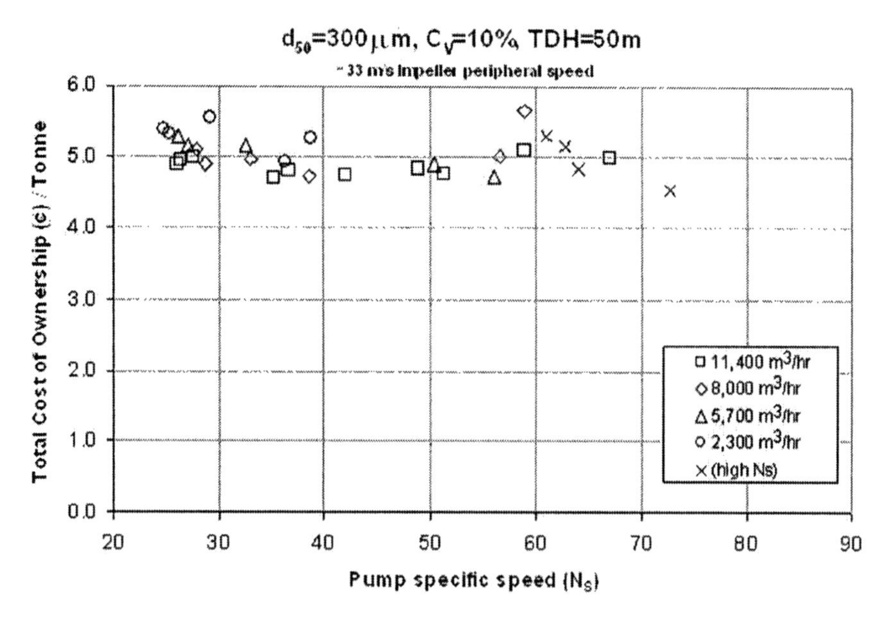

Figure 14.8. Cost per tonne versus specific speed

Table 14.3. Unit Cost for Different Slurry Duties

Solids Size	Concentration	Pump Head	Unit Cost
d_{50} μm	Cv (%)	TDH (m)	cents/tonne
150	20	50	2.5
600	20	50	3.6
300	10	50	5.1
300	40	50	1.5
300	20	35	2.1 (3.0)
300	20	65	3.6 (2.8)
300*	20	50	2.8

* Base Case
( ) Corrected to TDH of 50 m

14.4 Downtime Costs

If there is no back-up pump, when the pump stops, the plant shuts down. If the pump shut-down stops production, a downtime cost can be incurred. Downtime costs can be very large in a large mine, a figure of $100,000 (US) per hour not being uncommon. However, if the plant shuts down periodically for planned maintenance to refurbish screens or other equipment, then the pumps will not be seen to incur a downtime cost.

As shown in Figure 11.16, except in the case of low-specific-speed pumps, the pump suction liner has the shortest life, which determines when the pump must be stopped to replace parts. As shown in Chapter 11, the suction liner wear rate varies significantly with the duty and the design. Figure 14.9 (based on assuming a wear-out thickness) shows the expected suction liner life for the case of d_{50} = 300 μm, C_v = 20% and TDH = 50 m. Note that this figure is based on small front-clearing vanes, and an impeller fitted with deep clearing vanes would roughly double the lives shown, but with loss in efficiency of 2-5%.

Change-out time for suction liners varies with the type and size of the pump. In a small pump, the time to change parts could be as little as a couple of hours. In a large pump, 10 hours would not be excessive, and downtime costs could be as much as 1 million dollars per change-out. Given that the cost of operation shown earlier does not vary greatly with design, it would seem that low-specific-speed pumps should be considered whenever downtime cost is significant.

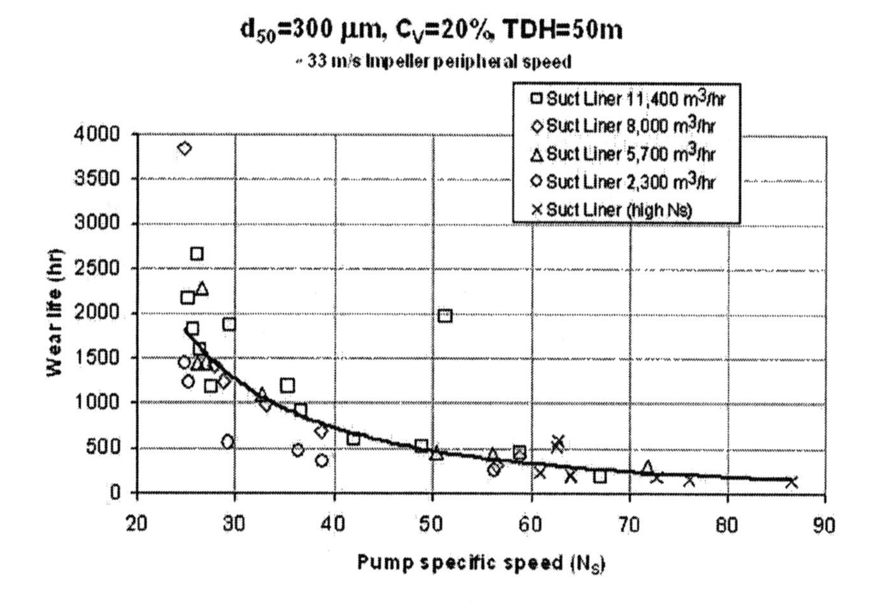

Figure 14.9. Suction liner life versus specific speed

Where downtime is significant, or the wear service class is heavy (class 4), deep clearing vanes should be used. For lighter services, the associated cost increase is likely to be greater than the wear-parts cost savings.

Until now we have not looked at the initial capital cost of the pumps, except to note that capital cost tended to be a small percentage of the total. Figure 14.10 is for different specific-speed pumps of various flow rates (and sizes). It shows that pump costs increase significantly with pump size and decrease with specific speed. The pump prices here should be taken to be typical, and it should be kept in mind that these do not include drive motors, gearboxes or bases. When these components are included the costs could rise to double or triple those shown.

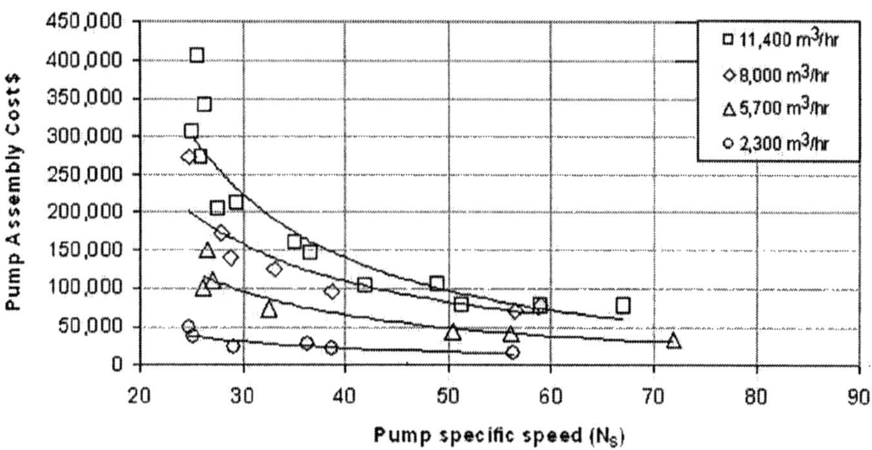

Figure 14.10. Pump cost versus specific speed

Lowering the specific speed of a design is one way to increase time between overhauls and reduce downtime cost. Lowering thc duty head per stage is another way. For the case in Table 14.3 with d_{50} = 300 µm, C_v = 20% and TDH = 65 m, it was shown that where corrected back to the TDH of 50 m, the unit costs were about the same. In other words, three pumps producing 33 metres of head will cost (per tonne) about the same to operate as two pumps producing 50 metres of head, except that the lower-head pumps will typically have 50% higher suction-liner life and consequently lower downtime costs. When downtime costs apply, then pumps should be employed that are slower-running with lower specific speed, a large-diameter impeller or lower head per stage. However, such choices involve higher initial capital costs. From an application view-point, using more pumps in series so as to reduce the head per stage will reduce the wear on the individual pumps. The number of wearing pumps (and parts), however, will increase. Here more capital, interest and rotating-assembly cost will be incurred. A comparison of total operating cost should be carried out to determine the best alternative.

14.5 Effect of System Size

The previous sections, and in particular Fig. 14.8 and Table 14.3, suggest that, for a given set of transport conditions, the costs per tonne are about the same regardless of the size of the pumps and the transport rate. These were for a base case with TDH of 50 m.

Values of pipeline head loss per unit length vary with the transport rate and pipeline diameter, typically being significantly lower in larger- diameter pipes. Detailed analyses have been presented in Chapters 3, 5 and 6. An example is shown in Table 14.4, based on a sand slurry with d_{50} = 300 μm and C_v = 20%.

This shows the minimum safe operating velocity, the transport rate and the horizontal pipe friction for different pipe diameters.

Table 14.4. Pipe friction in different pipe diameters
d_{50} = 300 μm C_v = 20%

Pipe Diameter	Flow	Velocity*	Solid Transport	Pipe Friction
	m^3/hr	m/sec	Tonnes/Hr	m/m
200	295	2.65	156	.0542
400	1700	3.8	900	.0363
600	4750	4.7	2510	.0294
* Velocity at deposition limit				

For a given length of horizontal pipeline, the head required is considerably lower for a larger pipe. This means that the heads used in Figures 14.8 and 14.9 would be lower for systems with larger transport rates and the same horizontal length, allowing the use of pumps with lower operating heads. The result will be considerable reductions in power, wear and capital cost per unit mass of solids transported. Thus, the larger-diameter pipelines and associated pumps are typically more cost-effective solutions.

14.6 Selection Procedure

As noted in Section 14.1, pump selection consists of determining the specific performance of each available pump and choosing the optimum one. It is straight-forward to determine the speed that possible selections would run at, to check the rotating assembly, ensure that the impeller passes the necessary size solids, and rank possibilities in order of efficiency.

Calculating or estimating the likely wear-parts costs and determining the pump with the lowest total operating cost (including any down-time) is more complicated. In a real world situation, operating conditions vary; parts wear out locally (as a result of operation away from the BEP) and are sometimes removed from service prematurely, making accurate predictions difficult. Accurate information on actual wear is hardly ever available for selection purposes. Where possible, use should be made of numerical modelling tools to establish peak wear rates of the three main wear components, impeller, suction liner and casing and establish a limiting life. In the absence of these, the data provided in Chapter 11 should be used.

In most practical cases, the size of solids is set along with the concentration of the mixture to be pumped. Depending on the solids size and concentration, the class of service can be established using Fig. 14.1. Given the solids size and solid concentration, Table 14.1 can then be used to select a maximum impeller peripheral speed, or preferably its equivalent maximum head per stage from Table 14.5. Where the system head is greater than the values shown in Table 14.5, more pumps in series should be considered.

Table 14.5. Maximum recommended head for different services

Service Class	1	2	3	4
Head per stage (m)	120	65	50	40

Even if downtime costs are not formally considered, the costs of changing out parts will be significant in class 4 services. On the other hand, in class 1 and 2, change-out costs will be small, the energy cost will be a large proportion of the total, and capital costs will be significant. In the cases of class 1 and 2 services, where the wear intensity is low, and where the head required is equal or less than the maximum for the respective classes, then designs of medium and high (50-60) specific speed should be considered primarily for the lower capital cost and higher efficiencies they make possible.

For class 3 services, medium-specific-speed (35-50) pump selections make the best compromise between capital, energy and other costs. The design with medium specific speed (35-50) is also likely to provide a pump with the lowest cost of wear parts.

For class 4 services, and where the head produced exceeds the service class limits, pump designs of lowest specific speed (20-35) should be used and/or consideration should be given to an extra pump in series.

Pumps selected for class 4 service should have deep clearing vanes fitted to the impellers, and in this case the wear rates predicted for the suction selection liner shown in Fig. 11.6 may be halved. As shown in Fig. 11.13, best suction-liner wear is achieved with values of the impeller ratio (impeller diameter/suction diameter) of 2.5 or greater. Generally designs with impeller ratios of less than 2 should be screened out for this service class. If downtime costs are significant, then pump designs of lower specific speed should be considered, in combination with lower-service-class head limits.

In the absence of specific results from numerical-wear models, the flow limits as percentages of Q_{BEP} which are shown in Table 14.1 should be observed. Suction-branch velocities should not be lower than the deposition-limit values for the same pipe diameter, as calculated in Chapter 5.

REFERENCES

Addie, G.R. & Sellgren, A. (2001), Semi-autogenous mill pumping cost considerations. *Proceedings,Third International Confenrence on Autogeneous and Semi-autogeneous Grinding Technology*, Vancouver, Canada, September.

Addie G.R., Dunens,E., Pagalthivarthi K., Sharpe J.,(2002) Oil sands tailings pump operating cost considerations. Proceedings, Hydrotransport 15,BHR Group, U.K., June pp 845-854

Addie, G.R., Kadambi, J.R. & Visintainer, R. (2005) Design and Application, Slurry Pump Technology, *9th International Smposium on Liquid- Solid Flows, 2005 ASME Fluids Engineering Summer Conference, Houston Texas.*

Sellgren, A. & Addie, G.R. (1993), The Economical Feasibility of Pumping Coarse Mine Residual Products,Proceedings *"Hydromechanisation 8", International Colloquium and Workshop, Germany.*

Sellgren, A.,Addie, G.R.,, Pagalthivarthi K.(2002) Wear and the total cost of ownership of slurry pumps, *Proceedings, ASME Fluids engineering Division Summer Meeting, Montreal, Canada, July 14-18.*

Sellgren, A., Addie, G.R., Visintainer,R., Pagalthivarthi, K.,(2005)Prediction of slurry component wear and cost, *Proceedings WEDA XXV and Texas A & M 37th Annual Dredging Seminar,* New Orleans U.S.A., June.

Chapter 15

PRACTICAL EXPERIENCE WITH SLURRY SYSTEMS

15.1 Introduction

Earlier chapters in this book were concerned with the basic principles of design, selection, application and operability of the various components of slurry systems. This chapter is devoted to some of the problems which can arise in operation. In draws on several decades of practical experience, and refers to case studies which are all real although they are not specifically identified. Some cases have been synthesized by compiling experience from several different installations, but they all are based on actual experience. Problems in operation usually arise when some element of the system - most commonly the pump or drive - is not correctly selected or installed, perhaps because it was on hand rather than being specifically selected for the duty. Sometimes problems arise simply from bad operating practice. Occasionally, as in the case of the notorious incident described in Section 15.8, problems arise from disturbances which were unexpected or were incorrectly analysed.

15.2 Pumps in Series

Considerations pertaining to pumps for series and parallel systems were covered in Section 12.4. It is normally important to ensure that the pumps are all identical, and for a series of pumps in a long pipeline, care needs to be taken in the sequence and timing of pump start-up. In this case variable-speed drives may be required on one or two of the pumps in the system. The start-up sequence and timing of a multiple-pump series system must be considered in light of the effect on the hydraulic grade line, and how this varies with time. If all the pumps are started simultaneously, there will be a brief initial period when the flow, and hence the frictional head loss, will be negligibly small. In a horizontal pipeline, this implies that the downstream pumps will, at least for a short time, be subjected to the combined zero-flow pressure generated by the upstream pumps. This can be much more severe than the normal steady-state pressure, and one of the writers has seen cases where three or more pumps 'blew up' because of this initial overpressure.

Figures 15.1 to 15.4 illustrate series pump installations with one, two and four pumps. Figure 15.1 shows a simple single-pump horizontal pipeline, with the normal steady-state grade line and the condition a short time after start-up. Figure 15.2 is also for a single- pump system but shows the effect on the grade line of having a mid-line pump location. The normal steady-state grade line is shown and also the condition a short time after start-up (before pipe friction becomes effective). Figure 15.3 is for a two-pump system and shows the cumulative effect a short time after starting both pumps, and also the steady-state pressure grade line when running only the downstream pump. Figure 15.4 indicates grade lines for a four-pump system with pumps at the suction end and along the pipeline. The grade lines shown are based on simultaneous start-up and on sequential start-up with time delays.

To avoid overstressing the line (and the pumps), it can be seen that pumps spaced along a pipeline must be started in a timed order arranged to minimize the maximum pressure. As well as limiting positive pressure, the start-up sequence should be such as to avoid creating any negative pressures in the line (which can produce cavitation or water hammer). Moreover, the overall start-up time should be limited in order to avoid problems with slurry settling or plugging.

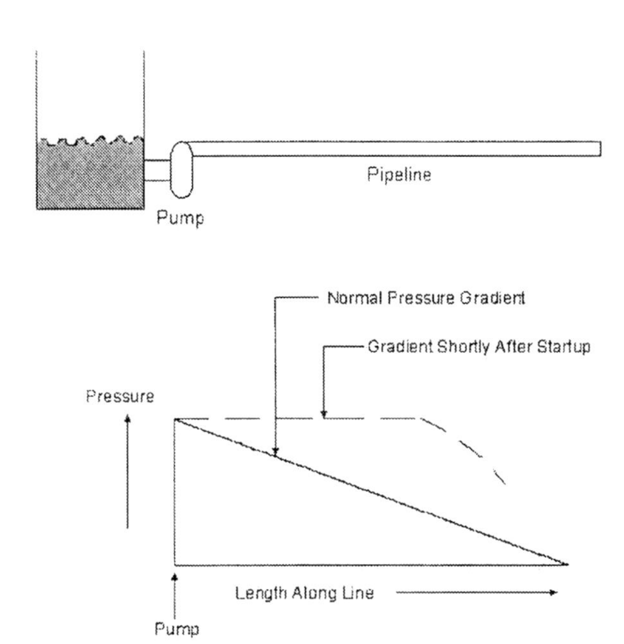

Figure 15.1. Start-up of pipeline with one pump at suction end

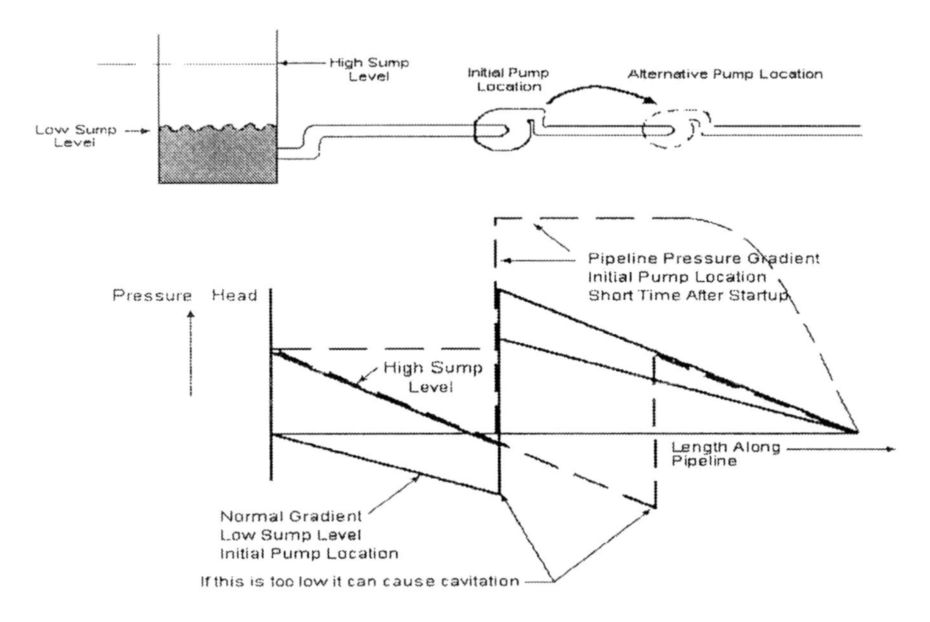

Figure 15.2. Start-up of pipeline with one pump at mid-line

For a series of six or more pumps spaced evenly along a horizontal line, the reduction of maximum pressure (without excessive start-up time) will not necessarily be achieved by a timed sequence of the type 1, 2, 3, 4.... Computer simulations have shown that, depending on circumstances, other timed sequences such as 1, 3, 2, 4... may be more effective.

For a line with pumps in series it sometimes happens that the first pump at the suction end is run at lower speed than other pumps. This condition is most common in dredging systems where, as shown in the Case Study of Chapter 8 and in Section 12.2, a submerged 'ladder pump' may be required to eliminate cavitation in the on-board pump. For convenience in stocking spares, the ladder and on-board pumps are often made identical. However, the ladder pump is sometimes run at lower speed, to reduce both the power of the ladder drive and the NPSH required by the ladder pump.

Figure 15.5 illustrates this case: identical pumps in series but driven at different speeds. In the specific case shown, the speed of the ladder pump is two-thirds that of the on-board pump (400 rpm and 600 rpm, respectively). Some general results from earlier chapters are used implicitly in Figure 15.5. As always for pumps in series, the total head developed at any discharge is the sum of the individual pump heads, leading to the 'Combined H-Q' curve shown. As indicated in Chapter 9, the net positive suction head required by a pump running at constant speed increases with discharge more than linearly (see Fig. 9.2 and 9.16). Similarly, if the pump speed is varied but the flow is maintained at the same percentage of the discharge at the best efficiency point (Q_{BEP}), the NPSH required follows the affinity law for developed head and thus varies with the square of the pump rotational speed. The curves in Figure 15.5 have been constructed on this basis.

Because of the lower speed and hence lower Q_{BEP} of the ladder pump, the NPSH required by the ladder pump exceeds that of the on-board pump at larger discharges. For the particular case in Figure 15.5, the NPSH required by the ladder pump is above that for the on-board pump if the discharge exceeds 133% of the ladder pump's Q_{BEP}. The NPSH available decreases as flow increases, and, as always, cavitation occurs when the NPSH available falls below the NPSH required.

We can now consider various operating conditions. For System A, characterised by high resistance, the discharge will be Q_1, with operation at the point labelled 'Duty 1'. The ladder pump is operating at 120% of its BEP flow, with less than optimal efficiency - 77% for the curves in Figure 15.5. The on-board pump operates at 80% of its BEP flow, at roughly the same reduced efficiency. For both pumps the NPSH available is well in excess of that required. This condition, with one pump operating above its BEP and the other below, but with neither close to cavitation, represents a reasonable compromise for this type of system.

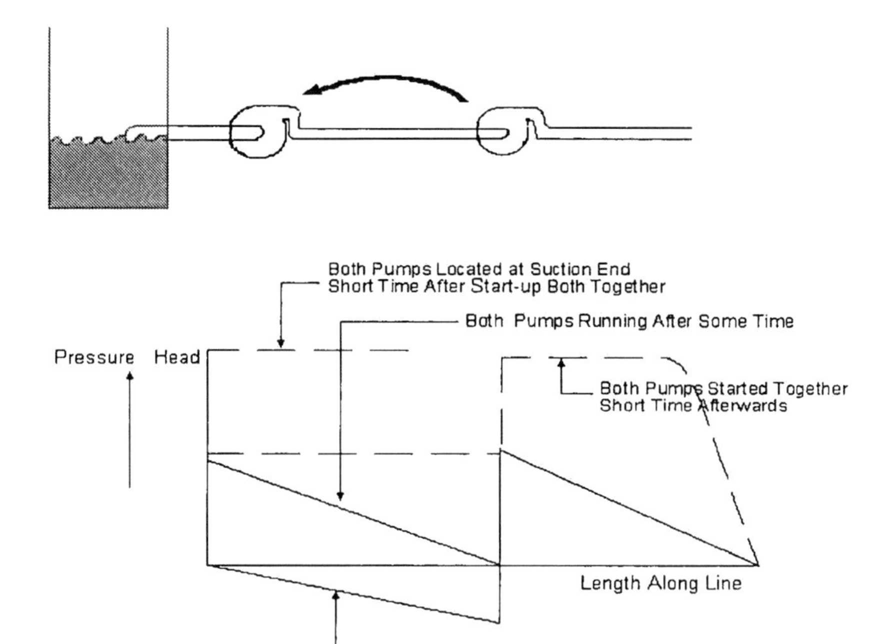

Figure 15.3. Start-up of pipeline with two pumps

Consider now the case where the system resistance is reduced, for example by pumping through a shorter line. The system curve has moved to that shown as System B on Figure 15.5, with discharge Q_2 and operation at the point labelled 'Duty 2'. At first sight, this might appear an easier duty. However, as is sometimes said of amatory porcupines, care is needed in addressing the task. The on-board pump is operating at 96% of its BEP flow with higher efficiency (nearly 80% for the pump of Fig. 15.5). However, the ladder pump is now operating at 144% of its BEP flow -- too far out on its 'overdischarge', with efficiency down to 67%. Moreover, the NPSH available to the ladder pump is just equal to the NPSH required. Thus Duty 2 represents the maximum possible discharge which can be obtained from the dredge without cavitating the ladder pump.

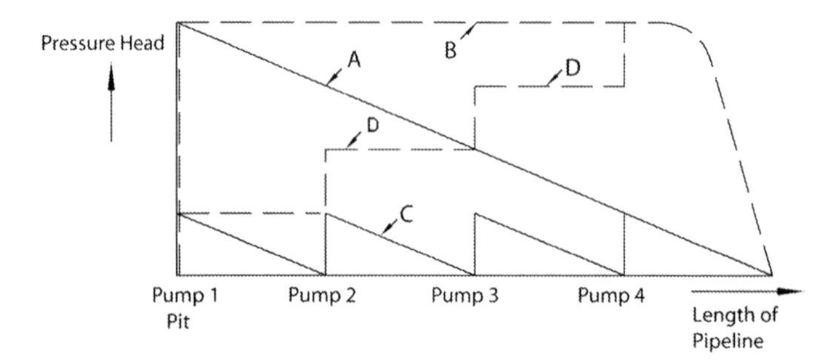

Figure 15.4. Start-up of pipeline with four pumps

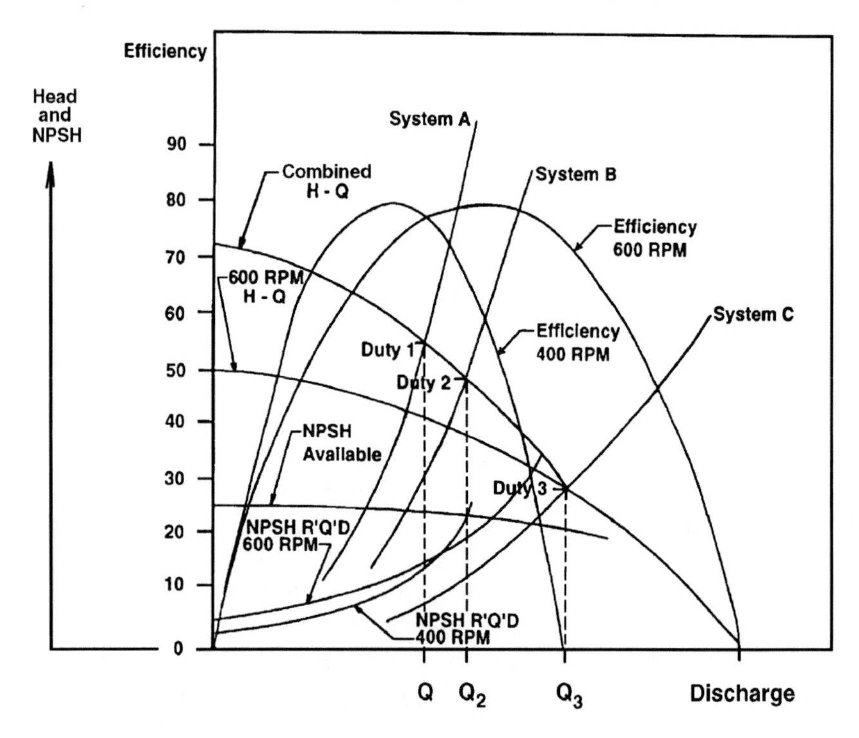

Figure 15.5. Pump and system characteristics for dredge with ladder pump operating at reduced speed

The curve for System C shows the effect of lowering the system resistance still further. It is assumed that sufficient NPSH is available to prevent cavitation, perhaps because the ladder pump is immersed more deeply. The discharge is now Q_3, with the system operating at Duty 3. At this discharge, the ladder pump ceases to generate any head. If the system resistance is reduced still further, so that the discharge is greater still, the ladder pump now represents a resistance in the system. When this condition has been encountered in the field, it has often been interpreted by the operators as failure or poor performance of the ladder pump. The analysis represented by Figure 15.5 shows that this is not the case: the condition arises from poor pump selection.

This example illustrates the importance of checking system operating points and suction conditions, especially when using, in series, pumps which are not identical or not driven at the same speed. It also shows that in some circumstances the NPSH required by a pump may not be lower if it is driven at lower speed. For a dredge, the best arrangement is to have a ladder pump with low shaft speed and low head (i.e. relatively high specific speed), while the on-board pump should have the same BEP discharge but should operate at higher rotational speed to give greater head (i.e. it should have lower specific speed than the ladder pump). It follows that the two pumps should not be of identical design. The ladder pump should be of a wider pattern (i.e. with larger value of the ratio b_2/D_2 defined in Section 9.2) and should possibly have an impeller of smaller diameter.

15.3 Pumps in Parallel

Systems with pumps operating in parallel are also subject to problems arising from mis-sizing or from difficulties in balancing and control. Figure 15.6 illustrates the case of three identical pumps running at the same speed and discharging in parallel into a common header. For this case, the combined pump characteristic is obtained by adding the individual discharges at the same developed head. With all three pumps in operation, the total system flow is A. However, each individual pump operates at flow B, corresponding to its best efficiency point. If one pump is taken out of service, the total system flow is reduced to C but each of the pumps in service moves out to point D. Because D is well above the BEP, the efficiency of the pumps is reduced. The reduction in efficiency and increase in flow cause each pump to draw substantially more power. With only one pump in service, the system operates at point E and the power drawn by this pump is increased further.

Figure 15.6 is schematic only. If the head-discharge curves of the pumps are flatter, as is normally the case for slurry pumps, the increase in power on taking one or two pumps out of service will be greater. Furthermore there is the likelihood of cavitation. If one of the pumps in the system has a variable-speed drive, conditions are similar to those shown in Fig. 15.6: as the speed of the variable-speed unit is reduced, the fixed-speed units will deliver more flow and draw more power.

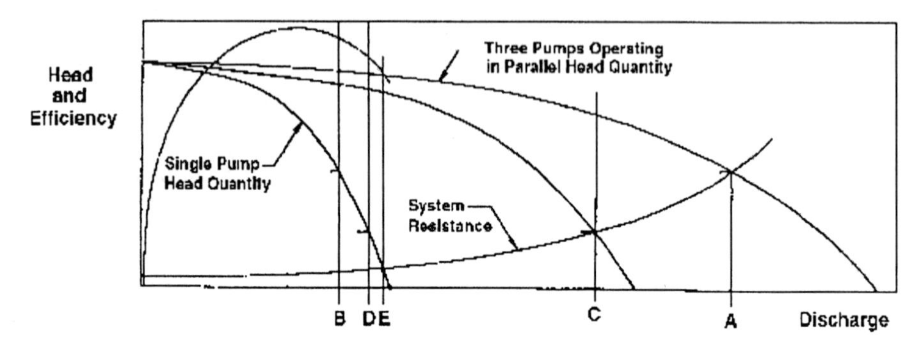

Figure 15.6. Pump and system characteristics for three pumps operating in parallel

It follows that, for a parallel system to operate as shown in Fig. 15.6, the pumps should have steep head-discharge characteristics and should be sized so that the BEP of each individual pump is close to discharge D. The drive units must be sized for the power corresponding to the maximum flow conditions at point E.

15.4 Suction Problems

The configuration of suction piping systems was covered in Section 12.2. The main consideration is usually to avoid cavitation by ensuring that the NPSH available exceeds that required by the pump. Intermittent cavitation can be particularly damaging if it leads to water hammer, a point discussed further in Section 15.8 below.

Where suction is from a pit, or where a dredge is operated without a ladder pump, it may be necessary to accept moderate cavitation for some operating conditions. Typically, these conditions occur at high solids loading. The loss in efficiency due to cavitation may then be offset by the reduction in specific energy consumption due to the increased concentration. For this type of application, it is essential to have a pump with good suction performance; i.e. not only with low NPSH requirement, but also with the ability to continue pumping when the NPSH available is significantly below that required at incipient cavitation. When operating under these conditions, the skill and vigilance of the pit operator or dredgeman are vital. A sudden blockage or rapid shift to excessively large solids concentration can worsen cavitation to the point where a large vapour cavity forms. Water hammer then results, possibly with disastrous consequences.

Sump and suction designs to prevent froth were discussed in Section 12.2. Entrainment of froth into the suction can be hard to distinguish from cavitation. The required NPSH increases with increasing flow while the available NPSH decreases with increasing flow, and therefore a small reduction in flow will often eliminate cavitation. Thus, a good test for cavitation is to increase the system resistance downstream of the pump by a valve or other means and look for an increase in pump discharge pressure or a decrease in noise. In this case, even a small reduction in flow can result in an immediate increase in head. If this behaviour can be repeated by use of the valve, it is very likely the problem is cavitation. However, if increasing the pump speed does not increase discharge, or if the head developed is unexpectedly low but the power drawn is still high, then it is likely that a pump is handling froth rather than cavitating.

15.5 Pumping Frothy Mixtures

In slurry pumping, frothy mixtures are encountered often, sometimes despite efforts to eliminate air entrainment. Common causes are excessive turbulence of the flow entering the pump, or a sump level that is too low. Other sources are chemical reactions such as in flotation processes, or the release of gases contained in the solids of the slurry.

When the volume of gas exceeds about 2% of the volume of mixture inside a centrifugal pump, the head, efficiency and NPSH performance of that pump is significantly affected. If larger amounts of air are present, the system performance can be severely limited. Depending on the circumstances, surging, water hammer or a combination of these may occur, possibly causing the whole system to cease pumping.

The effect of different admixtures of air on the performance of centrifugal pumps is also influenced by pump configuration. Representative results for air-water mixtures are shown in Fig. 15.7 (see Stepanoff, 1957, 1965). In this figure the parameter is the volumetric percent of gas, referred to the pressure at the pump suction, and measured relative to the volume of gas-free slurry. Pump head is expressed in metres of mixture.

Elimination of air in the pump may be as simple as the redirection of a return line to a sump or maintaining sufficient level in the sump. In the case of foamy slurries spray bars located above a flotation tank can often help. Gases contained in the solids (encountered, for example, when dredging some types of organic materials) can often be removed by the use of a vacuum-operated accumulator such as that shown in Fig. 15.8 (from Herbich, 1975).

Where froth is suspected in a trouble-shooting situation, or is known to occur and cannot be eliminated, it is necessary to obtain a clear understanding of conditions. The pumping equipment can then be sized to deal adequately with the frothy mixture. It must be appreciated here that the volume of mixture seen at atmospheric pressure at the top of a large tank will generally be different from that encountered by the pump at its inlet pressure. This inlet pressure can be either above or below atmospheric as a result of the combination of elevation differences and friction losses. For example, if there is 10% by volume of air in a slurry at the top of an open tank at zero gauge pressure and the pump operates with a vacuum of half an atmosphere, then the pump will encounter a mixture with 20% air. A situation of this type is illustrated diagrammatically on Fig. 15.9. The flow-rate axis of this figure is based on the volumetric flow of the mixture of water (or slurry) and air, and the head axis is based on pressure (Pascals or metres of water), not on height of mixture.

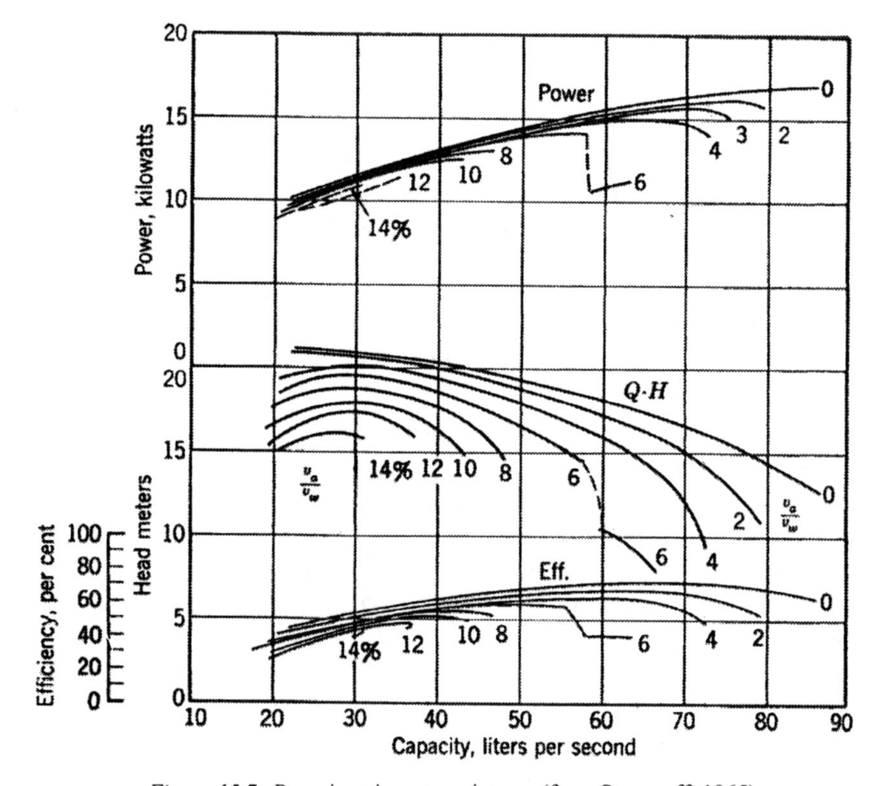

Figure 15.7. Pumping air-water mixtures (from Stepanoff, 1965)

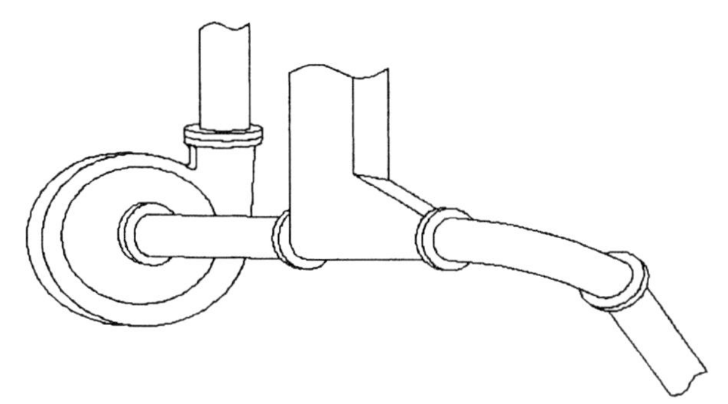

Figure 15.8. Vacuum accumulator (from Herbich, 1975)

Figure 15.9 shows the three system characteristics that must be considered. The first, indicated by a solid line, represents the case with no air. The second and third, indicated by dashed lines, represent the system behaviour for two fractional volumes of air, as found at the pump suction and the pump outlet. In practice the suction characteristic would not normally be measured, whereas the outlet characteristic can be obtained from the flow rate indicated on a magnetic-flux flowmeter on the discharge pipe and the head determined from the absolute discharge pressure.

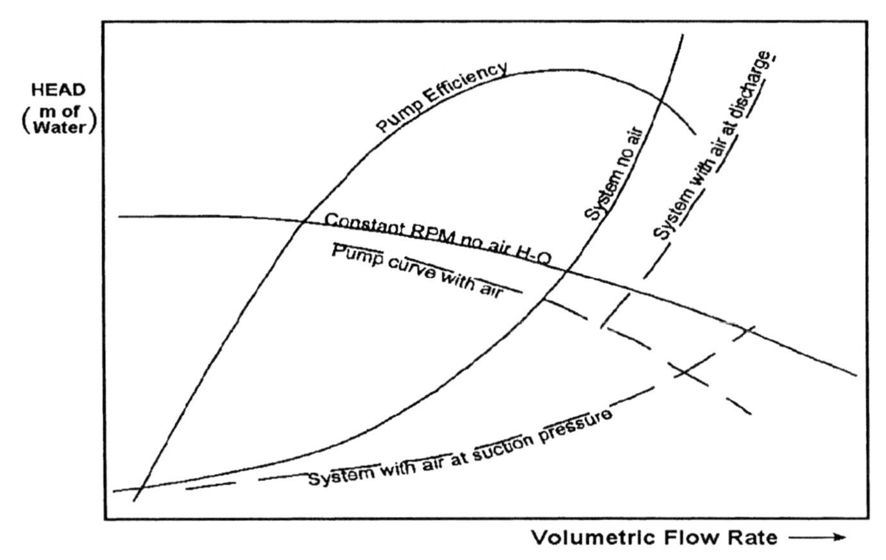

Figure 15.9. Diagrammatic illustration of operation of pump with air-slurry mixture

Figure 15.9 also shows the pump characteristics for two conditions. The solid line represents the H-Q curve with no air (the corresponding pump efficiency curve is also plotted). The dashed line represents the H-Q curve with air, reflecting conditions prevailing in the pump impeller (where the air fraction is intermediate between suction and outlet values). The intercepts of the dashed pump and system curves show operating conditions. Note that the heads at these intercepts are reduced compared to the intercept with no air, while the volumetric flow rates are increased. This increase is greater for suction conditions than for outlet conditions, because of the larger volumetric fraction of air at the lower pressure. The ratio of the volumetric flow rate with air to that with no air, mentioned above, is called the 'flow-rate froth factor', or simply the 'froth factor'. In new installations where various amounts of forth are expected, specifications often include froth factors of this type. They may be 110%, 150%, 200% or even 400%.

In an existing installation, if it is impossible to modify the sump and suction configuration to prevent entrainment of froth, consideration should be given to applying a froth factor, and to installing a larger, slower pump whose performance will be less impaired.

The experienced engineer trouble-shooting a malfunctioning pumping system will often suspect that entrained air is the villain. To investigate this point it is necessary first to estimate the pump input power, for example by use of an ammeter reading (see Section 12.6). A flowmeter and pressure gauge installed on the pump discharge line can be used to obtain values of flow rate and head. Together with the pump manufacturer's performance chart and an estimate of the fraction of entrained air (an initial guess can be refined by iteration), these values identify roughly where the pump operating point lies, and hence determine whether a significant amount of entrained air is present.

Froth can have a major effect on the suction performance of a pump. As noted above in connection with Fig. 15.9, entrained air shifts the operating point to larger volumetric flow rates. This produces an increase in the NPSH required, while reducing the NPSH available, so that cavitation might be expected. Cavitation in the strict sense, i.e. the formation and subsequent collapse of vapour pockets, is unlikely in these circumstances. Instead the entrained air bubbles expand rapidly in areas of low pressure within the pump, decreasing both head and efficiency.

15.6 Flow Instability

In the examples discussed up to this point, friction was the major component of the head to be supplied by the pump, and static lift was of only minor significance. Sometimes, however, the static head is the major component. This configuration often occurs with short-distance in-plant installations, such as cyclone feed circuits in mills. In typical feed circuits, the pump is fed from a sump and provides a vertical lift of some 20 to 30 metres to a bank of cyclones. The pressure required by the cyclones varies somewhat to suit the mill process, but this variation is a small proportion of the static lift, producing a very flat system characteristic. In conjunction with the relatively flat characteristic typical of slurry pumps, the result is an operating flow rate which is remarkably sensitive to variations in pump head. Specifically, a well-intentioned provision of more head than required tends to create significant difficulties.

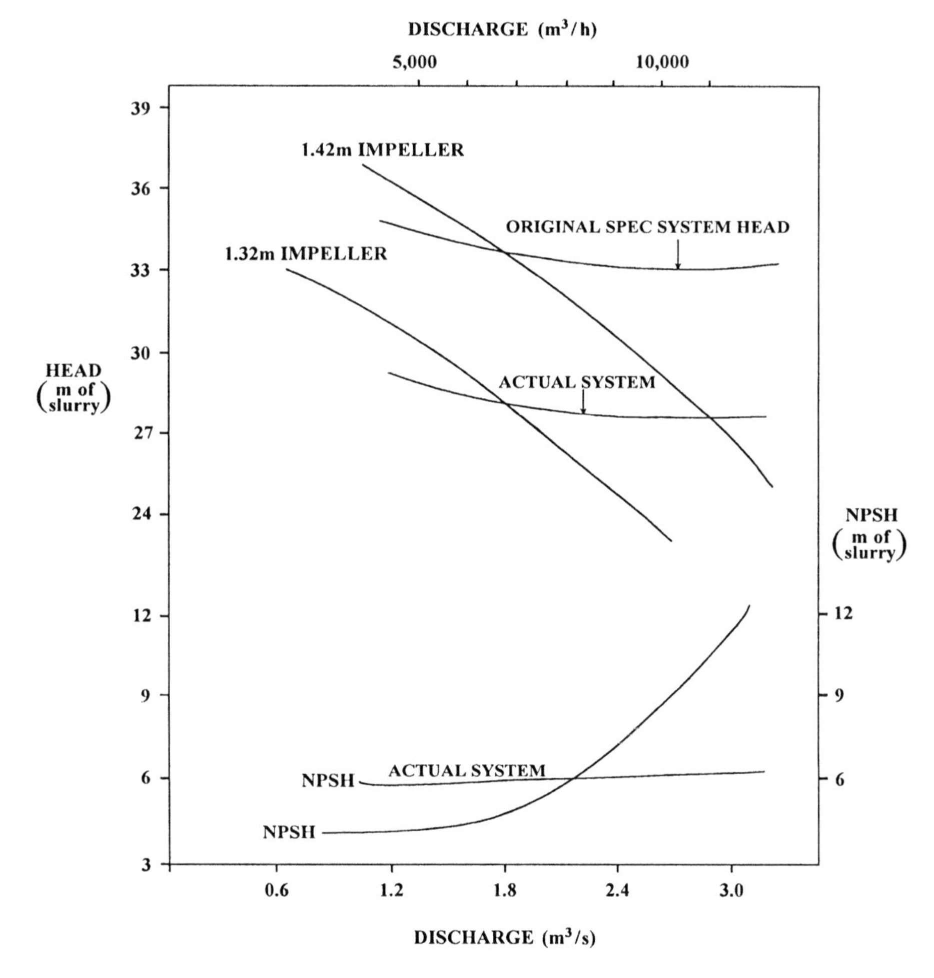

Figure 15.10. Pump and system characteristics for cyclone feed circuit

Changes in cyclone back-pressure can have a similar effect. Figure 15.10 shows head-discharge curves based on the cyclone feed circuit for a large copper producer in Chile. In this case the pump as supplied, with impeller diameter 1.42 m, gave the curve shown, which intercepts the originally-specified system head curve at the operating design flow of 1.8 m^3/s. Because of a process adjustment, there was a reduction of cyclone back pressure equivalent to about 5.8 m of slurry head. This change lowered the system head curve by the same amount, to the line marked "actual system". Note that both specified and actual system curves require less head as the discharge increases. In these circumstances the minor reduction in system head caused a large shift in the intercept with the pump curve, giving a flow rate near 3.0 m^3/s.

In this case, before the pump could reach the new flow rate, it drew down the sump, reducing the available NPSH and initiating cavitation. When the pump cavitated, the head that it produced dropped significantly, reducing the intercept flow-rate briefly, and allowing the sump to refill. The cycle was then repeated, giving an alternation of sump surge and pump cavitation that persisted for several weeks before the true cause was identified.

The solution was to trim the impeller (see Case Study 9.2), in this case from the original diameter of 1.42 m to 1.32 m (a change of only 7%), producing the new pump characteristic shown on Fig. 15.10. The result was to reduce the discharge at the intercept point to the original design value of 1.8 m^3/s.

We turn now from pumps operating at fixed speed to those with a variable-speed drive, and inquire what system parameter should be used to control pump speed. For settling slurries a logical choice is flow rate in the pipeline, which can be measured easily by means of a magnetic-flux flowmeter downstream of the pump. If the supply of slurry entering the sump on the suction side of the pump should diminish, additional make-up water can be supplied to this sump. One possible control method would be to have the pump speed controlled by the flow rate at discharge, and the make-up water to the sump controlled by the sump level.

For non-settling slurries, however, there is no need to maintain a minimum velocity in the pipeline. In one instance of this sort, a sump-pump-pipeline combination was installed to transport a non-settling slurry. The pump had a variable-speed driver, controlled by the technique mentioned at the end of the previous paragraph, using the signal from a magnetic-flux flowmeter located in the pipeline downstream of the pump. The system failed to deliver slurry at the required rate, and one of the writers was summoned to the site to rectify matters. It was found that the pump was running at the top of its speed range (about 900 revs/min) with the sump level surging violently. Perhaps understandably, the owner's representative swore that the pump was malfunctioning and should be removed forthwith.

It was observed that the sump seemed to be low most of the time, and that the flowmeter reading oscillated fairly wildly, indicating air entrainment. The first step was to control the pump speed manually. By slowing it down, stability returned to the system and the sump operated about 90% full. However, the next variation of slurry inflow to the sump caused the supply of make-up water to fluctuate to the extent that the sump either sucked dry or overflowed, and manual control of the pump speed was again required to regain stability.

A permanent solution to the problem was achieved by altering the control system. The normal technique of using the flowmeter signal to control pump speed was abandoned. This change was possible in the present case because the slurry was non-settling, so that low pipeline velocities would not be critical. The new control system used the sump level to adjust pump speed, making for a simple, reliable system in this case. Stability was achieved with the pump running smoothly in a narrow speed range near 600 revs/min.

15.7 Operation with a Stationary Bed

As noted in earlier chapters, for systems using centrifugal pumps operation near the limit of stationary deposition leads to instability, and hence is normally avoided. For a fixed throughput of solids, the operating velocity is kept above the deposit velocity either by selecting a relatively small pipe size or by pumping at low solids concentration.

Some systems, especially ones with a broad range of solids throughput, are known to operate at least part of the time with a stationary deposit of solids. To understand a system of this sort, it is best to consider a simplified example. This comprises a pipeline fed by two plant process lines with a slurry of near-constant solids concentration. (Substantial dilution of the slurry is to be avoided, either because of a limited supply of make-up water or to avoid separation problems at the tail end.) It is assumed that, at any given time, either both process lines are contributing to the pipeline, giving a pipeline flow Q_{max}, or only one is doing so, giving 0.5 Q_{max}.

In sizing the pipe, the conservative designer might choose to ensure that the minimum flow rate of 0.5 Q_{max} is clear of possible difficulties with deposition and stability, i.e. is on the rising right-hand limb of the system characteristic. As a result, the maximum flow Q_{max} lies much further up this limb of the characteristic. This design, which implies the use of a relatively small pipe diameter, is shown on the upper panel of Fig. 15.11.

As this figure is intended for illustrative purposes, specific numerical values have deliberately not been plotted. However, the form of

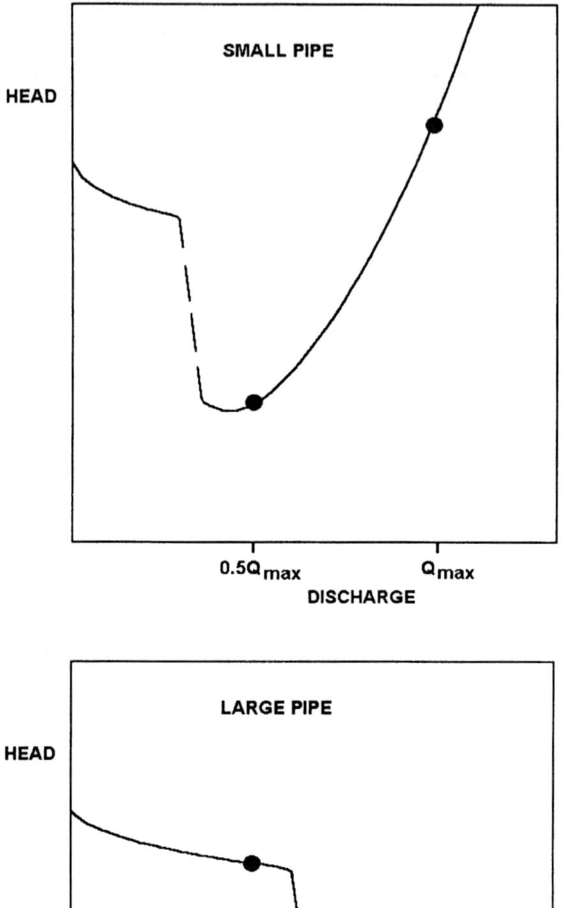

Figure 15.11. Operation with a stationary bed

characteristics are based on calculations for representative cases. Specifically, the right-hand portions of the curves are based on the heterogeneous flow model (see Chapter 6, specifically Section 6.6). The left-hand portions, representing the stationary-deposit condition, are based on Eq. 5.19. Both right-hand and left-hand limbs intercept the deposition locus (see Section 5.4) which is indicated by dashed lines on Fig. 15.11. As mentioned previously, the upper panel on this figure refers to a relatively small pipe, sized so that a flow rate of 0.5 Q_{max} is on the rising limb of the system characteristic. As a result, the full flow Q_{max} lies high on this curve, requiring large head and power.

The head and power required at Q_{max} can be greatly reduced by using a larger pipe (about 30% larger in this case), for which the Q_{max} operating point is only slightly to the right of the low point in the system characteristic. This design is shown on the lower panel of Fig. 15.11. It gives a major reduction of power (proportional to the product of Q and head) at Q_{max}, but moves the operating condition at 0.5 Q_{max} to the stationary-bed zone, as indicated on the figure. For the conditions shown on the lower panel of the figure, the power requirement for 0.5 Q_{max} is actually larger than for Q_{max} (by about 30%). There is a slight negative slope to the system characteristic at 0.5 Q_{max}, but, with appropriate pump selection and an adequate power supply, stable operation at this point can probably be achieved. For this system the overall specific energy consumption will depend on the time split between operating at Q_{max} and 0.5 Q_{max}. If an equal time at each flow is taken for illustration, it is found that the overall energy consumption with the large pipe is only about 60% of that with the small pipe.

The above result would appear to favour the large-pipe option with deposition during part of the operating cycle. It must be noted, however, that different slurries or other design conditions can lead to other results (Addie & Hammar, 1994). Moreover, only two discrete operating flow rates have been considered here, but there are intermediate flows (especially those in the range of the dashed line on the figure) where stable operation may be impossible. In addition, difficulties associated with deposition are much more severe if the pipe is not horizontal (see Chapter 7, specifically Case Study 7.1). Therefore, it is strongly recommended that detailed pilot-plant testing be conducted before implementing any proposed design involving stationary-bed operation.

15.8 Transient Conditions

Start-up of a system with several pumps in series was discussed in Section 15.2. Even for single pumps and relatively short lines, start-up can create special problems which may damage both the pump and the pipeline.

When a centrifugal pump is started by a synchronous electric motor against an open discharge valve, the torque varies approximately as the square of the rotational speed, provided that the specific speed of the pump is sufficiently low (less than 3500 in customary US units, i.e. $N_s < 3500$ or $n_s < 0.01$). Slurry pumps normally lie in this range. Mechanical friction in the bearings and glands absorbs a fraction of the drive torque, the proportion absorbed being typically lower for large pumps. The instantaneous head and flow and the torque imparted by the pump to the fluid (see below) depend on the characteristics of the pump at its instantaneous speed of rotation. The difference between the instantaneous drive torque and delivered torque represents the torque available to accelerate the impeller and fluid up to design speed. Most synchronous motors are able to deliver sufficient starting torque to get the pump up to speed within an acceptable time, but this point should be checked for each individual installation. If the drive is oversized, and the inertia of the impeller is large compared to that of the motor - which is normally the case - it is advisable to calculate the angular acceleration of the rotating assembly and hence determine the torque on the pump shaft. It has been known for shafts to fail in torsion from this cause, especially in pumps which start and stop often.

The variations in power and torque during start-up depend on the condition of the pipeline, specifically on the proportions of friction and static lift in the system characteristic, and whether the line is full or empty at start-up. The possible conditions are illustrated by Fig. 15.12. For System A, the pump is required to deliver to a pipe with a large static lift, which is to be started full. For this case, flow commences only when the impeller has accelerated to a speed sufficient to generate a head which will overcome the static requirement. Figure 15.13 shows the corresponding variation during start-up of the torque absorbed by mechanical friction and imparted to the fluid. The initial torque arises from mechanical friction, so that the pump starting condition corresponds to point B rather than point A. Around point C, at a speed which depends on the pump size but is typically in the range 10-20% of full speed, the torque imparted to the fluid takes over as the dominant component. Thereafter, the torque depends on the system.

For the extreme case of a system dominated by static lift, there is no flow until the pump reaches the speed corresponding to point D. The torque follows the curve CD, corresponding to shut-off conditions with no flow. Thereafter, the discharge increases rapidly and therefore the torque also increases, along the curve DE. At the other extreme, where there is no static lift so that the system resistance arises solely from friction, the flow increases steadily as the impeller comes up to full speed. The torque therefore increases steadily, along the curve CE. Intermediate cases, with different proportions of friction and static lift, will be represented by curves within the hatched region on Fig. 15.13. The discussion above shows that, if a synchronous motor is used, the impeller will be accelerated up to full rotational speed most rapidly in a system containing a large proportion of static lift.

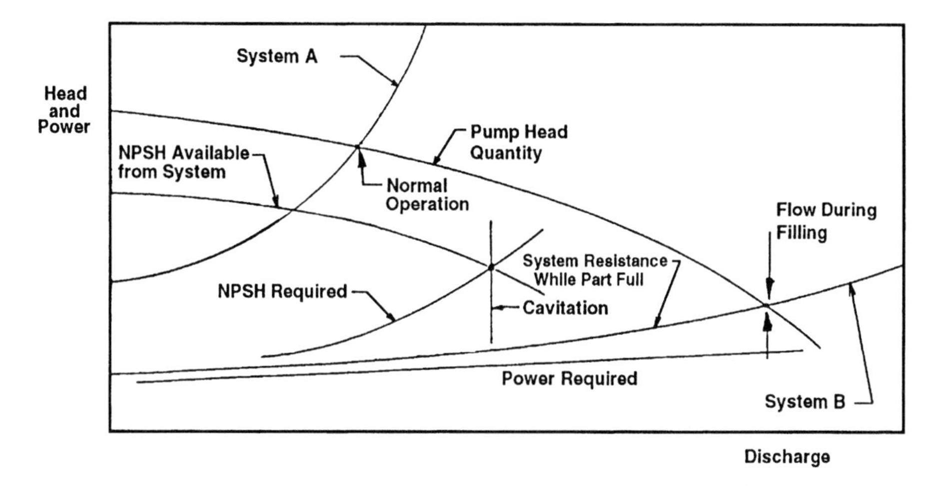

Figure 15.12. Pump and system characteristics during pipeline filling

Consider next the case where the system is not completely full on start-up. The initial system characteristic is now represented by System B in Fig. 15.12. The static lift and friction components correspond to the part of the system which does contain fluid on start-up. The pump will run up to design rotational speed as discussed above, so that the head-discharge characteristic runs up to the System B head-discharge characteristic. With the partially filled line, the initial discharge and power will be greater than during normal steady operation and, as shown by Fig. 15.12, the difference can be very large. Slurry pumps, with flat H-Q characteristics, will experience greater increase in flow and power than will pumps with steeper characteristics. It is not uncommon to find electric motor drives burned out from the high power required during system filling.

The NPSH required by a pump increases with discharge. Given a constant static suction head, increasing friction losses in the suction piping cause the NPSH available to decrease with increasing Q. Figure 15.12 illustrates a case where the flow during line filling exceeds that at which the NPSH available falls below the NPSH required, so that the pump cavitates. In an extreme case like that of Fig. 15.12, cavitation will be severe enough to generate a vapour pocket. This condition is likely to lead to difficulties with water hammer, a point discussed below.

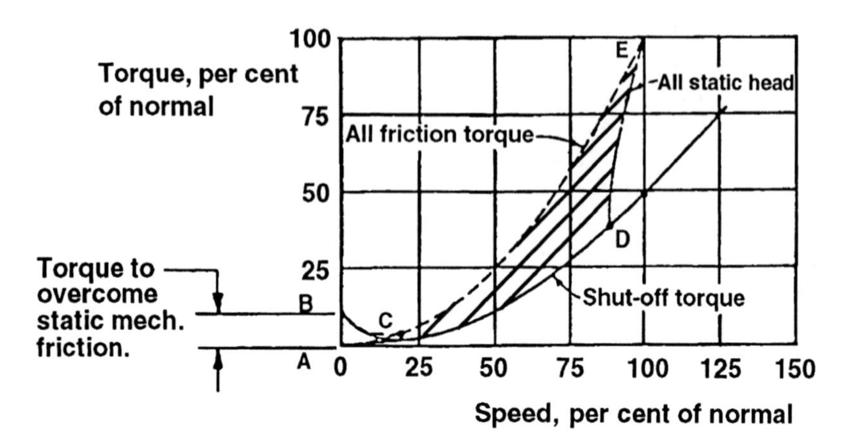

Figure 15.13. Variation of torque with speed for pump starting up on a system initially full

It was shown in Chapter 13 that fluctuations in the size and concentration of solids in a slurry can also lead to fluctuations in operating conditions, particularly when the pipe is long and the pumps are sized for operation below the equivalent fluid velocity. For example, a sharp increase in the concentration of the slurry arriving at the pump causes a surge in flow. This is analogous to the kind of start-up condition shown on Fig. 15.12, and therefore causes a power surge, possibly leading to cavitation and water hammer. Sudden reduction in slurry concentration causes a reduction in flow so that, if the system is operating close to the deposit velocity, the solids can settle out. The resultant problems in restarting the line were discussed in Chapter 13.

Sudden power failure is always a risk, and can present major problems in field applications. If the line is running at high velocity, sudden loss of torque to the pump can cause the pressure in the pump to fall sharply, to the point where a vapour pocket forms and water hammer results.

A particular case of the problems caused by power failure is provided by a pump delivering to a system dominated by static lift. Here power failure tends to cause the fluid to run backwards through the pump. (This tendency can be stopped by installing a quick-acting check valve, but the result may well be destructive water-hammer pressures.) If the flow is permitted to run in reverse, the ensuing changes in discharge, speed and torque are illustrated on Fig. 15.14. In an extreme case, the reverse flow has been known to reach 120% to 150% of the design delivery discharge; i.e. the pump moved from the design point A to, say, point B. Although the pump is now running at more than full speed in reverse, the torque remains positive.

Provided that the drive train - most critically, the gearbox - is able to withstand this condition, no damage results. In a specific case, the reverse flow has been known to last for some 30 minutes. Even though it was not actually damaging, the condition was nevertheless alarming. Furthermore, the sump overflowed badly. A branch was therefore installed on what was normally the discharge side of the pump, so as to divert the reversed flow into standby storage. In order to avoid damaging transients, the diverter valve in this branch was set to open slowly, at some delay time following power failure.

On the next occasion when power failed, the slurry again ran backwards, so that the pump again reached a point like B on Figure 15.14. The diverter valve now opened, so that reverse flow through the pump ceased. Static head in the pump sump may even have caused a slight positive flow through the pump. However, the impeller - which in this case was large - had sufficient inertia that it was still running at high speed in reverse. Thus, the diversion of the reverse flow had the effect of moving the pump operation from B to, say, point C on Fig. 15.14. The torque between the shaft and the impeller therefore changed sign: in effect, the fluid was driving the impeller forward rather than the usual condition. In this specific assembly, the shaft was connected into the impeller by a threaded stud. With the torque reversed, this stud unscrewed. It was sufficiently long that, as it unscrewed, the shaft pushed the gearbox into the motor. The cost is not recorded, and the thoughts of the engineers in the darkened pump house may best be left to the imagination.

The term 'water hammer' is applied to elastic pressure transients in a liquid or slurry. There is an extensive literature on such pressure transients, and for more detail than can be provided here the reader is referred to the classic text of Parmakian (1955) or to subsequent works by Wylie and Streeter (1978), Sharp (1981), and Chaudhry (1987).

In general, a pressure wave moves through a fluid at a characteristic sonic velocity or 'celerity' which depends on the fluid properties. In a pipeline the elasticity of the walls reduce the celerity of the disturbance, denoted by a. A typical value of a for a steel pipe filled with water (or an aqueous slurry) is about 1000 m/s. As given by the elastic-column model and verified experimentally, the rise in pressure generated by a change in velocity is

$$\Delta p = - a \, \rho_f \, \Delta V \tag{15.1}$$

where ΔV is the change in velocity. In terms of head, the transient rise is

$$\Delta H = \frac{-a}{g} \bullet \Delta V \qquad (15.2)$$

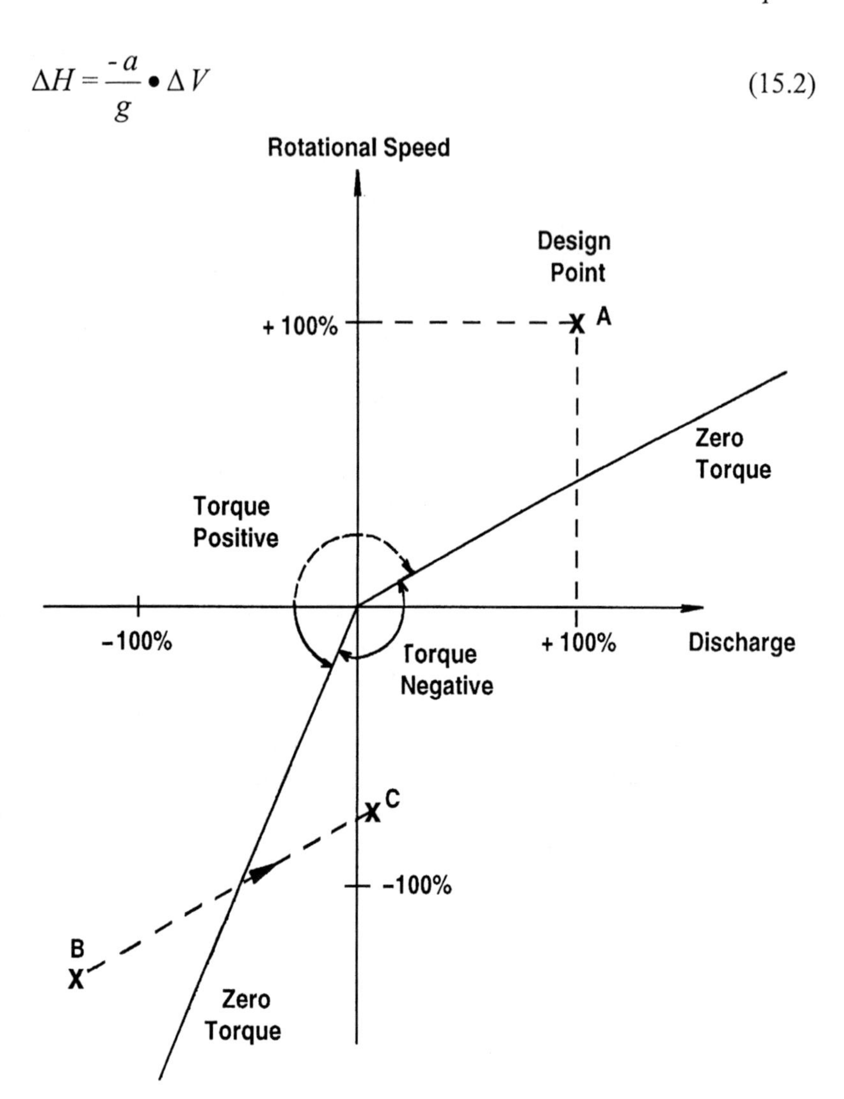

Figure 15.14. Variations in discharge, rotational speed and torque

Because the ratio a/g is approximately 100 (in both SI and ft.-lb-sec. units), a head transient in a pipeline is about 100 times the velocity change.

Equations 15.1 and 15.2 apply to any rapid change in operating conditions. For example, rapid closure of a valve in a pipe causes the pressure to rise on the upstream side and to fall on the downstream side. If the pressure on the downstream side falls below the vapour pressure of the fluid, a vapour cavity will form. Equation 15.1 shows that this happens for quite modest flow velocities. Subsequent collapse of the vapour cavity causes the valve to experience a water hammer in the reverse direction, as it is struck by the column of fluid moving back to fill the cavity.

Rapid opening and closing of valves is always to be avoided. However, vapour pockets can be caused by other types of maloperation, of which one of the commonest is sudden gross cavitation in a pump. Conditions which can cause this to occur have been identified in earlier sections of this chapter. Carstens & Hagler (1964) showed how vapour pockets caused by poor pump operation can lead to severe water hammer. As implied by Eq. 15.1 and 15.2, the pressure transients generated by closure of a vapour pocket can be very large indeed. For example, column closure at the relatively modest velocity of 5m/s (16.4 ft/sec) causes a pressure peak of about 50 bar (i.e. 500 m or 1640 ft. of head). The initial disturbance sets up wave fronts which travel in either direction along the pipe at velocity a. In a pipe of constant section, pressure waves pass by each other without changing amplitude or wave form. However, when they reach the end of the pipe or a change in section, the pressure waves are reflected. The maximum pressure transients thus arise at points of initiation and reflection, or where different wave fronts meet. Thus it can happen - and all too often does happen- that maloperation of the pump at the feed end of a long line can set up a pressure wave which travels along the line until it meets the first change in section. This is typically the next pump, which may suffer more or less severe damage. In extreme cases, the water hammer explodes the pump shell.

REFERENCES

Addie, G.R. & Hammar, J.R. (1994). Pipeline head loss tests of settling slurries at low velocities. *Proc. Intern'l Mechanical Engineering Congress (Mech '94)*, Perth W. Australia.

Carstens, M.R. & Hagler, T.W. (1964). Water hammer resulting from cavitating pumps. *Proc. Am. Soc. Civil Engrs.*, Vol. 90, No. HY6, pp. 161-184.

Chaudhry, M.H. (1987). *Applied Hydraulic Transients, 2nd Ed.*, Van Nostrand Reinhold Inc., New York.

Herbich, John B. (1975). *Coastal & Deep Ocean Dredging*, Gulf Publishing Company, Houston, TX.

Parmakian, J. (1955). *Waterhammer Analysis*, Prentice Hall, republished 1963, Dover, New York.

Sharp, B.B. (1981). *Water Hammer: Problems and Solutions*, Edward Arnold, London.

Stepanoff, A.J. (1957). *Centrifugal and Axial Flow Pumps, 2nd Ed.*, John Wiley & Sons, New York.

Stepanoff, A.J. (1965). *Pumps and Blowers, Selected Advanced Topics: Two Phase Flow*, John Wiley & Sons, New York.

Wylie, E.B. and Streeter, V.L. (1978). *Fluid Transients*, McGraw-Hill, New York.

Chapter 16

ENVIRONMENTAL ASPECTS OF SLURRY SYSTEMS

16.1 Introduction

One of the themes running through this book is that understanding of the behaviour of slurry systems has advanced greatly over the last few decades, so that hydraulic conveying systems can now be designed for much higher efficiency, achieved in part by better control and by operating at higher solids concentration. Reduced consumption of energy and water represent not only cost savings but also improvements in environmental performance. Hydraulic transport systems can show greatly reduced impacts on human health and the environment, by reducing emissions including dispersion of the material being transported. This is a serious practical concern. Where a particulate material is transported dry, for example by trucks or belt conveyors, it can become dispersed so that workers and the surrounding environment, along with any local population, are exposed. For the specific case of dry mining and transport of uranium ores, which is still common practice at mines in Namibia and Australia for example, it has been shown that worker exposure to fugitive dust emissions at the mine site is a major component of the human health risks associated with nuclear power generation (Sollberg-Johansen 1998). This extreme (and surprising) case illustrates the importance of integrated environmental management in mining and mineral processing.

This chapter reviews some of the environmental advantages of hydraulic conveying compared with other approaches to handling and transporting bulk solids. Hydraulic handling represents a way to integrate processing and transport in mineral processing systems. The advances in understanding of slurry behaviour mean that systems can be designed not just for energy efficiency but for flexibility, and this assists in integrated system design.

16.2 The benefits of integrated processing

In the past decade, approaches to assessing and managing the environmental performance of industrial processes and products have generally moved to a life cycle basis; i.e. rather than simply assessing the local impacts of an individual operation or plant (including fugitive dust emissions and noise) it is now common to include in addition the significance of materials and energy over their complete supply chains starting from primary resource extraction (see Clift 2001). This is the so-called 'cradle-to-grave' approach, embodied in an analytical approach known as Life Cycle Assessment (LCA). Applying the life cycle approach to mining and mineral processing means allowing for all the off-site environmental impacts of supplying energy, such as electricity and transport fuels, allowing for consumption of materials (including water) and for the impacts of waste disposal (including tailings).* Mining and mineral processing clearly come into the supply chains of most manufactured products; therefore applying LCA to products has in turn directed more attention to the environmental performance of the minerals sector.

Integrated environmental assessment underpins the 'pollution prevention' approach of avoiding the production of pollutants rather than trying to trap them so that they are not released into the environment (Allen & Shonnard 2002). In many cases, it has been possible to develop pollution prevention approaches which reduce costs as well as reducing environmental emissions; these 'win-win' developments go by the name of 'clean technologies' (Clift 2001). Although not introduced as such, many of the developments in slurry transport set out in this book are examples of clean technology. For example, improving system design to operate at higher solids concentration can reduce consumption of energy and/or water; i.e. it reduces environmental impacts and resource use as well as capital and operating costs.

Systematic development of clean technologies requires an integrated approach to design of processing systems. This approach is starting to become generally accepted in mining and mineral processing (e.g. Stewart & Petrie 1996; Villas Bôas & Barreto 1996). As a general principle, integrating process systems to minimise the number of distinct operations tends to improve both economic and environmental efficiency (Allen & Shonnard 2002). Handling bulk materials in slurry form aids process integration. Given that the great majority of mineral processing operations treat the material in slurry form, there is immediately a

---

* It should also strictly mean allowing for the resources used and environmental impacts for producing plant and equipment; however, except in very exceptional cases, these turn out to be small by comparison with the impacts of plant operation (Allen & Shonnard 2002; Clift 2001).

presumption in favour of slurrying the material as soon as possible. In addition to reducing material losses, this can sometimes have additional processing advantages. An example is provided by the processing of Albertan oil sands: replacing belt conveyors by slurry lines for transport from the mine to the processing plant enabled some pre-treatment in the transport line, so that some of the processing carried out at the mineral processing plant was eliminated, leading to increased throughout and economic savings (Cymerman *et al.* 1993).

A study by Sellgren & Addie (1998) illustrates the environmental and economic benefits which can be realised by taking an integrated approach to mining and mineral processing systems. The study refers specifically to the management and disposal of coarse waste rock and fine tailings, a problem which arises in a range of applications including coal washing (Williams 1996) and processing of oil sands (Cymerman *et al.* 1993). The conventional practice, shown schematically in Figure 16.1 (a), is to provide separate handling systems and disposal areas for the coarse rock and the fine tailings. However, it is possible in principle to integrate these into a single transport and disposal system, shown schematically in Figure 16.1 (b).

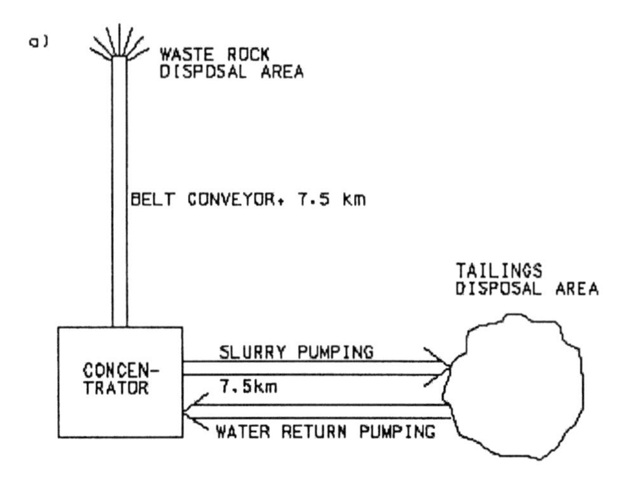

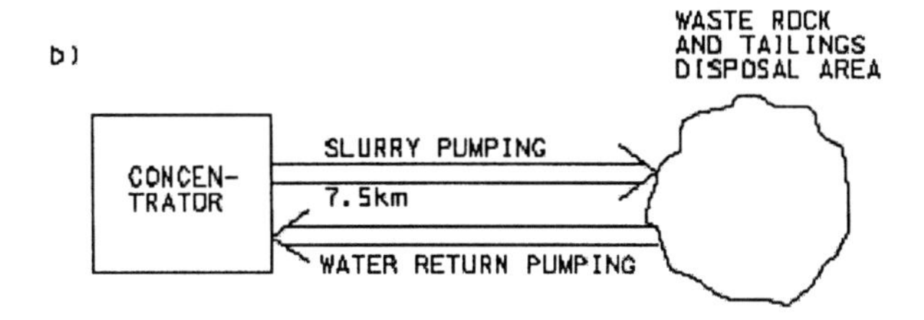

Figure 16.1. Schematic comparison of conventional handling of waste rock and tailings with an integrated system based on slurry transport of coarse and fine particles, from Sellgren and Addie (1998) a) Conventional system. b) Integrated system.

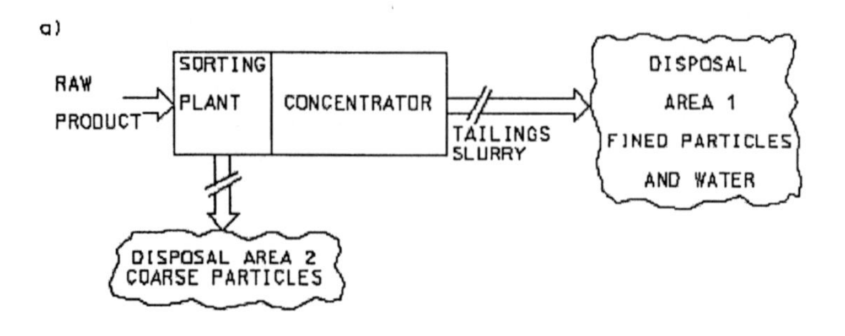

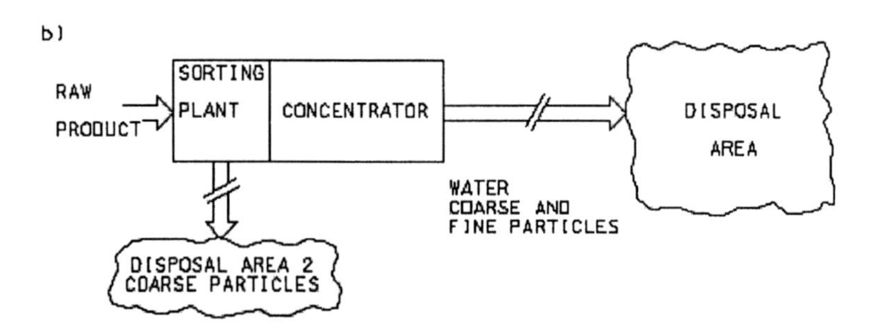

Figure 16.2. Specific case considered by Sellgren and Addie (1998): a) Conventional system. b) Integrated system.

The specific cases considered by Sellgren & Addie are shown in Figure 16.2. The conventional system used a belt conveyor for the rock and hydraulic conveying for the tailings, in each case moving the material 7.5 km from the concentrator to disposal. The rock was up to 75 mm (3 inches) in size, while the tailings were below 2 mm with a median size of 100 μm. The quantities of the two streams were taken to be equal. The feasibility of the system in Figure 16.2 (b) had been confirmed by comprehensive tests at the GIW Hydraulic Testing laboratory, which had shown that the mixture could be conveyed satisfactorily by centrifugal pumps at concentrations well above 35% by volume (i.e. 60% by weight). The integrated system had a capital cost less than half that of the conventional system while the energy consumption, which is the dominant component of operating cost, was some 20% lower in the integrated system. In this example, all the energy used is electrical power drawn from the distribution grid, so that the reduced operating cost translates directly into reduced environmental impact. In other cases, the comparison is not so straightforward; examples are developed in Section 16.3. Furthermore the mixed material had much improved disposal properties: low porosity and hence low water content, good strength and stiffness enabling use of much steeper slopes, little risk of particle separation which could lead to migration of the fine material. Clearly the integrated system represents a 'clean technology' in which both economic cost and environmental impact are reduced.

16.3 Energy consumption and associated emissions

Earlier chapters showed how the specific energy consumption can be calculated for solids transported as slurries. These relationships enable comparisons to be made with other possible transport modes. In Eq. 4.9, the specific energy consumption (SEC) in kWh/tonne-km was shown to be

$$SEC = 2.73\ i_m \,/\, S_s\, C_{vd} \qquad (16.1)$$

Eq. 16.1 refers to the energy delivered to the slurry by the pump and its drive system. If the pump has efficiency η_P, and it driven by an electrical motor with drive efficiency η_D, then the electrical energy actually delivered in order to convey the slurry (in kWh/tonne-km) is

$$SDEC = 2.73\ i_m \,/\, \eta_P\, \eta_D\, S_s\, C_{vd} \qquad (16.2)$$

However, in an integrated slurry transport system the conveying water will typically be returned from the slurry pipe delivery, be it processing plant or tailings pond, back to the feed end. Therefore we must allow for an additional delivered energy consumption required to pump the returned water:

$$SDEC' = 2.73\ i'_w \,/\, \eta_P\, \eta_D \qquad (16.3)$$

where i'_w is the friction gradient in the return water line. The maximum volume of return water corresponds to the volume of water in the slurry; i.e. $Q_M(1-C_{vd})$. If the return pipe is the same size as the slurry line and the friction gradient is proportional to the square of the mean velocity in the pipe (see Chapter 2), then

$$i'_w = i_w \left(1 - C_{vd}\right)^2 \qquad (16.4)$$

where i_w is, as usual, the friction gradient in the slurry pipe for water alone moving at the mean slurry velocity.

Combining Eqs. (16.2), (16.3) and (16.4), the total specific delivered energy consumption including the return water system is given by

$$SDEC = \frac{2.73}{\eta_P \, \eta_D} \left[\frac{i_m}{S_s \, C_{vd}} + i_w \left(1 - C_{vd}\right)^2 \right] \qquad (16.5)$$

To give representative figures for these parameters, Table 16.1 shows the results for two case studies considered in earlier chapters:

Case Study 13.1*: Coarse sand

Case Study 6.2: Washed coarse coal

It may be noted that these slurries both represent materials which are not well 'tailored' for hydraulic transport: they are both rather coarse, and require relatively high conveying velocities – particularly high in the case of the sand slurry (see Case Study 6.1 (a)). The figures in Table 16.1 are calculated assuming that, both for the slurry and the return water, the pump efficiency is 85% (i.e. $\eta_P = 0.85$) and the drive efficiency is 90% (i.e. $\eta_D = 0.9$). As might be anticipated, the energy required to return the conveying water is a minor but nevertheless significant part of the energy consumption for the whole system.

We now explore the environmental comparison between the hydraulic transport system and possible alternatives. In addition to belt conveyors, considered in the example in Section 16.2, bulk solids are commonly moved by truck. In the examples in Table 16.1, the tonnages to be transported are too large for trucking to be feasible, at least for Case Study 13.1, but nevertheless the comparison with trucking will be pursued for the sake of illustration. Specific energy consumption alone is not a sufficiently informative basis for comparison. Instead, the option might be compared in terms of the associated emissions of carbon dioxide, the gas making the largest contribution to global climate change. Table 16.1 shows representative emissions per tonne of payload per km for standard European trucks with 40-tonne capacity (SimaPro 2003) with 50% load factor; i.e. empty on the return trip. These figures are for trucks driven on metalled roads, so that emissions for the rough terrain frequently encountered in mining operations will be substantially higher. The power drawn by the conveying pumps is assumed to be taken from the electricity distribution grid. The associated carbon dioxide emissions then depend on the mix of primary energy sources used to generate electricity. Two different examples are considered here, summarised in Table 16.2:

i. A high-carbon energy economy, illustrated by the energy mix used in the UK in 1994;

ii. A lower carbon energy economy, illustrated by the energy mix used in 1994 in mainland Europe South of the Baltic (i.e. the UCPTE system); carbon emissions per kWh delivered are much lower due to the smaller proportion of coal used and the higher

* This case follows from Case Studies 5.1, 6.1 and 13.1, where the pump efficiency was found to be 84.5%, close enough to the value of 85% assumed in Table 16.1

proportion of essentially carbon-free generation, primarily nuclear and renewable (including hydropower).

Table 16.1. Energy consumption and associated carbon dioxide emissions for slurry systems.

Pump efficiency, $\eta_P = 0.85$

Drive efficiency, $\eta_D = 0.9$

	Case Study 13.1		Case Study 6.2	
Material	Sand		Coal	
Throughput (tonne/hr)		4000		577
Pipe diameter, m		0.55		0.44
Specific gravity, S_s		2.65		1.4
C_{vd}		0.24		0.25
Mixture velocity, V_m (m/s)		7.35		3.01
i_m (m. water/m. pipe)		0.082		0.0298
i_w (m. water/m. pipe)		0.053		0.0137
SEC kWh/tonne-km				
Slurry		0.353		0.232
Return water		0.084		0.021
TOTAL		0.44		0.253
Delivered energy consumption, kWh/tonne-km				
Slurry		0.46		0.303
Return water		0.11		0.028
TOTAL		0.57		0.33
Associated carbon dioxide emissions, gms/tonne-km				
UK: 680gm/kWh		390		220
UCPTE: 120gm/kWh		70		40
Compare emissions from 40 tonne diesel truck, gms/tonne-km		90-95		90-95
No. trucks per hour		100		14

Table 16.2. Energy mix for electrical power generation (figures from SimaPro 2003)

	UK	Mainland Europe (UCPTE)
Coal	59%	16%
Lignite	-	7%
Oil	8%	10%
Gas	3%	7%
Nuclear	27%	44%
Renewables (primarily hydro)	3%	15%
Carbon dioxide emission (gms) per kWh delivered	680	120

Note: The carbon emission figures allow for generating efficiency and for energy losses in production and transport of the fuels and in the electrical distribution grid; i.e. they are assessed on a full fuel cycle basis. The generating mix in both the UK and UCPTE systems has shifted since 1994, most significantly for the UK where the proportion of coal has decreased and gas has increased leading to a significant reduction in carbon dioxide emissions per unit of delivered energy. Nevertheless the 1994 figures are used here as representative of relatively high- and relatively low-carbon energy economies.

Table 16.1 shows that for the coal (Case Study 6.2), the hydraulic transport system using a generating mix like UCPTE is associated with much lower carbon emissions, even compared with the low estimates for the emissions from trucks. For the sand, the carbon dioxide emissions are somewhat lower for hydraulic transport, although this comparison is purely illustrative because the number of truck movements required would be impractical.

The comparison proves to be different if electrical power is drawn from a high-carbon economy like the UK (1994) mix: in this case, the estimated carbon dioxide emissions are higher for the hydraulic system. However, for coal the two options would be comparable if the off-road conditions were to double the fuel required by the trucks. For the sand, the hydraulic system with this background generating mix gives higher carbon dioxide emissions (although, as noted above, this comparison is of hypothetical interest only).

This kind of comparison will be specific to each particular case, but the comparison in the form of Table 16.1 can be developed easily provided data are available for the background electricity supply system. For a relatively low-carbon energy economy like UCPTE, hydraulic transport will evidently show lower carbon dioxide emissions than trucking for any practical values of the specific energy consumption. For a high-carbon economy, like the UK was in 1994, hydraulic transport will lead to higher carbon dioxide emissions than trucking unless SEC is less than a range around 0.1 to 0.2 kWh/tonne-km – but other factors such as the scale of operation (as in Table 16.1) or integrated process system design (see Section 16.2) may still favour a hydraulic system.

16.4 A cautionary tale – Velenje and its lakes

We end this chapter, and the book, with a true story of the environmental disasters which can be used by thoughtless use of technology, and how more sensitive use of technology has gone a long way towards restoring the damage. The story illustrates the concepts of unsustainable and sustainable engineering (see Azapagic *et al.* 2004). Hydraulic conveying is a key part of the story.

Velenje is an industrial town with a population around 40,000, in North Central Slovenia, the northernmost republic of what was once Yugoslavia. Velenje is on the edge of one of the thickest lignite deposits in the world, which has been mined since 1875. All the lignite mined is now used in the Šoštanj Thermal Power Plant (ŠTPP) which provides 750 MW electrical output corresponding to about 30% of electricity demand in Slovenia (Vrtačnik 2001). However the lignite has relatively low calorific value (10 MJ/kg), with 1.4% sulphur and 17 to 20% ash. Therefore the plant generates great quantities of ash and, prior to installation of flue gas desulphurisation plant in 1991 – 2001, emitted quantities of sulphur oxides plus nitrogen oxides which devastated the countryside, particularly to the North East of Velenje into Austria. Mining of the lignite brought its own problems. It has a very high water content: in effect, the lignite is an aquifer. The lignite is relatively close to the land surface, and the overburden is unconsolidated sand and gravel. It has been mined so that the overburden collapses behind thc cutter as the lignite is removed. This has so far lowered the land surface over an area of about 6 km^2 (more than 2 square miles) and has already displaced some 2,000 people from villages over the lignite body; most of them have moved away from the area. The subsidence has led to the formation of 'artificial' lakes, which already cover 2 km^2 (0.8 square miles) and are growing. Lake Velenje is the largest of these; it is about 20 m deep.

The impact on the water system in the area has also been disastrous. Velenje is in the Šalek valley, where the main watercourse is the river Paka. Flow in the Paka is seasonally variable, between 100 m^3/s in November and March and 0.2 m^3/s in January and August (Šterbenk *et al.* 2004). At full capacity, the cooling towers of the ŠTTP evaporate 0.4

m^3/s. Fortunately other tributaries join the Paka in the Šalek valley, and the lakes act as balancing reservoirs. Water from the Paka was also used to transport ash from the power plant. Until 1983, it was discharged into Lake Velenje. A substantial proportion of the ash is calcium oxide. As a result, the lake became strongly alkaline (pH ≈ 12), to the point where it could sustain no life. From 1983 the ash was contained by a dam (see Figure 16.3) but the conveying water was still discharged into the lake so that the pH remained high. The flow of water through Lake Velenje is small: the mean residence time of water in the lake is 2½ years. With such small movement, the upper layers of the lake reacted with carbon dioxide in the atmosphere so that the pH at the surface fell to below 11, but with the formation of calcium carbonate which deposited in the lake. The alkaline lake water discharged into the River Paka where more calcium carbonate formed. Partly treated sewage from the town of Velenje and the effluent from a leather tannery situated close to the ŠTPP plant were also discharged into the Paka. The water quality in the Paka was very good when it entered the municipality of Velenje but, unsurprisingly, very bad downstream of Šoštanj.

In short, Velenje and the ŠTPP are an excellent example of the kind of elephantine man-made disaster which happens when industrial development is pursued without regard to its environmental consequences.

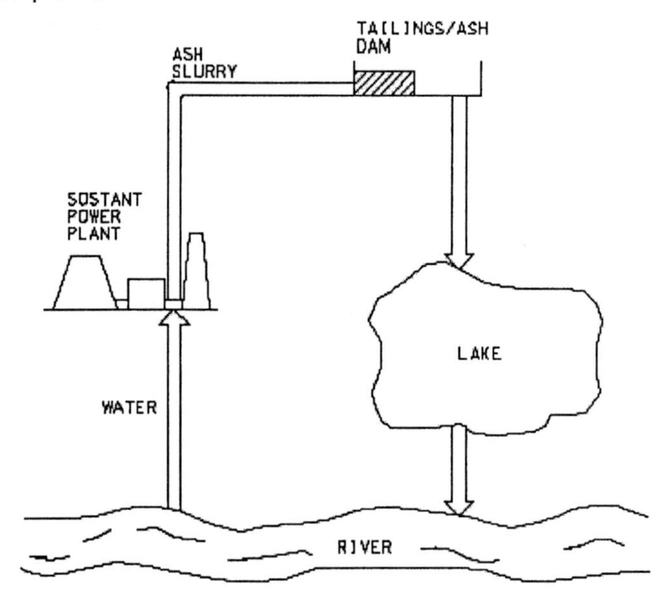

Figure 16.3. Former ash disposal system at Šoštanj

Systematic monitoring of water quality in the River Paka began in 1987. However, serious efforts to rectify the environmental disaster began in 1992, after Slovenia had separated from Yugoslavia and was already contemplating joining the European Union (which actually happened in 2004). The leather tannery closed down. Following extensive public consultation, a programme to improve water quality in the Paka and Lake Velenje was developed. In 1994, the ash transport system was modified as shown in Figure 16.4. Most importantly, a closed-loop system was introduced so that the strongly alkaline conveying water is returned to the power plant and not discharged into Lake Velenje. In spite of the low throughflow, the lake now recovered rather rapidly: its pH reduced to 8 in about three years, largely due to neutralisation through reaction of calcium oxide and hydroxide with carbon dioxide in the atmosphere to precipitate limestone. Along with the other lakes in the Šalek valley, Lake Velenje is now sufficiently healthy to have fish and crayfish populations and to be a resource for recreation including fishing, bathing and water skiing (Vrtačnik 2001). Water quality in the Paka has improved greatly; further improvement now depends on further upgrading of the municipal waste water treatment plant (Šterbenk *et al.* 2004).

Flue gas desulphurisation (FGD) plant was installed on the two main units of the power plant in 1995 and 2001. The FGD plant uses limestone slurry to react with sulphur (and nitrogen) oxides in the flue gases, producing a solid rich in gypsum (calcium sulphate) which tends to form a strongly cemented mass when wet. It is taken by truck to the ash disposal site and used to form barrier dams.

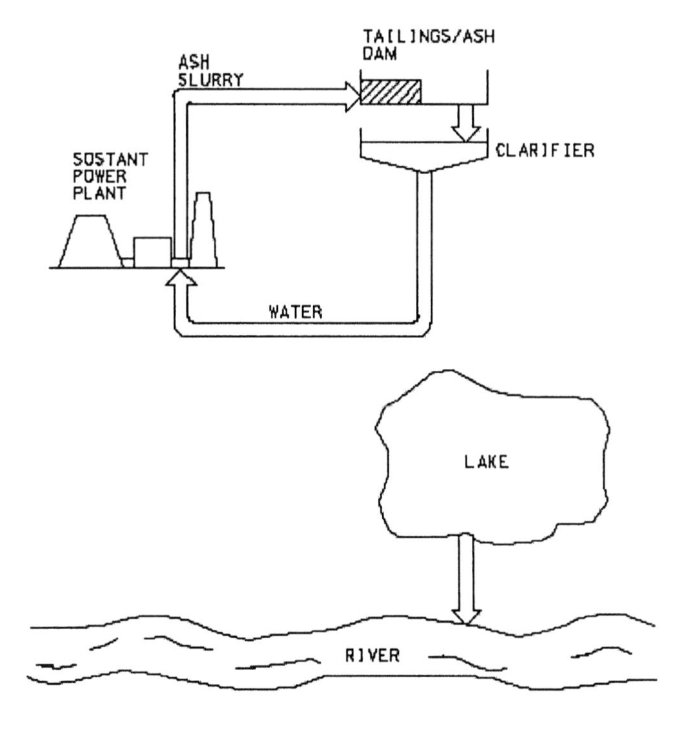

Figure 16.4. Modified ash disposal system at Šoštanj

The overall story since 1994 has been of successful remediation of the Šalek valley, which had been one of the most polluted areas in Slovenia. It shows what environmental damage can be caused by thoughtless use of technology, and how different the consequences can be if the technology is used sensitively. Hydraulic conveying around Lake Velenje at first contributed to ruining the local environment, but then played a major role in keeping an important power plant in operation while restoring much of the environmental damage. At the most obvious level, it illustrates the importance of closed-loop conveying systems. At a more general level, it illustrates how even a well-honed technology needs to be deployed with careful consideration of all its consequences, sometimes in consultation with people who might be directly or indirectly affected: the elephant cannot be allowed to trample anything in its way.

Finally, for the reader who may be wondering, the Slovenian for 'elephant' is *slon*.

REFERENCES

Allen, D.T. & Shonnard, D. R. (2002) *Green engineering: Environmentally conscious design of chemical processes*, Prentice-Hall, Upper Saddle River, NJ.

Azapagic, A., Perdan, S. & Clift, R., eds. (2004) *Sustainable development in practice: case studies for engineers and scientists*, John Wiley & Sons, Chichester.

Clift, R. (2001) Clean Technology and Industrial Ecology, Chapter 16, pp 411–444 in *Pollution: causes, effects and control*, 4th ed., ed. R. M. Harrison, Royal Society of Chemistry, London.

Cymerman, C. J., Leung, A. & Maciejewski, W. (1993) Oil sand hydrotransport at Syncrude Canada Ltd., *Proc 18th International Technical Confce on Coal Utilization and Fuel Systems*, Florida.

Sellgren, A. & Addie, G.R. (1998) Effective integrated mine waste handling with slurry pumping, *Proc. 5th International Confce on Tailings and Mine Waste*, Fort Collins, Colorado; Balkema, Rotterdam.

SimaPro (2003) Life Cycle Inventory Database version 5.1, PRé Consultants, Amersfoort.

Sollberg-Johansen, B. (1998) *Environmental life cycle assessment of the nuclear fuel cycle*, PhD Thesis, University of Surrey.

Stewart, M. & Petrie, J. G. (1996) Life cycle assessment for process design – the case of mineral processing, pp. 67-81 in *Clean Technology for the Mining Industry*, ed. M. A. Sánchez, F. Vergara and S. H. Castro, University of Concepción, Chile.

Šterbenk, E., Drev, A. R. & Bole, M. (2004) Water management of a small river basin toward sustainability – the example of the Slovenian river Paka, pp. 349-356 in *Sustainable Development of Energy, Water and Environmental Systems*, ed. N. H. Afgan, Ž. Bogdan and N. Duić, A. A. Balkema Publishers, Lisse.

Villas Bôas, R. C. & Barreto, M. L. (1996) Clean technology solutions for the mineral industries: the need of P^2 (pollution prevention) solutions, pp. 43-52 in *Clean*

Technology for the Mining Industry, ed. M. A. Sánchez, F. Vergara and S. H. Castro, University of Concepción, Chile.

Vrtačnik, J. (2001) *Ecological Remediation of Šoštanj Thermal Power Plant, 1987 - 2000*, Termoelektrarna Šoštanj d.o.o., Šoštanj, Slovenia.

Williams, D. J. (1997) Effectiveness of co-disposing coal washery wastes, *Proc. 4th International Confce on Tailings and Mine Waste*, Fort Collins, Colorado; Balkema, Rotterdam.

Appendix

WORKED EXAMPLES AND CASE STUDIES IN U.S. CUSTOMARY UNITS

Example 2.A - Flow of Water in a Pipe

Water at room temperature of 68°F flows at 1900 USgpm through a standard 8-inch steel pipe. Calculate:

(a) the mean velocity and pipe Reynolds number;
(b) the Moody friction factor;
(c) the hydraulic gradient;
(d) the thickness of the sub-layer;
(e) the shear velocity;
(f) the hydraulic power required.

(a) Standard (schedule 20) 8-inch pipe has an internal diameter of 8.125 inches. Therefore the pipe cross-sectional area is

$$\frac{\pi}{4}(8.125/12)^2 = 0.36\ ft^2$$

and so the mean velocity V is

$$(1900/450)/0.36 = 11.75\ ft/s$$

This mean velocity is towards the low end of the range typically used for settling slurries. For water at 68°F, ρ = 1.937 slugs/ft^3 and μ = 0.000021 lb s/sq ft. Therefore the Reynolds number is

$$Re = \frac{\rho VD}{\mu} = \frac{1.937(11.75)(8.125/12)}{0.000021} = 7.35 \ x\ 10^5$$

Because Re is dimensionless, the same value is obtained whatever system of units is used. Note that Re is high and the flow is well into the turbulent range.

(b) For commercial steel pipe, the equivalent sand-grain roughness is typically 0.00015 feet, as noted previously. Therefore the relative roughness is

$$\varepsilon/D = 0.00015\,/(8.125/12) = 2.23 \ x\ 10^{-4} = 0.000223$$

Referring to Fig. 2.10, flow conditions are in the "transitional rough" range where the friction factor depends on the relative roughness and (weakly) on Re. From Fig. 2.10 or Eq. 2.33

$$f = 0.0152$$

Values for the Moody friction factor in the range 0.01 to 0.02 will prove to be fairly typical for water alone pumped at conditions commonly used for settling slurries.

(c) From Eq. 2.35, the hydraulic gradient is

$$i = \frac{f\,V^2}{2gD} = \frac{0.0152(11.75\,)^2}{2\,(32.2)(8.125/12)} = 4.84x\,10^{-2}\ ft\ head\ per\ ft\ pipe$$

Values of the order of a few metres head per hundred metres of pipe are again typical of water pumped at conditions appropriate for a settling slurry.

(d) From Eq. 2.29, the wall shear stress is

$$\tau_o = \rho\, f\, V^2/8 = 62.4(11.75\,)^2(0.0152)/8 = 0.508\ lb/\ ft^2 = 0.0035\ lb/\ inch^2$$

The thickness of the viscous sublayer is estimated from Eq. 2.27 as

$$11.6\mu/\sqrt{\rho\tau_o} = 11.6\ (0.000021)/\sqrt{1.937\ x\ 0.508} = 0.000245\ ft$$

This value can be compared with the equivalent sand-grain roughness of 0.00015 ft. As expected for a flow well into the 'transitional rough' range, the sub-layer thickness and the equivalent roughness are of comparable magnitude. The value for sub-layer thickness is typical and worth noting.

(e) From Eq. 2.30, the shear velocity is

$$U_* = V_m\sqrt{f/8} = 11.75\sqrt{0.0152/8} = 0.51\ ft/s$$

and $U_* / V_m = \sqrt{f/8} = 0.044$

This value of U_*, of the order of V/20, is typical.

(f) The power required from the pumps is given by Eq. 2.37 as

$$\frac{P}{L} = \frac{8\rho Q^3 f}{\pi^2 D^5} = \frac{8(1.937)(4.23)^3(0.0152)}{\pi^2(8.125/12)^5} = 12.7\ ft\ lb/s\ ft$$

or 125 HP/mile, which again gives a typical order of magnitude for water alone.

Example 2.B - Calculation of Terminal and Hindered Settling Velocities

The calculation of settling velocities is poorly suited for US customary units. Instead, particle size and other characteristics should be converted to SI units, and the procedure given in Chapter 2 can then be followed directly.

Case Study 3.1 - Non-Newtonian Slurry of Phosphate Slimes

A phosphate-slimes clay-water slurry is to be pumped over a horizontal distance of 2296 ft, using a pipe of internal diameter 12 inches. The slurry will be taken from a pond in which the mixture specific gravity is 1.13. Tests have been carried out with this material at the GIW Hydraulic

Laboratory using a pipe of internal diameter 7.98 inches. Test data for i_m and V_m appear in Table A.1, together with values of $8V_m/D$ and τ_o (the wall shear stress equals $\rho_w g i_m D/4$). Fig. 3.10 shows the values of i_m plotted *versus* V_m.

Table A.1. Data for phosphate-slimes slurry in 7.98 inch pipe

Run	V_m (ft/s)	$8V_m/D$ (s^{-1})	i_m	τ_o (psf)
1	1.75	21.0	0.1004	1.04
2	5.00	60.1	0.1130	1.17
3	6.55	78.8	0.1150	1.19
4	8.49	102.1	0.1189	1.23
5	10.63	127.9	0.1218	1.26
6	12.49	150.3	0.1237	1.28
7	14.52	174.7	0.1273	1.32
8	16.79	202.0	0.1348	1.40
9	18.51	222.6	0.1472	1.53

The preliminary design of the pumping system calls for a discharge Q_m of 4755 USgpm in the 12 inch pipe. However, this value is not yet definite, and Q_m values of 3170 USgpm and 6340 USgpm are also to be considered.

(a) Find the values of i_m for the three values of Q_m noted above. On dividing by the pipe area of 0.7854 ft^2, the required velocities are found to be 8.99 ft/s, 13.49 ft/s and 17.97 ft/s. Also calculate the required head and pressure at the pump for each case.
The first step is to scale the data points from the 7.98 inch pipe to the 12 inch one. For the laminar-flow points the appropriate scaling laws are given by Eq. 3.11 and 3.12, which show that i_m scales inversely with the diameter ratio while V_m scales directly with this ratio. Points from test runs 1 to 7 have been scaled on this basis and are shown on Fig. 3.10. For runs 8 and 9 the flow is turbulent. The hydraulic gradient can still be scaled by Eq. 3.11, but V_m must now be scaled by Eq. 3.13. This requires evaluation of U_*, i.e. $\sqrt{(\tau_o/\rho)}$.

For example, run 8 with τ_o = 1.4 psf has U_* = 0.797 ft/s, for which Eq. 3.13 gives a scaled-up value of V_m equal to 17.6 ft/s. The laminar-turbulent transition point is obtained by projecting the turbulent line back to intercept the laminar line. For the test data in the 7.98 inch pipe this intercept occurred to the right of point 7, at a velocity of about 16.1 ft/s. For the larger pipe the intercept lies between points 4 and 5, with a slightly reduced velocity, about 15.4 ft/s.

Figure 3.10 shows that conditions will be laminar for the two lower flows to be investigated, and turbulent for the highest flow. The values of i_m can be taken directly from the figure, and are listed in Table A.2. As i_m is expressed in feet of water per foot of pipe, $\Delta p/\Delta x$ equals $\rho_w g i_m$ or 0.4336 i_m, and the required pressure at the pump is the product of this quantity and the pipe length.

Table A.2

Qm (USgpm)	3170	4755	6340
V_m (ft/s)	8.99	13.49	17.97
i_m	0.0760	0.0795	0.0915
$\Delta p/\Delta x$ (psi/ft)	0.03298	0.03448	0.03963
Pump pressure (psi)	75.7	79.2	91.1
Pump head (ft water)	174.5	182.4	210
Pump head (ft slurry)	154.4	161.4	185.8

(b) Find the power which must be supplied to the motor driving the pump, assuming motor efficiency of 95% and pump efficiency of 75%. Also find the specific energy consumption, both per ton of mixture and per ton of delivered solids on a dry-weight basis, assuming S_s = 2.65.

The output power from the pump, in HP, equals QH/3960 with H in ft of water and Q in USgpm, and the power required by the motor will equal the input power divided by product of the efficiencies, giving QH/2821. For example at Q_m of 4755 USgpm H is 182.4 ft of water, and the power required by the motor is 307 HP. To find the specific energy consumption in this case, we also require the

transport in tons per hour. On the basis of mixture specific gravity of 1.13, this is 4755(10)(1.13)(60)/(1.2)(2000) or 1343 tons/h, so that the ton-mile/hour is 583.9, and the specific energy consumption is 307/584 or 0.525 HPh/ton-mile. On a dry-solids basis the useful throughput tonnage is based on $Q_m S_s C_v$. The relative density of the mixture, i.e. 1.13, equals 1 + (S_s -1)C_v from which, with S_s = 2.65, C_v is equal to 0.0788, giving a dry-solids throughput of 252 tons/h for SEC = 2.81 HPh/ton-mile. The same result could be obtained from the pump output. For HP input to the pump motor, it must be divided by (0.95)(0.75), giving 2.81 HPh/ton-mile, as before. The various results, for all three values of Q_m, are given in Table A.3.
As the solids concentration is fixed, the specific energy consumption varies only with i_m, in a direct proportionality. The flat curve for laminar flow, plotted on Fig. 3.10 shows a very small rise in i_m as Q_m goes from 3170 to 4755 USgpm, and the rise in SEC for this step is less than 5%. On the other hand, the equal increase of Q_m from 4755 to 6340 USgpm requires an increment in SEC of more than 15%, as a result of the shift from laminar to turbulent flow.

Table A.3

Q_m (USgpm)	3170	4755	6340
Power required (HP)	197	308	473
SEC for mixture (HPh/ton-mile)	0.495	0.525	0.597
SEC for dry solids (HPh/ton-mile)	2.69	2.81	3.23

(c) Suppose now that testing had stopped after run 6, so that no turbulent-data flow points were available. What effect would this have on the results found above?
In this case the transition to turbulent flow must be predicted from the parameters of the laminar-flow rheogram. The first step is to prepare a logarithmic plot of τ_o *versus* $8V_m/D$ and find the slope n′. As it happens, the data from the laminar test runs for this material were used in plotting Fig. 3.4, from which it can be seen that all points lie close to a line with n′ = 0.110. The ratio (3n′+1)/4n′ is

thus 3.02, which is multiplied by the values of $8V_m/D$ to give du/dy as shown on Fig. 3.5. All points on that figure except the lowest one can be considered as obeying a Bingham formulation of the rheogram, with τ_y = 1.1 psf and η_B = 4.2 10^{-4} lb-s/ft^2.

If the power law model is considered appropriate, the locus of laminar-turbulent intercepts will correspond to a constant f evaluated (from Eq. 3.19) as 0.0142. On this basis the intercept velocity V_T should be almost independent of pipe diameter, with a predicted value of 18.1 ft/s in the prototype pipe (D = 12 in). (For the test pipe, with D = 7.98 in, the predicted V_T is slightly larger, 18.6 ft/s). It will be recalled, however, that for a power-law fluid there will be a 'faired' transition to fully turbulent flow, with the transitional points lying above the intercept rather than passing through it. Comparison with the points plotted on Fig. 3.10 show that this is the case here, implying that the intercept-locus method under-predicts friction loss in this area and hence is not appropriate for the present example.

A more conservative approach is to use Eq. 3.22, based on a Bingham plastic. From Fig. 3.5 showed the Bingham parameters of this material are τ_y = 1.1 psf and η_B = 4.2 x 10^{-4}lb-s/ft^2. with V_m in ft/s and D in ft.

In the present instance, the flow at Q_m = 3170 USgpm is clearly laminar, and that at Q_m = 4755 USgpm coincides with the transition point predicted by Eq. 3.22. Thus neither of these points will be influenced by the omission of the turbulent flow data. However, for Q_m = 6340 USgpm, the value of i_m would be altered drastically. On the basis of 0.000388 V_m^2 it is predicted to be 0.1415 ft water/ft pipe, i.e. some 55% in excess of the value obtained previously on the basis of scale-up of the turbulent-flow data (and more than 70% above the result of extrapolating the laminar flow line to this velocity). Compared to the change of about 5% found previously for a velocity change of 4.49 ft/s in the laminar-flow region, it is now seen that the same change is now found to account for a 78% increase in i_m, and hence in specific energy consumption. Nothing could give a clearer indication of the differing effects of laminar and turbulent flow of non-Newtonian slurries.

Case Study 5.1 - Preliminary Pipe Sizing

A pipe is to be designed to convey solids at 4410 tons/hour (on a dry-weight basis). The material is described only as <coarse sand', and it is estimated that the largest volumetric solids concentration which can be fed to the pumps is C_{vd} = 0.24. Preliminary estimates are required of suitable combinations of pipe size, C_{vd} and V_m. For convenience it will be assumed that pipes are available with inside diameter D in increments of 1.97 inches (50 mm).

First, the tons per hour is converted to ft^3/s of solids, Q_s, by multiplying by 2000, then dividing by the estimated solids weight of 165 lbs/ft^3 and by 3600 s/hour, giving Q_s = 14.8 ft^3/s. The mixture discharge Q_m equals Q_s/C_{vd} and hence the minimum Q_m, corresponding to the maximum C_{vd} of 0.24, equals 61.7 ft^3/s or 27800 USgpm.

The next step is to select a pipe size which gives a reasonable velocity for this value of Q_m. For example, a pipe of D = 23.6 inches has a cross-sectional area of 3.04 ft^2, and hence the velocity in this pipe must be at least 61.7/3.04 or 20.2 ft/s in order to satisfy the maximum - C_{vd} condition. This velocity seems to be in the right 'ball park', but must be compared with the deposition velocity V_{sm}. The particle size distribution of the coarse sand may not be known, but it is safe to assume that it contains grains of the 'Murphian' size, which is about 0.03 inches (0.7 mm) for this pipe diameter. For this particle and pipe size combination, the nomographic chart gives $V_{sm} \approx 19.0$ ft/s (a version of this chart which shows U.S. units is included in this Appendix as Fig. A.1). It should be recalled that the value of V_{sm} estimated from the chart tends to be high. This point can be checked by means of Eq. 5.11, using a reasonable value of f_w, say 0.012. The result is a revised value of V_{sm} of only 15.3 m/s, so that the minimum operating velocity of 20.2 ft/s provides a generous margin against deposition.

Other pipe sizes should also be investigated. The next smallest pipe (D = 21.7 inches), is found to have a minimum operating velocity of 24.1 ft/s, based on C_{vd} = 0.24. This can be compared to V_{sm} values of 17.7 ft/s from Fig. A.1 and 14.6 ft/s from Eq. 5.11. Using the latter number as the upper limit for deposition, it is seen that the minimum operating velocity would be 65 percent in excess of the deposition limit, producing stable operation at the cost of high frictional losses. This pipe size could merit further investigation of the sort illustrated in the case studies of subsequent chapters, but it is obvious that no smaller pipe would be suitable, and the 21.7 inch pipe appears at present to be considerably less attractive than the 23.6 inch size.

Turning to a larger pipe, with D = 25.6 in, it is found that for C_{vd} =

0.24 the minimum velocity is 17.3 ft/s, while the V_{sm} values are 20.0 ft/s from the nomographic chart and 15.9 ft/s from Eq. 5.11. The minimum operating velocity of 17.3 ft/s is less than 10 percent in excess of V_{sm} by Eq. 5.11, which is not a sufficient margin. A margin of 30 percent should be adequate, giving $V_m = 20.7$ ft/s. Increasing the velocity to this value requires a lower solids concentration, i.e. 0.20 instead of 0.24. This combination (D = 25.6 in, $V_m = 20.7$ ft/s, $C_{vd} = 0.20$) merits further investigation, but any larger pipe size (which would involve higher operating velocity and lower delivered concentration), would probably not be suitable.

It is worth noting that this case study shows that a preliminary selection of appropriate pipe sizes can be made on the basis of continuity and deposition velocity, even though very little is known about the solids in the slurry. In the present instance, it is found that to transport 4410 tons of coarse sand per hour, only pipes of internal diameter 23.6 in and 25.6 in appear to be attractive.

Case Study 5.2 - Transporting Clay Balls

The dredging industry is normally concerned with pumping sand slurries for which heterogeneous flow is expected (the type of flow dealt with in Chapter 6), but on occasion very coarse particles must be transported. Cases of well-defined coarse-particle transport arise when dredging cohesive clays. The cutter head of the dredge shaves slices off the surface of the clay, and after passing through the pump these emerge as fist-size lumps, with a typical dimension of roughly 4 inches. The clay balls are then transported through the discharge pipe to a disposal area.

The following data, for a typical application of this sort, were provided by Ing. K. Oudmaijer of the Netherlands (personal communication). The clay had a specific gravity of 1.79 and the diameter of the balls could be taken as 4 inches. Ing. Oudmaijer measured μ_s by wiring a group of balls together, placing them (under water) on a steel plate and inclining the plate until slip occurred. He obtained a value of 0.31. The carrier fluid was seawater with a specific gravity of 1.02. The pipe was steel, with internal diameter 27.6 inches, and the pressure drop was measured over a horizontal reach of pipe 6070 ft in length. The measured pressure gradient was 0.0256 psi/ft (580 Pa/m) at $V_m = 15.1$ ft/s and $C_{vd} = 0.0714$.

The friction factor of the pipe is not known; for commercial pipe at the Reynolds number of this application a value of 0.012 might be expected. However, it should be recalled that the actual friction losses in dredge lines are raised considerably because of dents in the pipe and misaligned joints

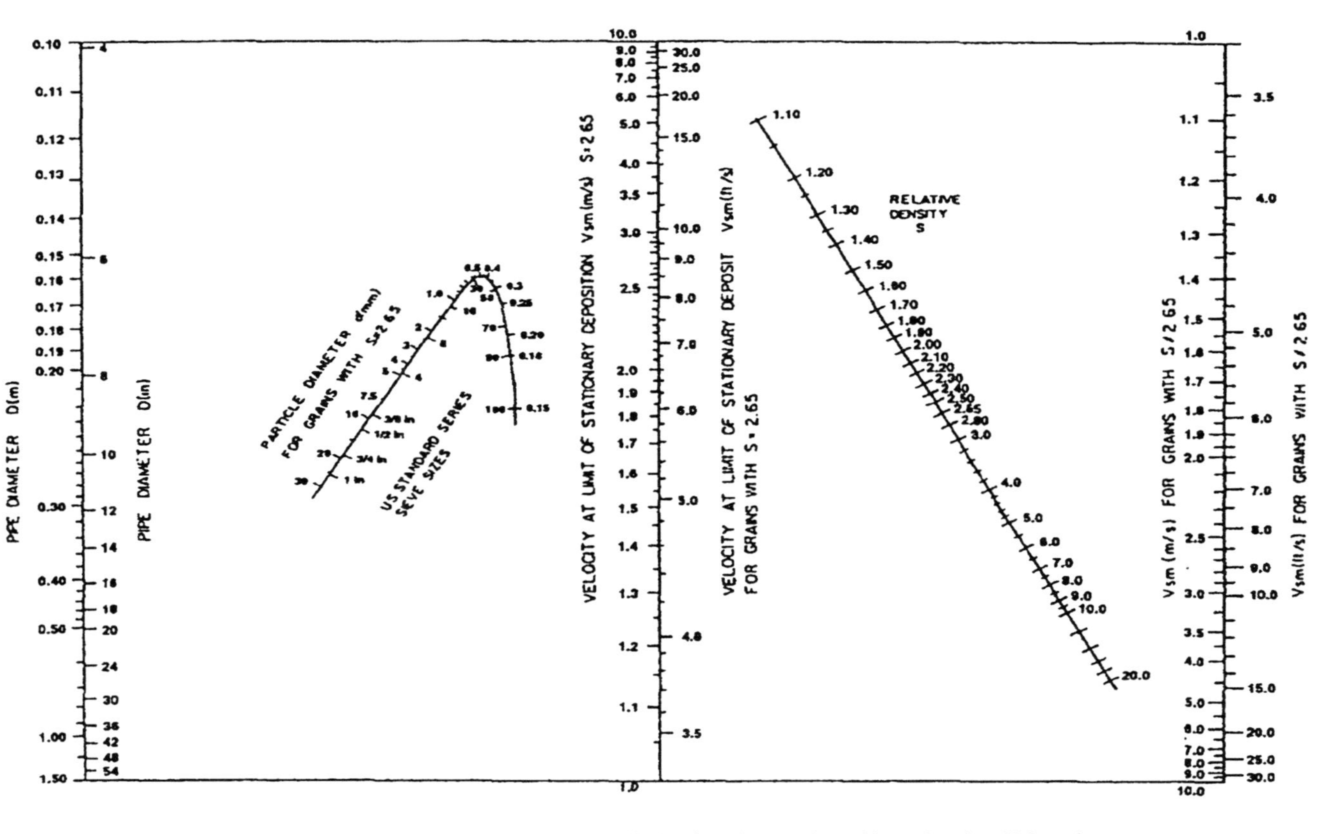

Figure A.1 Nomographic chart for maximum velocity at limit of stationary deposition, showing U.S. units

between lengths. Thus a higher friction factor must be employed, and a value of 0.020 is suggested.

The pressure gradient is to be calculated and compared with the observed value. The first step is to find the plug flow gradient i_{pg}, using Eq. 5.8, with $\mu_s = 0.31$, $S_s = 1.79$ and $S_f = 1.02$. The value of the loose-packed concentration, C_{vb}, was not measured, and a typical value of 0.6 will be employed, giving

$$i_{pg} = 2(0.31)(1.79 - 1.02)(0.6) = 0.286$$

The relative concentration C_r equals C_{vd}/C_{vb}, i.e. 0.0714/0.60 or 0.119. In order to use Fig. 5.4, it will be necessary to know the relative velocity V_r, which is based on the velocity at the deposition limit V_{sm}. The latter quantity can be estimated from the nomographic chart (see Fig. A.1). It will be recalled that with this chart the first step is to use the left-hand panel to find V_{sm} for $S_s = 2.65$, correcting later for different density. The pipe diameter of 27.6 inches is readily located on the vertical axis at the left of the chart, but it is seen that the particle diameter of 4 inches lies off the plotted portion of the 'demi McDonald'. However, the large-particle branch of this curve is virtually straight, and the numbers follow an essentially logarithmic scale on which the distance from 1 inch to 4 inches will be the same as that from 1/4 inch to 1 inch. The line starting at D = 27.6 inches and passing through this extrapolated point intersects the central vertical axis at a value of about 7.9 ft/s. As the relative density is not 2.65, the right hand panel of the chart is also needed. For this purpose, the relative density is taken as 1.79/1.02 or about 1.75. A line starting from 7.9 ft/s on the central axis is projected through this value of 1.75 on the inclined axis to give the corrected V_{sm} on the right-hand vertical axis, about 5.1 ft/s.

The relative velocity V_r is $V_m/5.1$, and for the operating velocity V_m = 15.1 ft/s, V_r is approximately 3.0. Extrapolating from Fig. 5.4, with $C_r \approx 0.12$, gives a value of ζ if about 0.1. When multiplied by i_{pg} (found previously to be 0.286) this gives the solids effect ($i_m - i_f$) as 0.029 feet of water per foot of pipe. The fluid friction gradient i_f is calculated as $S_f f V_m^2/(2gD)$, or 0.031, from which i_m is predicted to be 0.060 feet of water per foot of pipe. The equivalent pressure gradient is 0.0260 psi/ft (590 Pa/m), which is very close to the observed value of 0.0256 psi/ft (580 Pa/m).

Alternatively, the simplified form of Eq. 5.13 can be used. For the present case V_{sm} is 5.09 ft/s, as found above, and 0.55 V_{sm} equals 2.80 ft/s. As μ_s has been evaluated at 0.31, the right hand side of Eq. 5.13 is multiplied by the ratio 0.31/0.40 or 0.775, and the solids effect is given by

$$i_m - i_f = (S_m - 1)(0.775)(V_m/2.80)^{-0.25}$$

For the specific case referred to above, (S_m - 1) equals (1.75-1)(0.0714), or 0.0536, and V_m = 15.1 ft/s, giving

$$i_m - i_f = 0.027$$

This number is a reasonably good approximation of the solids effect of 0.029 found above using Fig. 5.4. The resulting pressure gradient is now predicted to be 0.0252 psi/ft (570 Pa/m), which is again very close to the observed value.

Case Study 6.1 - Effect of Particle Size and Grading on Sand Transport

In Case Study 5.1 a preliminary pipe selection was made for a system to transport 4410 tons/hour of coarse sand. One of the pipe sizes selected was D = 25.6 inches, for which the combination V_m = 20.7 ft/s and C_{vd} = 0.20 gives the required solids flow rate.

In that case study the mean size and grading of the particles was not specified; and thus the hydraulic gradient and the specific energy consumption could not be calculated. The present case study will investigate these points.

The first material to be considered is a clean sand (negligible fraction smaller than 40 μm) with S_s = 2.65, d_{50} = 0.70 mm (0.03 inches) and d_{85} = 1.00 mm (0.04 inches). We will consider an aqueous slurry (S_f = 1.00) in the pipe with D = 25.6 inches. The conditions found in Case Study 5.1 (C_{vd} = 0.20, V_m = 20.7 ft/s) will be investigated initially, and the first requirement will be:

(a) Find the hydraulic gradient i_m and the specific energy consumption for this case. For these particles the associated velocity w can be read directly from Fig. 6.5, giving 0.127 m/s or 0.417 ft/s for the 0.70 mm particle and 0.148 m/s or 0.486 ft/s for the 1.00 mm one. When multiplied by cosh(60d/D), these become 0.418 ft/s and 0.488 ft/s, giving a ratio of 1.17. From Eq. 6.15, the associated value of σ_s is 0.069, and when this is substituted into Eq. 6.16, M is given as 1.79. As this is larger than 1.7, M = 1.7 will be used. (This result is not surprising, since d_{85}/d_{50} is less than 1.5, indicating a narrow particle grading.)

With M evaluated, we can turn to V_{50}, equal to

$w\sqrt{\frac{8}{f_w}}\cosh(60d/D)$, based on d_{50}. For $f_w = 0.012$ (found from the Stanton-Moody diagram for commercial steel pipe of this size and velocity range), the resulting value of V_{50} is 10.8 ft/s. (If the simplified approach of Section 6.6 had been used, M would remain at 1.7 and V_{50} would be 11.4 ft/s). On the basis of mixture velocity V_m = 20.7 ft/s, and $(S_m - S_f) = (1.65)\ 0.20 = 0.330$, Eq. 6.3 gives the solids effect $(i_m - i_f)$ as $(0.330)(0.22)[1.921]^{-1.7}$ or 0.0239 (ft water/ft pipe). With no fines, the fluid is water and $i_f = i_w$, given by $0.012\ (20.66)^2/(64.4)(25.6/12)$, or 0.0373. The value of i_m is the sum of i_w and the solids effect or 0.0612 (ft water/ft pipe). The specific energy consumption is proportional to i_m/S_sC_{vd}, which is 0.115, giving SEC of 0.613 HPh/ton-mile.

(b) Next, it is of interest to vary some of the quantities and as input, to find the effect on i_m and SEC. The initial case to be investigated is a variation of V_m only, with C_{vd} kept constant. The second case involves adjusting C_{vd} as well as V_m, so that the solids throughput Q_s is maintained at a constant value.

The effect of varying V_m alone could be obtained by differentiation, but it is probably clearer simply to change V_m by about 10% in each direction, repeating the calculations for, say, V_m = 22.6 ft/s and 18.7 ft/s. As i_w varies with V_m^2 (ignoring any changes in f_w) and $(i_m - i_w)$ varies with $V_m^{1.7}$, the calculations are straightforward. For V_m = 22.6 ft/s, i_m = 0.0653 and SEC = 0.656 HPh/ton-mile. For V_m = 18.7 ft/s, i_m = 0.0589 and SEC = 0.594 HPh/ton-mile.

If the same values of V_m are employed, but C_{vd} is adjusted to maintain Q_s, the values shown in Table A.4 are obtained. As with the figures for the constant-C_{vd} case, this table indicates a significant increase in SEC when V_m is increased to 22.6 ft/s. As before, the specific energy consumption is diminished when V_m is decreased to 18.7 ft/s, but once again, the change is smaller for a decrease in velocity. Thus it would appear that the lower velocity of 18.7 ft/s could be attractive. On the other hand, the lower velocity is less than 20 percent above the deposition limit (V_{sm} = 15.9 ft/s by Eq. 5.10, see Case Study 5.1), and considerations of system stability may be involved. In order to resolve these questions, it will be necessary to know the pump characteristic as well as the system characteristic. Pump characteristics will be discussed in Chapter 9, and the intersection of pump and system characteristics in Chapter 13.

(c) It is also of interest to find the effect on i_m of changes in particle grading. As an initial example, suppose that the coarse sand specified above is replaced by a fine, clean, narrow-graded sand with d_{50} = 0.20 mm (0.008 inches). From Fig. 6.5, w for this sand is 0.08 m/s or 0.28 ft/s. In the pipe of 25.6 inches diameter the resulting value of V_{50} from Eq. 6.1 is 6.8 ft/s. For V_m = 20.7 ft/s, and C_{vd} = 0.20, the solids effect as calculated from Eq. 6.3 is 0.0110, giving i_m = 0.0483 ft water/ft pipe. (If the simplified approach had been used, the value of i_m would be 0.0498 ft water/ft pipe.) Clearly, i_m is considerably less than the value of 0.0612 found in part (a) for the coarse sand. It may be compared with the result of the crude 'equivalent fluid' model, by which i_m is estimated as the product of S_m and i_w. As S_m = 1.330, and i_w = 0.0373, the equivalent fluid model gives i_m = 0.0496, rather close to the result of the new model for this specific instance of fine-sand slurry flow.

Consider next a coarse sand with d_{50} = 0.70 mm as before, but with a broader particle grading so that d_{85} = 1.50 mm (0.06 inches). For the latter size, w = 0.578 ft/s, and for the 25.6 inch pipe cosh(60d/D) is 1.01. Substitution into Eq. 6.15 shows that the value of σ_s for this material is 0.146. When this is substituted into Eq. 6.16, M is found to be 1.38. (The simplified approach gives M = 1.0/ln (1.50/0.70 = 1.31.) The value of V_{50} remains at 10.8 ft/s and for V_m = 20.7 ft/s and C_{vd} = 0.20, the solids effect becomes $(0.330)(0.22)[1.921]^{-1.38}$ or 0.0295. Adding i_w of 0.0373 gives i_m = 0.0668 ft water/ft pipe, which is about 9% higher than the value obtained previously for the narrow-graded particle distribution with the same median size. (A similar increase, about 11%, is obtained if the simplified approach is used throughout.)

Table A.4

V_m (ft/s)	C_{vd}	i_m	SEC (HPh/ton-mile)
18.7	0.221	0.0619	0.564
20.7	0.200	0.0612	0.615
22.6	0.183	0.0636	0.699

Case Study 6.2 - Transport of Coal Slurry

The material for this study is coal, with S_s = 1.4, μ_s = 0.44, C_{vb} = 0.60. The coal is clean with few fines, and d_{50} = 2.0 mm (0.08 inches)and d_{85} = 2.8 mm (0.11 inches). As the ratio of d_{85}/d_{50} is less than 1.5, the particle grading is narrow, and M will be 1.7.

The pipe to be used has D = 17.3 inches and f_w = 0.013. Solids concentration C_{vd} can be taken initially as 0.25. The minimum in the curve of constant C_{vd} is to be investigated, comparing the velocity at this point with V_{sm} and V_{50}, and finding the minimum specific energy consumption and how this is influenced by changes in C_{vd}.

(a) The limit of deposition will be determined first. For d = 2.0 mm (0.08 inches) and D = 1.44 ft, Fig. A.1 shows that V_{sm} would be 13.4 ft/s for material with S_s = 2.65, and for S_s = 1.4, the right-hand side of that figure gives V_{sm} = 6.2 ft/s. Equation 5.10 gives a value of 6.4 ft/s, but the smaller number should be taken, i.e. V_{sm} = 6.2 ft/s. (Note that for these 2.0 mm coal particles the nomographic chart shows a smaller value than Eq. 5.10; the reverse of the condition found in Case Study 5.1 for 0.7 mm sand.)

(b) To obtain V_{50}, it is sufficient to use the simplified approach of Eq. 6.11, which gives V_{50} = 8.7 ft/s. For this narrow-graded material M = 1.7, and with $(S_s - 1)C_{vd}$ = 0.4 (0.25) = 0.10, the expression for solids effect (Eq. 6.3) becomes:

$$i_m - i_w = 0.022\left(\frac{V_m}{8.7}\right)^{-1.7} = 0.870\,V_m^{-1.7}$$

Also, i_w is $\dfrac{0.013V_m^2}{(64.4)(17.3/12)} = 0.000140\,V_m^2$. The sum of these two terms is i_m, and differentiation gives the following relation for the velocity at minimum i_m:

$$1.7(0.870)\,V_m^{-2.7} = 2(0.000140)\,V_m$$

from which V_m equals 10.2 ft/s. At this point i_m is calculated to be 0.0313 ft water/ft pipe and i_m/S_sC_{vd} is 0.089, giving SEC = 0.474 HPh/ton-mile. (This is the quantity which must be provided by the pumps to the slurry; the energy which must be supplied to the pumps will, of course, be larger).

(c) In this case, the value of V_m at the minimum point is comfortably above the deposition limit of 6.2 ft/s, so no problem with deposition should arise. However, stability problems would occur near the minimum, as shown in Chapter 13, and thus the operating velocity should be larger than 10.2 ft/s. It is also worth noting that in the present study, the flow is definitely heterogeneous near the minimum point. At the minimum point V_m/V_{50} equals 1.17 and the stratification ratio, which is given by 0.5 $(V_m/V_{50})^{-1.7}$ equals 0.38, implying that slightly over one-third of the submerged weight of solids is carried by intergranular contact, with fluid support accounting for the rest.

(d) Finally, it is of interest to consider the effect of a change in concentration on the minimum point. As shown in the discussion following Eq. 6.5, the velocity at minimum i_m varies *cæteris paribus* with $C_{vd}^{0.27}$. Thus, if C_{vd} increases from 0.25 to 0.30 the value of V_m at the minimum point will increase from 10.2 ft/s to 10.7 ft/s, not a large change. It can readily be calculated that the equivalent value of i_m is 0.0345 ft water/ft pipe. This gives SEC = 0.437 HPh/ton-mile, or about 92% of the value obtained previously for $C_{vd} = 0.25$.

Example 7.1

A broadly-graded granular material with S_s=2.65 has a grading curve that passes through the points listed below. (Here, as elsewhere in this book, particle diameters are expressed in μm.) An aqueous slurry of this material having C_v=0.25 is to be pumped through a horizontal pipe with internal diameter 11.8 inches. The throughput velocity V_m will be 10.2 ft/s. Estimate the hydraulic gradient i_m using the method of Section 7.2.

Percent Finer	Diameter (μm)
20	40
44	150
50	200
70	400
85	1400
95	4500

We can begin with the hydraulic gradient for water, i_w, which is given by $fV^2/(2gD)$. With a typical f of 0.013, i_w=0.0226. From the grading table it is seen that the -40 μm fraction, X_f, is 0.20, giving S_f=1.0+0.25(0.20)(1.65), which is 1.082. For these granular fines, at relatively modest concentration, it can be assumed that ν_r and α are both equal to 1.0, placing the upper boundary of X_p at 150 μm, and giving X_p as 0.44-0.20=0.24. Also, i_f can be estimated as equal to $S_f.i_w$ or 1.082(0.0226).

By Eq. 7.1, i_e is given by 1.082(0.0226)[1.0+0.25(0.24)(2.65-1.082)] = 0.0268 (ft water/ft pipe). The lower limit of the heterogeneous fraction is 150 μm, as seen above, and its upper limit is 0.015D, i.e. 4500 μm. From the grading data provided, these boundaries of X_h represent 44% finer and 95% finer, with an average of 69.5%, which gives the representative diameter for this fraction, d_h, as about 400 μm. Substitution into Eq. 7.3 (with ν_r=1.0 and S_{fp}=1.182) gives V_{50}=8.9 ft/s. When this value (together with relative densities and other properties) is substituted into Eq. 7.2, it is found that Δi_h=0.0349 (ft water/ft pipe).

Finally, there is the 5% of the solids that travel as fully-stratified load, producing an additional component of hydraulic gradient, Δi_s, in accord with Eq. 7.4. Here X_s = 0.05, S_{fph} = 1.392 and B' is taken as 0.5. The deposition limit V_{sm} for this fraction can be estimated from Fig. A.1 for d = 4.5 mm as 8.9 ft/s, producing Δi_s = 0.0129. Summing i_e, Δi_h and Δi_s [Eq. 7.5] gives $i_m \approx$ 0.068 (ft water/ft pipe). [It may be of interest that the calculations for the same example using the simplified single-fraction approach of Section 6.5 give i_m some 30% higher than that obtained with the present method.]

Example 7.2

The coal slurry used for plotting Fig. 7.3 is to be pumped through a horizontal pipe with internal diameter 17.3 in. The volumetric concentration C_v will be 0.26, giving S_m = 1.13 and ρ_m = 2.19 slugs/ft^3. For a pressure gradient ($\Delta p/\Delta x$) of 0.89 psf/ft, find j, τ_o, U_*, μ_{eq}, V_m and Q. The lowest line on the figure indicates that, for the design concentration, μ_{eq} equals 6.6 (10^{-5}) $U_*^{-1.6}$ in SI units or 9.3(10^{-6}) $U_*^{-1.6}$ with μ_{eq} in lb.s/ft^2 and U_* in ft/s.

As $\Delta p/\Delta x$ equals $\rho_m gj$, j is given by 0.89(2.19)(32.2) = 0.0126 ft of slurry per ft of pipe. From Eq. 2.17, $\tau_o = (\Delta p/\Delta x)D/4$ i.e. $0.89(17.3)/48 = 0.321\,\text{psf}$. Furthermore, U_* equals $\sqrt{\tau_o/\rho_m} = \sqrt{0.321/2.19} = 0.383\,\text{ft/s}$. Using U_*, the value of μ_{eq} can be obtained graphically from Fig. 6.9 or by means of the fit equation, which gives μ_{eq} = 0.00205 Pa.s or 4.3(10^{-5})lb.s/ft^2. As the points in this region are essentially on the fit line, there is no need to distinguish U_{*f} from

U_*. Substitution into Eq. 6.24 gives V_m = 9.8 ft/s (corresponding to f = 0.0122). Flow rate is the product of V_m and $\pi D^2/4$, giving Q = 16.0 ft^3/s.

Example 7.3

For the pipe of the previous example, it has been decided to use a design flow rate of 21.2 ft^3/s. Find the corresponding values of U_*, μ_{eq}, $\Delta p/\Delta x$ and j. From the previous example, it can be expected that the fit line from Fig. 6.9 will also apply here, and there is no need to distinguish U_* and U_{*f}. It will be necessary to substitute V_m, pipe diameter, and b into Eq. 6.24. The value of V_m is readily found by dividing flow rate by pipe area, yielding 13.0 ft/s. Substitution then gives

$$13.0 = 2.5\, U_*\, \ell n\left(\frac{2.19(17.3/12)\, U_*^{2.6}}{9.3(10^{-6})}\right)$$

On iteration it is found that the required value of U_* is 0.479 ft/s. The corresponding value of μ_{eq} is 3.02(10^{-5}) lb.s/ft^2, and that of f is 0.011. The shear stress equals $\rho_f U_*^2$, *i.e.* 0.502 psf, from which $\Delta p/\Delta x$ is 1.39 psf/ft and j is 0.0198 ft of slurry per ft of pipe.

Worked Examples of Vertical Hoisting from Chapter 8

1. In an iron-ore mine the ore is ground to -40 μm in a sub-surface facility, and then pumped vertically 2625 feet to the surface. The pipe has a diameter of 7.87 inches. The concentration by volume, C_{vd}, is 20% and S_s is 4.9. Determine the pressure requirement to pump the slurry to the surface at a velocity of 6.56 ft/s.
   Solution: According to the discussion leading to Eq. 8.9 it follows that:
   $$(p_1 - p_2)/2624 = \rho_w\, g\,(1 + (S_s - 1)\, C_{vd})(1 + f_w V_m^2 / 2gD)$$
   The friction factor for water flow, for a smooth pipe, f_w, is 0.014.

$$(p_1 - p_2) = 2624 \times 1.94g\,(1 + (4.9 - 1) \times 0.20)$$
$$\times (1 + 0.0014 \times 6.56^2/(2g \times 7.87/12))$$
$$(p_1 - p_2) = 295900\ lb/\ ft^2 = 2055\ psi$$

2. Centrifugal slurry pumps are used to pump a sand slurry of d_{50} = 1.5 mm (0.06 inches) out of a quarry. The pipe is vertical with length 328 feet and diameter 3.94 inches. Tests have shown that the settling velocity of the largest particles is approximately 1.48 ft/s. Select the operating velocity and calculate the head requirement in feet of slurry.

   Solution: Following the guidelines given in connection with Eq. 8.10, the velocity V_{all} is 4 times 1.48, i.e. 5.9 ft/s. At this velocity, $V_m^2/2g$ is 0.541 ft. With the friction factor f_w taken as 0.016 for smooth-pipe conditions, the head is obtained from Eq. 8.10 as

$$328\,[1 + \frac{0.016(0.541)}{3.94/12}] = 336.5\ ft\ of\ slurry$$

   or 336.5 S_m in ft of water.

3. Consider the upward vertical transport of coarse ore particles with diameter 75 mm (3 inches) in a pipe flow with velocity V_m = 8.2 ft/s. A delivered volumetric solids fraction of 0.05 (5%) is required. Estimate the resident volumetric solids fraction if v'_t . v_t = 2.95 ft/s.
   Solution: Eq. 8.4 gives

$$C_{vi} = 0.05/(1 - 2.95/8.2) = 0.078\ (7.8\%)$$

Case Study 8.1 - Inclined Flow in a Dredge Ladder

Case Studies 5.1 and 6.1 dealt with flow of sand slurries in a horizontal pipe of D = 25.6 inches. This pipe will now be considered as the pressure line carrying the slurry from a suction dredge to a disposal area. However, in this example the friction factor will be not be re-evaluated to reflect dredge-line conditions. What will be considered here is the analysis of conditions in the inclined suction pipe located on the dredge ladder. This pipe (which has length 59.0 ft in the present case) conveys the slurry from the cutter head to the main pump on board the dredge. The angle of the ladder can be adjusted according to the depth to be dredged. For simplicity we will consider only an angle of 30° to the horizontal. The dredging is taking place in fresh water, and it is assumed initially that the size of the suction pipe is the same as that of the discharge pipe and that the sand is as specified in Case Study 6.1, with d_{50} = 0.70 mm (0.03 inches). As in that study the required transport rate of 4410 tons/hour (on a dry-weight basis) is satisfied by $V_m = 20.7$ ft/s and $C_{vd} = 0.20$.

(a) The first point is to investigate the effect of the incline on the deposition velocity in the suction pipe. Figure 8.2 shows that an inclination angle of 30° gives the largest increase in deposition velocity. At this angle the upper limit of the envelope for experimental points has $\Delta_D = 0.33$, equivalent to an increase in V_{sm} of $0.33\sqrt{2(32.2)(1.65)(25.6/12)} = 4.9\ ft/s$. As shown previously in Case Study 5.1, Eq. 5.11 gives a value of V_{sm} for the horizontal pipe of 15.9 ft/s. If the maximum increase of 4.9 ft/s is added, the resulting deposition limit velocity on the slope could be as large as 20.8 ft/s. However this is rather pessimistic, as the typical experimental data (rather than the envelope) show $\Delta_D \approx 0.30$, for V_{sm} of about 20.3 ft/s on the slope. Thus the operating velocity of 20.7 ft/s, although well clear of deposition in the horizontal discharge line, is very close to the deposition point on the inclined suction pipe. This deposition-limit condition is especially severe for the 0.7 mm particle, which is in the 'Murphian' size range for V_{sm}. If, say, the particle diameter were only 0.20 mm (0.008 inches), V_{sm} would be 10.5 ft/s for the horizontal pipe, and roughly 15.4 ft/s on the incline, well below the operating velocity of 20.7 ft/s.

(b) The next step is to consider the pressure drop in the suction pipe. For fully- stratified flow, detailed calculations would be required. However as noted in Section 8.5, for heterogeneous flow a reasonable approximation can be obtained from the formula of Worster and Denny (Eq. 8.11). In the present case, for the 0.03 inch (0.70 mm) sand slurry at $V_m = 20.7$ ft/s, this formula gives the excess gradient, Δi (30°), as $\Delta i\ (30°) = 0.0239 \cos 30° + (1.65)(0.20) \sin 30°$ where 0.0239 represents the solids effect in the horizontal pipe, as obtained in Case Study 6.1. Thus Δi (30°) equals 0.0207 + 0.1650 or 0.1857. The clear water gradient i_w was found in Case Study 6.1 to be 0.0373, giving a total value of 0.223, and on multiplying this by the suction-pipe length of 59.0 ft, the drop is found to be 13.2 ft of water. (The greater part of this arises from lifting the submerged particles from the bottom to the water level, this is represented by the term in $\sin\theta$ in Eq. 8.11). When the velocity head is added in, together with minor losses, it is estimated that the pressure at the suction of the pump (which is located close to the water line) will be sub-atmospheric to the extent of about 22.8 ft of water. In Chapter 9, it is seen that only a limited sub-atmospheric pressure can be tolerated at the suction of a pump, otherwise cavitation will occur. In the present instance, the low pressure at the pump suction may well induce cavitation.

(c) As shown above, although the design conditions are well suited for operation in the discharge pipe, conditions are very near the deposition limit in the suction pipe, and the pressure drop in this pipe is probably enough to cause the pump to cavitate. This difficulty would not be alleviated by a change of particle size alone; for example if the vacuum at the suction side of the pump is 22.8 ft of water for the 0.03 inch (0.70 mm) particles, a particle size of 0.008 inches (0.20 mm) would reduce it only very slightly; to about 22.1 ft, and cavitation would probably still occur in this case.

For water, and for any slurry that behaves as an equivalent fluid (i.e. $i_m = S_m\, i_w$, see Eq. 2.47), a normal method of eliminating cavitation problems is to have suction piping larger than that on the pressure side of the pump. Thus a larger suction pipe, say D = 27.6 inches, might well have been installed along with the pressure-pipe diameter of 25.6 inches. In this case the slurry of 0.008 inch (0.2 mm) particles (which, as seen in Case Study 6.1, behaves much like an equivalent fluid) will show satisfactory performance. The vacuum at pump entry is reduced from the figure of 22.1 feet of water given above for the 25.6 inch pipe to about 19 feet for the 27.6 inch pipe, which may be sufficient to eliminate cavitation. Despite the success of this strategy for a slurry of fine sand, the results of applying it to a coarse-sand slurry would be disastrous. For the 0.03 inch (0.7 mm) particles in the 27.6 inch pipe, Eq. 5.11 gives V_{sm} = 16.5 ft/s for a horizontal pipe, and the equivalent value on a 30° incline will be 21.6 ft/s. However, for the required discharge the mean velocity in the 27.6 inch suction pipe is much less than that in the 25.6 inch discharge line (17.8 ft/s *versus* 20.7 ft/s). Thus deposition will take place immediately in the 27.6 inch suction line. This deposit implies that the flow will become stratified. In the lower portion of the pipe the velocity will be zero, or even negative (at large angles of inclination reverse flow may actually be observed near the bottom of a pipe) and the required total flow rate Q_m can only be maintained if the velocity in the upper portion of the pipe is much larger than V_m, which in turn will cause extremely high velocity heads and friction losses. Moreover, the stratification in the suction line indicates that the *in situ* concentration of solids here will be much larger than that in the discharge line. Hence the term in $\sin\theta$, which shows the effect of submerged weight of solids, will rise abruptly when deposition occurs.

It follows from the points outlined above that for slurries of particles of coarse-sand size and above, use of an inclined suction pipe larger than the pressure line is highly likely to produce *both* deposition and cavitation. Such enlarged suction pipes are manifestly not to be recommended for coarse-particle slurries.

Case Study 9.1 - Sizing a Water Pump

A pump is required for a clear-water flow of 39,600 USgpm and 164 feet of head. It is to be operated at 400 rpm because of an existing driver. Estimate the impeller diameter, the expected efficiency (assuming operation at best efficiency point) and the required net positive suction head.

a) From the speed, head and flow the design specific speed of the pump (at the best efficiency point) will be

$$N_s = \frac{400\sqrt{2.5}}{50^{0.75}} = 33.64 \text{ in SI units}$$

$$N_s = \frac{400\sqrt{39{,}600}}{164^{0.75}} = 1737 \text{ in US units}$$

An impeller vane outlet angle of 25° is an appropriate selection. This is used in Fig. 9.8, together with the SI specific speed, to obtain a value of 0.463 for the head coefficient, which is defined as

$$\psi = \frac{gH_{BEP}}{U_2^2}$$

Here H_{BEP} is the head at the best efficiency point, and U_2 is the tangential velocity of the impeller outlet, equal to $\frac{\pi DN}{60}$.

Rearranging gives

$$D = \frac{60}{\pi N} \times \sqrt{\frac{gH}{\psi}}$$

$$= \frac{1}{\pi} \times \frac{60}{400} \times \sqrt{\frac{32.2(164)}{0.463}} = 5.099 \text{ ft.} = 61 \text{ inches}$$

(b) From Fig. 9.10, we can interpolate from the solid lines that a water pump of this specific speed and of this size (flow), should achieve an (ideal) efficiency of 92 or more percent but that well designed slurry pump will be 4.5% or so less, i.e. 87.5% efficient.

(c) From Fig. 9.14, the sample slurry pumps shown indicate that it is reasonable to expect a σ value of 0.15.

$$\sigma = \frac{NPSH}{H}$$

Therefore,

$$NPSH = 0.15(164) = 24.6 \textit{ feet of water}$$

To operate properly, the pump should therefore be provided with at least 25 feet of NPSH.

Case Study 9.2 – 'Trimming' a Pump Impeller

A pump with characteristics shown on Fig. A.2 (US-unit version of Fig. 9.2) is to be operated at 500 rpm on water at 12,672 USgpm against a system head of 147.6 feet. Calculate the reduced impeller diameter necessary, neglecting any slip losses. Determine the power needed, and the NPSH required.

(a) For impeller turn-down, as noted at the end of section 9.2, the head equation is

$$\left(\frac{H_1}{H_2}\right)=\left(\frac{D_1}{D_2}\right)^2$$

where D is the impeller diameter and 2 and 1 refer to initial and final values. For the discharge relation, the equation in customary use in the United States will be employed, i.e.

$$\frac{Q_2}{Q_1}=\frac{D_2}{D_1}$$

Using Fig. A.2 as a guide we may find (by trial and error), the ratio of D_1/D_2 which transforms the design Q and H to a point on the 500 rpm head characteristic. This ratio was found to be 1.091, giving

$$\mathrm{H}=147.6\times(1.091)^2=175.8\ \mathrm{ft}$$

and

$$Q=12672\,(1.091)=13825\ US\ gpm$$

where the values of 175.8 ft and 13,825 USgpm refer to the point on the 500 rpm head line.
Using the same ratio we can calculate the trimmed impeller diameter as

$$D_2=\frac{46}{1.091}=42.16\ inches$$

In this case, the turndown diameter ratio is 0.916, which is well above the value of 0.8 regarded as the limit where significant slip losses reduce head and efficiency.

(b) Thus negligible slip will occur and the efficiency can be taken to be the same as that found by interpolation at the 500 rpm take-off point on Fig. A.2. This efficiency is 82.8%. In this case, the pump input power for the water-only duty is

$$\frac{12{,}672 \bullet 147.6}{3960 \bullet 0.828} = 570 \text{ BHP}.$$

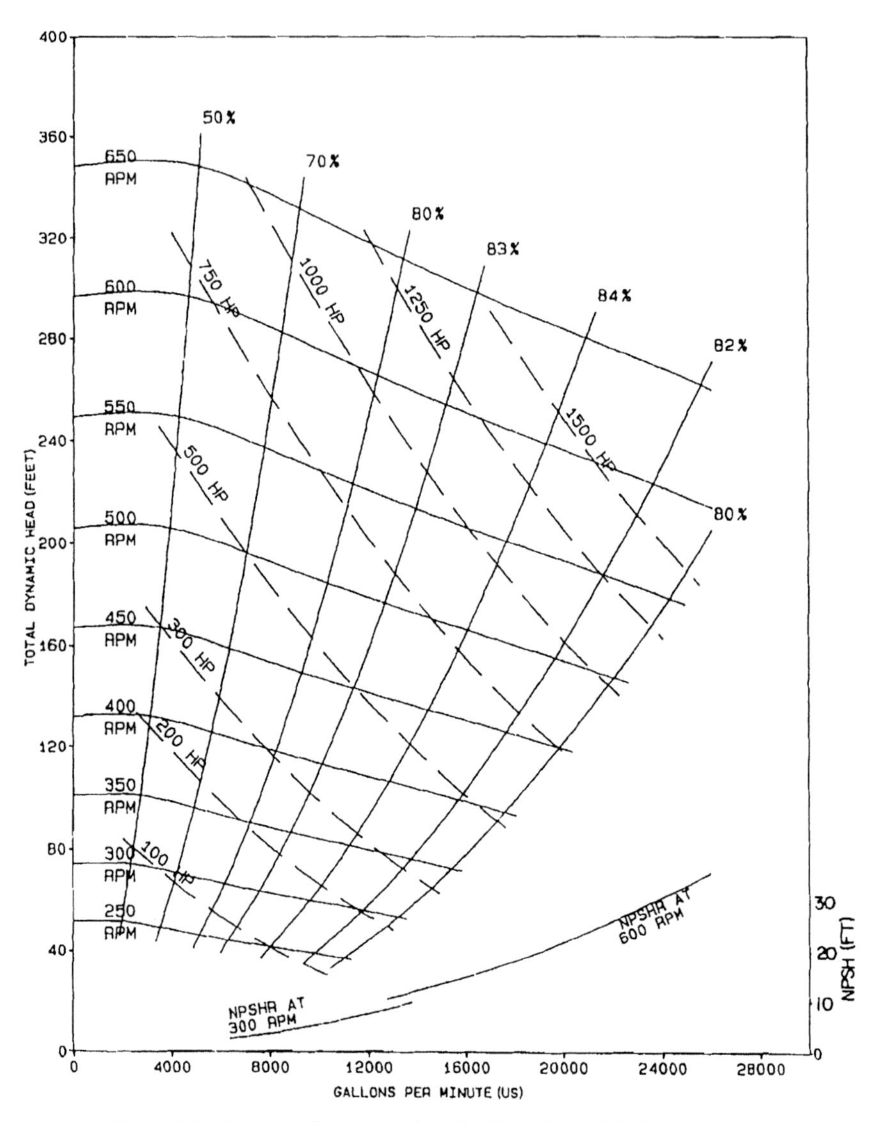

Figure A.2. Pump performance chart for Case Study 9.2 (U.S. units).

(c) The pump suction geometry and speed are the prime factors governing the suction capability of the pump. Provided the turndown is kept small, as in this case, the NPSH at the duty for all practical purposes will be that of the so called take off point of 13,686 USgpm flow, 172 feet head at 500 rpm.

Figure A.2 only provides us with NPSH for 600 rpm. Figure 9.13 shows a plot of σ versus Q/Q_{BEP} derived from data for the same series of pumps. The take-off point is at 82% of the BEP flow. This corresponds to a σ of 0.07 on Fig. 9.13. The NPSH at the duty is therefore equals 0.07 (147.6) or 10.3 ft.

Worked Examples from Chapter 10

Example 10.1

A sand slurry is to be pumped using a large pump with a 52 inch impeller. Determine the reduction in head R_H if the solids concentration by volume C_{vd} = 0.20, the solids specific gravity S_s = 2.65 and the average particle size d_{50} = 0.016 inches (400 μm).

The impeller diameter of 52 inches is 1320 mm, which can be used with Fig. 10.9. It follows directly from this figure that for d_{50} = 400 μm S_s = 2.65 and C_{vd} = 15%, R_H equals 3.5%. The adjustment from C_{vd} = 15% to 20% gives: 3.5(20/15) for R_H = 4.7%.

Example 10.2

Determine R_H for a pump with an impeller diameter of 31.5 in (800 mm) pumping an ore product (S_s = 4.0) with d_{50} = 0.020 in (500 μm), X_h = 0.28 and delivered volumetric concentration of 20%.

Figure 10.9 shows that for d_{50} = 500 μm, D = 800 mm, S_s = 2.65, X_h = 0 and C_{vd} = 15%, R_H is 6.4%. Correction factors for C_{vd} = 20%, S_s = 4 and X_h = 0.28 are; respectively:

$$\frac{20}{15} = 1.33$$

$$\left[\frac{4-1}{1.65}\right]^{0.65} = 1.475$$

$$(1 - 0.28)^2 = 0.52$$

R_H is then $6 \cdot 1.33 \cdot 1.47 \cdot 0.52 = 6.2\%$.

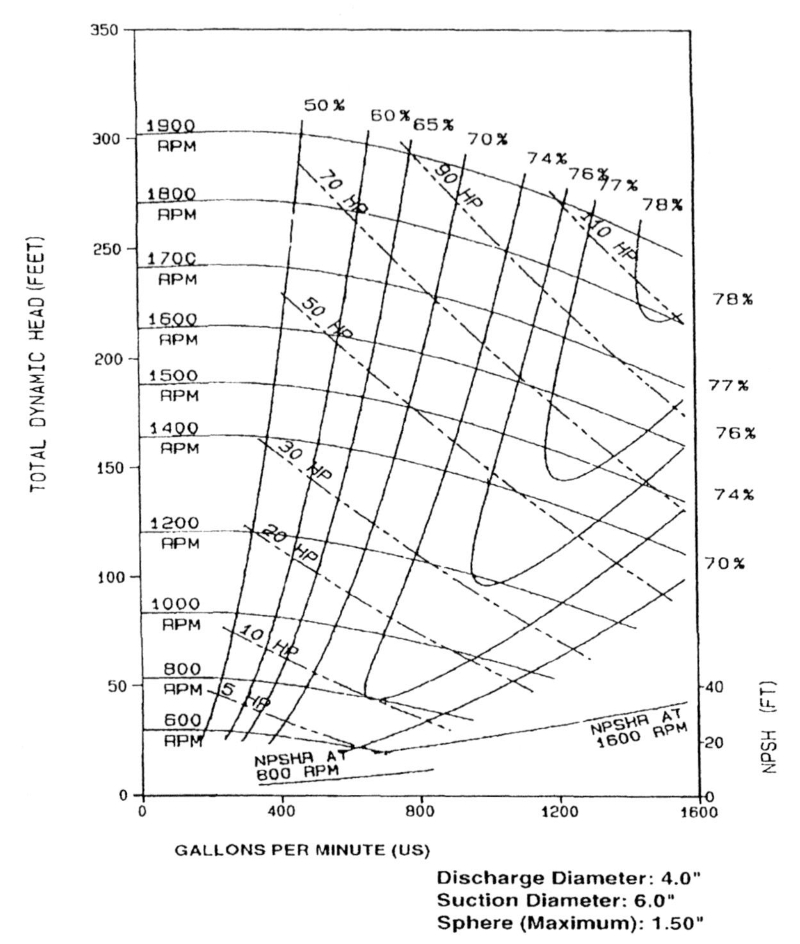

Figure A.3. Pump characteristic curves for Example 10.3 (U.S. units)

Example 10.3

A smaller pump is to be used to pump a slurry of the same solids as in Example 10.1, but at a volumetric concentration of 10%. The pump has an impeller diameter of 15 inches (380 mm), and its clear-water performance characteristics are shown on Fig. A.3.

Following Example 10.1 with a pump impeller diameter 380 mm, the diagram gives R_H = 7%. Correction for C_{vd} = 10% means that R_H = $7 \cdot (10/15) = 4.7\%$, say 5%. Assuming that R_η equal to R_H, it is also 5%. The head required for this application is 148 ft of slurry at a flow rate of 990 USgpm. The pump has a 4 inch discharge branch, 6 inch suction branch.

The running speed and power requirement for pumping this slurry are to be determined.

The head ratio H_r was defined in Section 10.1 as $1\text{-}R_H$, and for the present example this equals 0.95, *i.e.*

$$H_r = \frac{H_m}{H_w} = 0.95$$

In order to be able to produce the required 148 ft head of slurry, the pump must thus be capable of producing a head of water, H_w, equal to

$$H_w = \frac{148}{0.95} = 156 \text{ ft of water}$$

For this water head, and the discharge of 990 USgpm, the pump characteristics shown on Fig.A.3 indicate that the pump must run at 1500 rpm, and that the clear water efficiency η_w will be 75%. With R_η assumed to be equal to R_H, then

$$\eta_r = 1 - R_\eta = 0.95 = \frac{\eta_m}{\eta_w}$$

Thus

$$\eta_m = \eta_r \eta_w = 0.95(0.75) = 0.713 \text{ or } 71.3\%$$

The slurry specific gravity, S_m, is 1 + 0.10 (2.65 - 1) or 1.165, and by Eq. 9.1 P_m equals $S_m P_w$. For the calculated values of η_m and other quantities the power requirement is found to be

$$P_m = \frac{1.165(148)(990)}{(3960)(0.713)} = 60.5 \text{ HP}$$

Case Study 13.1

Case Studies 5.1 and 6.1 were concerned with the transport of 4410 tons per hour of coarse sand (S_s = 2.65) as a slurry in water, at delivered concentrations up to C_{vd} = 0.24. In this Case Study the design will be completed by examining system operability. Because the earlier studies all represent stages in the design, they will be summarised here.

Case Study 5.1 was the preliminary selection of pipe size, not needing information on the particle size beyond noting that the slurry has settling characteristics. Three pipes sizes were identified as suitable for more detailed study, keeping $V_m > V_{sm}$ to avoid deposition:

Condition	D (in)	V_{sm} (ft/s)	C_{vd}	V_m (ft/s)
A	21.7	14.6	0.24	24.1
B	23.6	15.3	0.24	20.2
C	25.6	15.9	0.20	20.7

In Case Study 6.1 the particle size distribution was introduced, and used to calculate the friction gradient and specific energy consumption. It was shown that, for the 25.6 inch pipe with 20% solids of the type specified (Condition C above) the friction gradient is 0.0612 ft water/ft pipe with specific energy consumption (SEC) of 0.617 HPh/ton-mile. It was also found that increasing C_{vd} and decreasing V_m reduced the specific energy consumption. A revised operating point (Condition D) was therefore chosen for further analysis.

D	=	25.6 inches	(V_{sm} =15.9 ft/s)
C_{vd}	=	0.221	
V_m	=	18.7 ft/s	
i_m	=	0.0619 ft water/ft pipe	
SEC	=	0.563 HPh/ton-mile	

The feasibility of operating at this point was not investigated in Case Study 6.1.

We now pursue this study further by examining system operability. Attached figures show the system characteristics and SEC curves for three pipe sizes: 25.6, 29.5 and 21.7 inches. The curves are all calculated using the approach illustrated in Case Study 6.1, with values for the deposit velocity (V_{sm}) obtained as in Case Study 5.1. We consider first the pipe size which appeared to be attractive in the earlier case studies : D = 25.6 inches. Figure A.4 shows the system characteristics (expressed in terms of i_m) and SEC curves. The full curves are drawn for constant delivered concentration. The broken curves refer to constant throughput: throughput can be kept constant by increasing mixture velocity while decreasing delivered concentration. Conditions C and D considered in earlier case studies can be located on this figure.

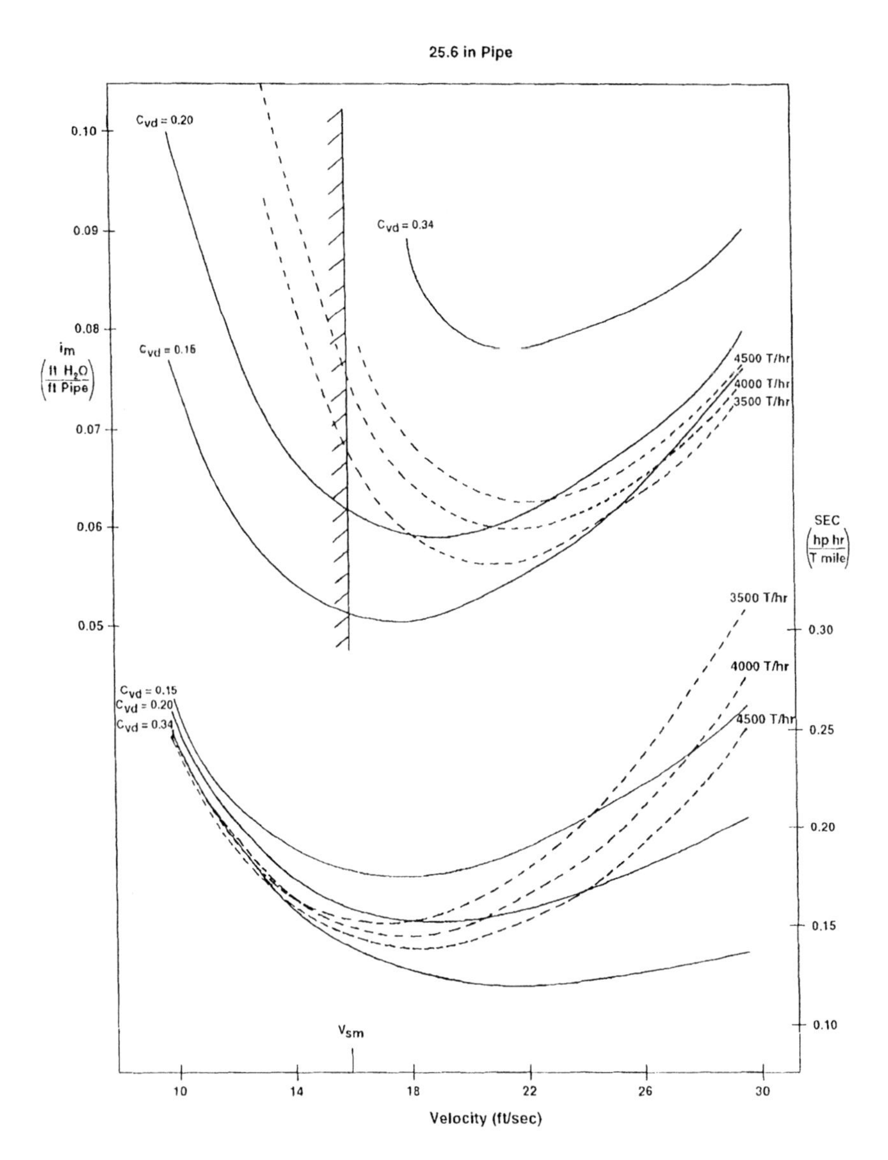

Figure A.4. Friction gradient (i_m) and specific energy consumption (SEC) for transport of coarse sand in 25.6 in pipe (U.S. units)

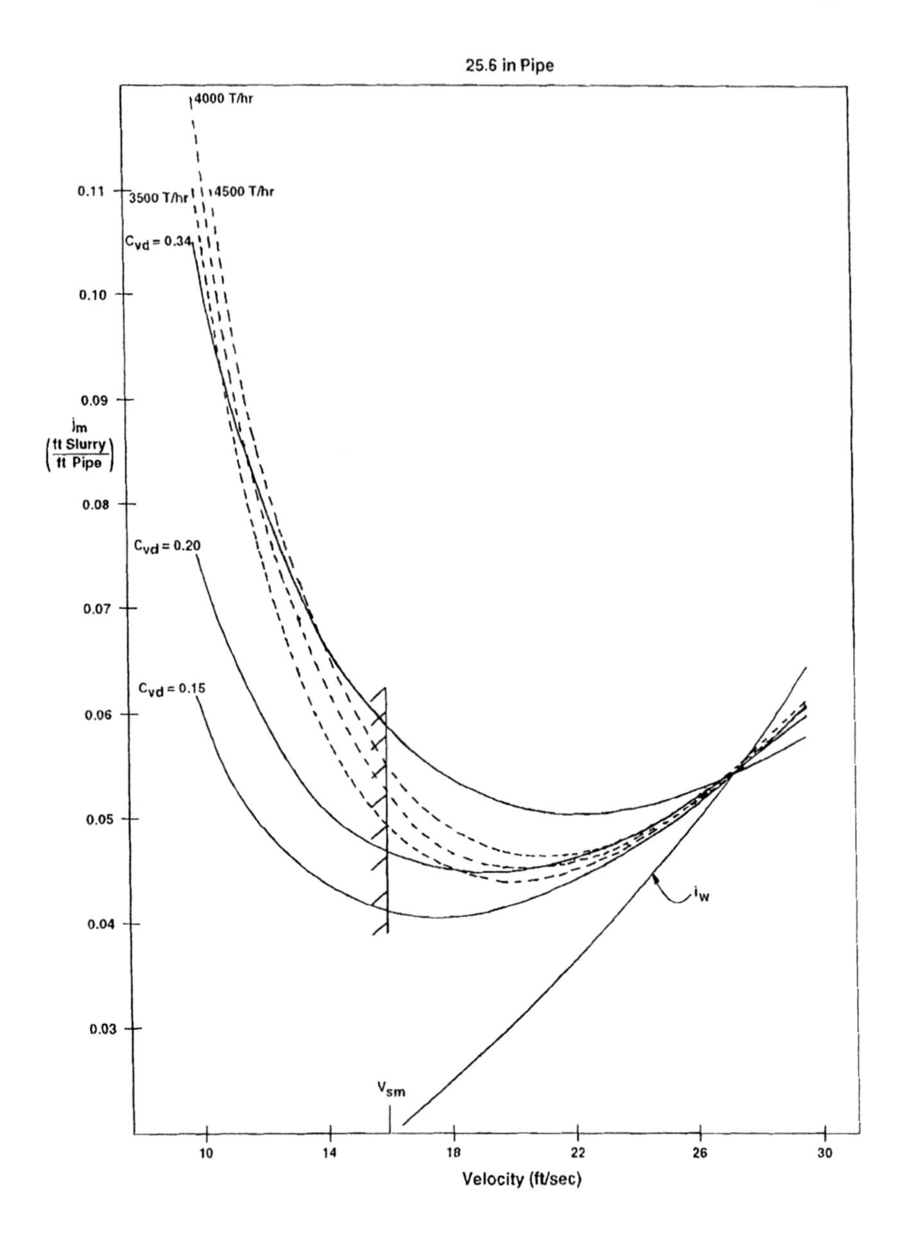

Figure A.5. Friction gradient in terms of head of slurry (j_m) for coarse sand in 25.6 in pipe (U.S. units)

Only now that the system characteristics are plotted is it evident that Condition D is too far back on the system curves to represent a feasible operating point for this material - the system characteristic shows i_m

decreasing as V_m increases, so that operation will be unstable. Even Condition C is marginal. Given that the booster pump will have variable - speed diesel drive, operation at C might be feasible if solids consistency and throughput do not fluctuate too widely. However, if the variations are significant - for example, throughput variations from 3800 ton/hr to 5000 ton/hr (see the first figure) - then operation will become unstable.

To achieve truly flexible operation, it is preferable to operate around the 'standard' velocity, where $i_w = j_m$. Figure A.5 shows the same system characteristics, but now plotted in terms of j_m (i.e. i_m/S_m in ft slurry/ft pipe). The recommended operating velocity is therefore around point E, i.e. at V_m around 24.9 ft/s, with C_{vd} . 0.168 and with SEC increased to 0.820 HPh/ton-mile. Thus, once operability is considered, it is seen that lower concentrations should be used, with correspondingly higher energy consumption.

We consider now whether going to a larger pipe would improve matters. Figures A.6 and A.7 show the curves of i_m, SEC and j_m for a pipe of 29.5 inch diameter. To achieve operation at the standard velocity, it is now necessary to reduce the concentration further still: V_m = 25.7 ft/s, with C_{vd} = 0.121. Again because of the cost of pumping the conveying water, the specific energy consumption is high: SEC is 0.997 HPh/ton-mile at a throughput of 4410 tons per hour. Furthermore, increasing the pipe size will increase the capital cost. Thus the 29.5 inch pipe is not recommended.

A smaller pipe size should also be considered. The final two figures (A.8 and A.9) show the curves of i_m, SEC and j_m calculated in the usual way for a pipe 21.7 inches in diameter. Condition A from Case Study 5.1 is indicated. It is now seen that this condition is close to the 'standard' velocity. It therefore represents a flexible but stable operating point. Furthermore, because the delivered concentration is relatively high, at about 24% solids, the specific energy consumption is quite attractive, at 0.689 HPh/ton-mile.

Thus, of the three pipe sizes considered in this case study, the smallest would be preferred. In addition to its favourable specific energy consumption, it will have the least capital cost. It is left to the reader to carry out the comparison for the intermediate size, D = 23.6 inches. A full design would now proceed to consider pump selection again, making reference to Chapter 14.

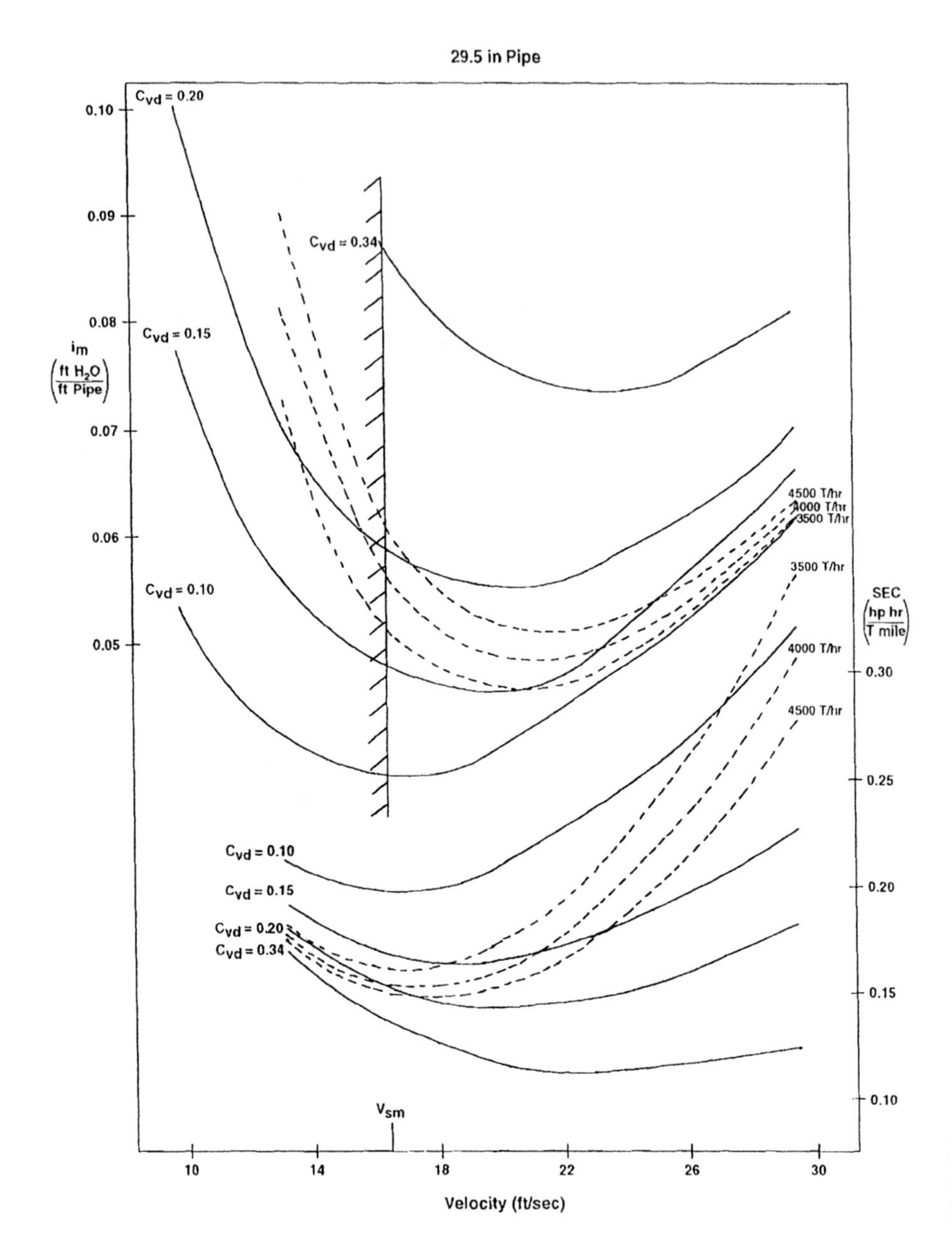

Figure A.6. Friction gradient (i_m) and specific energy consumption (SEC) for transport of coarse sand in 29.5 in pipe (U.S. units)

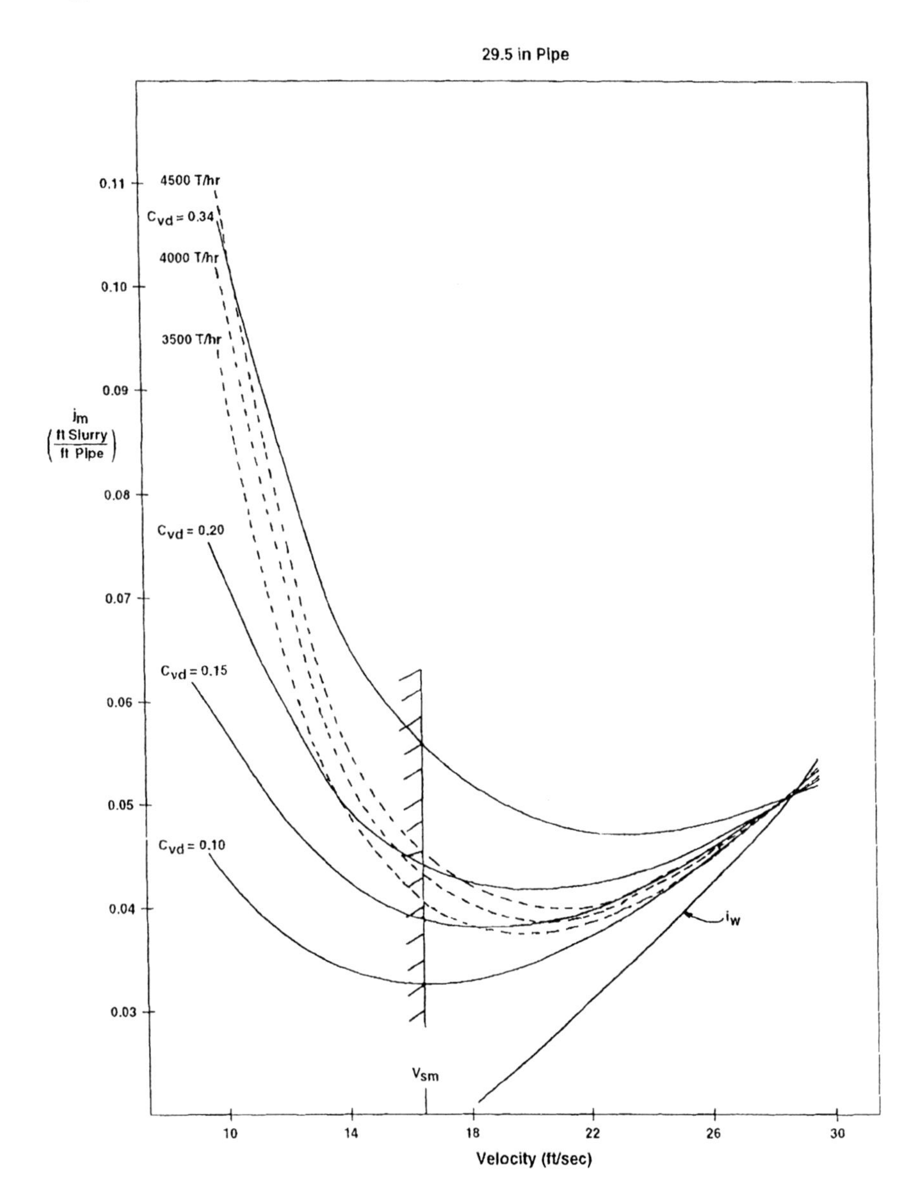

Figure A.7. Friction gradient in terms of head of slurry (j_m) for coarse sand in 29.5 in pipe (U.S. units)

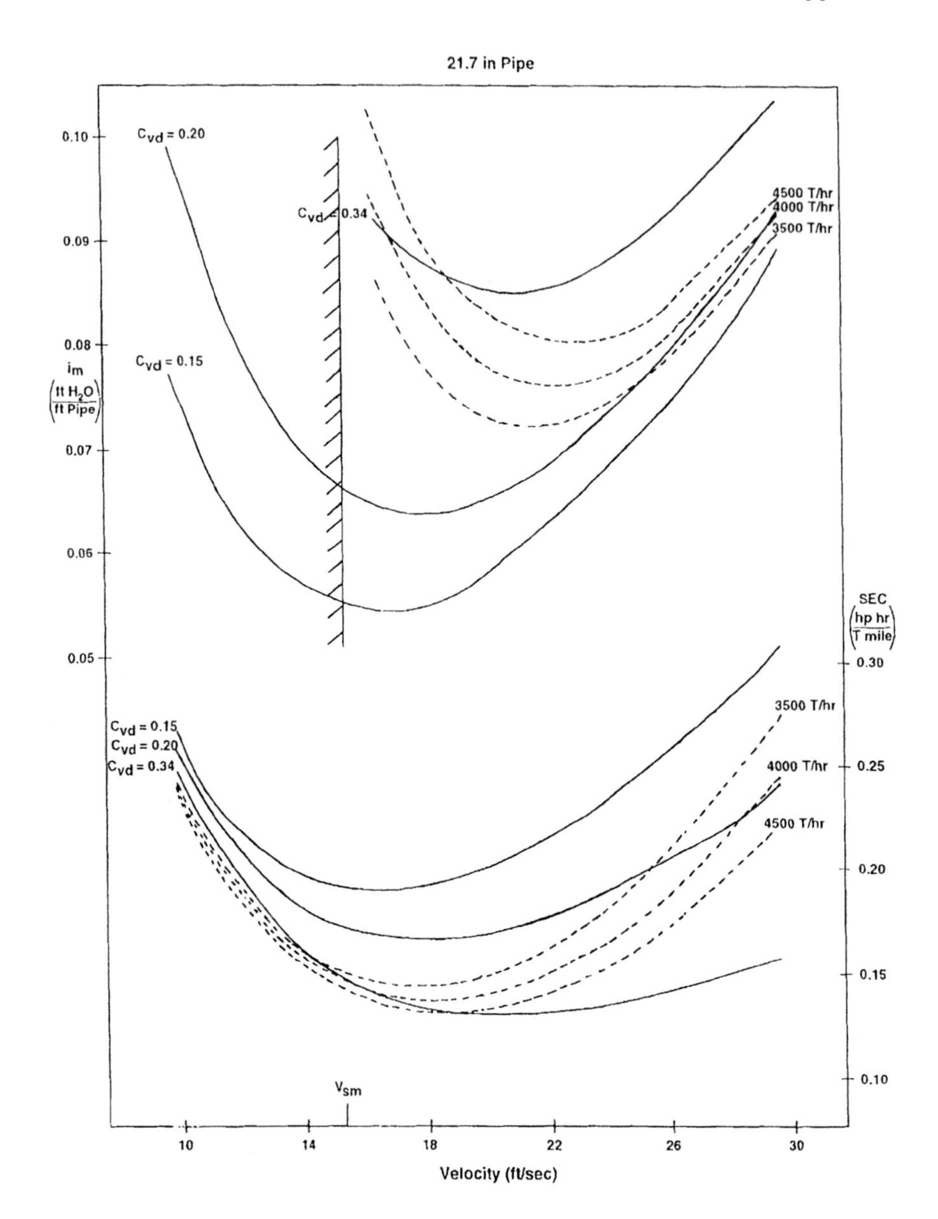

Figure A.8. Friction gradient (i_m) and specific energy consumption (SEC) for transport of coarse sand in 21.7 in pipe (U.S. units)

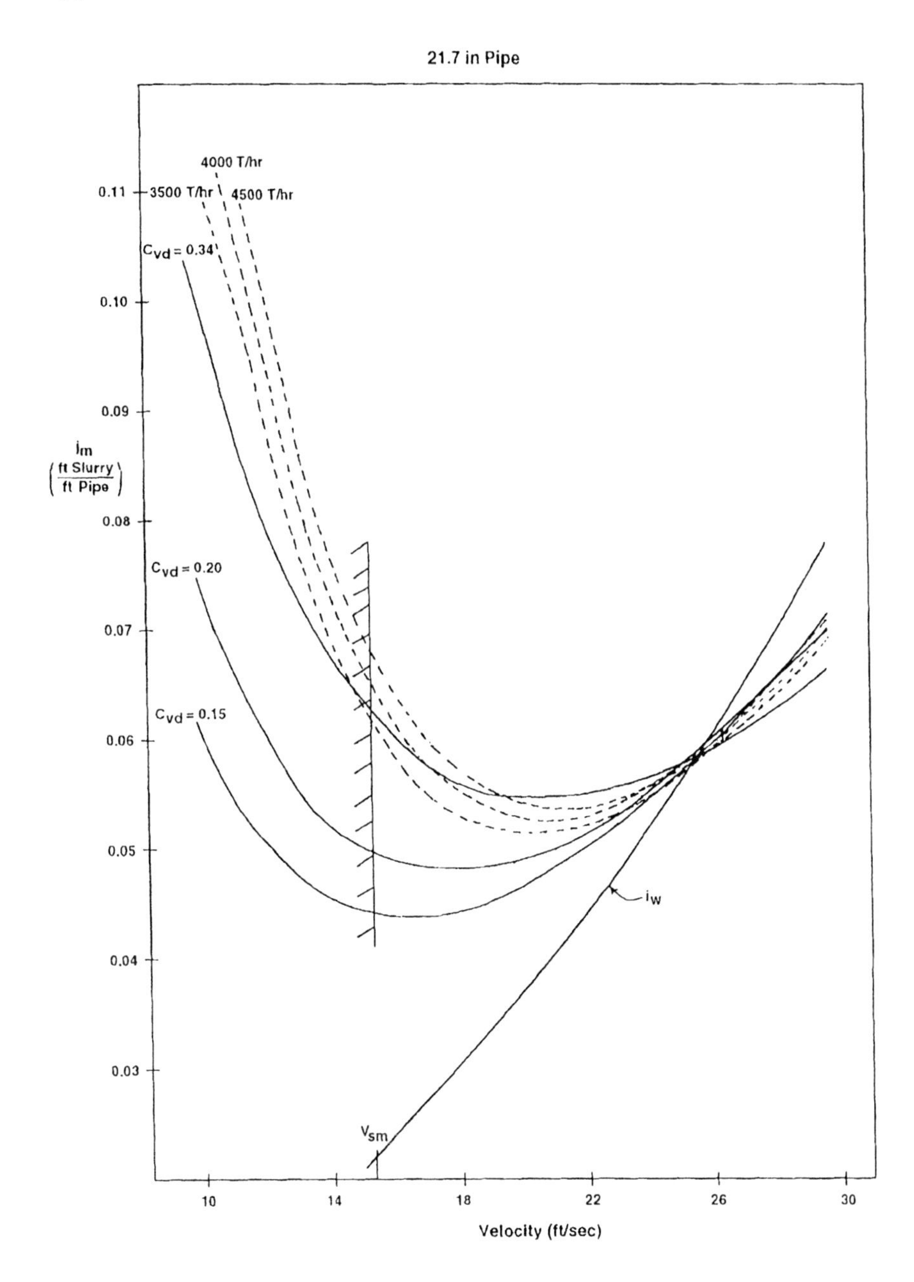

Figure A.9. Friction gradient in terms of head of slurry (j_m) for coarse sand in 21.7 in pipe (U.S. units)

The conclusions illustrated by this case study are general: for slurries with settling characteristics, considerations of energy consumption together with operational stability favour use of high solids concentration and small pipe diameter. Also, as indicated throughout this chapter, stable operation rather than deposition controls the operating velocity for this type of slurry.

On the negative side, the mixture velocity can be too high. For example, Condition A, as selected above, has a design mixture velocity of 24.1 ft/s, and the pipe will experience substantial wear. As noted in Chapter 11, for some applications, e.g. conveying of phosphate matrix, wear is accommodated by rotating the pipe at intervals, usually through 120°, so as to expose new parts of the pipe wall to the stratified solids which cause the abrasion.

Index

CPSIA information can be obtained at www.ICGtesting.com
Printed in the USA
LVOW082006090212
267952LV00002B/2/A